Electronic Nose Technologies and Advances in Machine Olfaction

Yousif Albastaki
University of Bahrain, Bahrain

Fatema Albalooshi
University of Bahrain, Bahrain

A volume in the Advances in Computer and Electrical Engineering (ACEE) Book Series

IGI Global
DISSEMINATOR OF KNOWLEDGE

Published in the United States of America by
IGI Global
Engineering Science Reference (an imprint of IGI Global)
701 E. Chocolate Avenue
Hershey PA, USA 17033
Tel: 717-533-8845
Fax: 717-533-8661
E-mail: cust@igi-global.com
Web site: http://www.igi-global.com

Library of Congress Cataloging-in-Publication Data

Names: Albastaki, Yousif, 1962- editor. | Albalooshi, Fatema A., 1979- editor.
Title: Electronic nose technologies and advances in machine olfaction / Yousif Albastaki and Fatema Albalooshi, editors.
Description: Hershey, PA : Engineering Science Reference, [2018] | Includes bibliographical references.
Identifiers: LCCN 2017022953| ISBN 9781522538622 (hardcover) | ISBN 9781522538639 (ebook)
Subjects: LCSH: Olfactory sensors--Handbooks, manuals, etc. | Gas detectors--Handbooks, manuals, etc. | Odors--Handbooks, manuals, etc. | Intelligent sensors--Handbooks, manuals, etc.
Classification: LCC TP159.C46 E535 2018 | DDC 681/.754--dc23 LC record available at https://lccn.loc.gov/2017022953

This book is published in the IGI Global book series Advances in Computer and Electrical Engineering (ACEE) (ISSN: 2327-039X; eISSN: 2327-0403)

British Cataloguing in Publication Data
A Cataloguing in Publication record for this book is available from the British Library.

All work contributed to this book is new, previously-unpublished material. The views expressed in this book are those of the authors, but not necessarily of the publisher.

For electronic access to this publication, please contact: eresources@igi-global.com.

Advances in Computer and Electrical Engineering (ACEE) Book Series

Srikanta Patnaik
SOA University, India

ISSN:2327-039X
EISSN:2327-0403

MISSION

The fields of computer engineering and electrical engineering encompass a broad range of interdisciplinary topics allowing for expansive research developments across multiple fields. Research in these areas continues to develop and become increasingly important as computer and electrical systems have become an integral part of everyday life.

The **Advances in Computer and Electrical Engineering (ACEE) Book Series** aims to publish research on diverse topics pertaining to computer engineering and electrical engineering. **ACEE** encourages scholarly discourse on the latest applications, tools, and methodologies being implemented in the field for the design and development of computer and electrical systems.

COVERAGE

- Digital Electronics
- Power Electronics
- Computer science
- Applied Electromagnetics
- Sensor Technologies
- Computer Architecture
- Circuit Analysis
- Computer Hardware
- Analog Electronics
- Chip Design

IGI Global is currently accepting manuscripts for publication within this series. To submit a proposal for a volume in this series, please contact our Acquisition Editors at Acquisitions@igi-global.com or visit: http://www.igi-global.com/publish/.

The Advances in Computer and Electrical Engineering (ACEE) Book Series (ISSN 2327-039X) is published by IGI Global, 701 E. Chocolate Avenue, Hershey, PA 17033-1240, USA, www.igi-global.com. This series is composed of titles available for purchase individually; each title is edited to be contextually exclusive from any other title within the series. For pricing and ordering information please visit http://www.igi-global.com/book-series/advances-computer-electrical-engineering/73675. Postmaster: Send all address changes to above address.

Titles in this Series

For a list of additional titles in this series, please visit: www.igi-global.com/book-series

Free and Open Source Software in Modern Data Science and Business Intelligence Emerging Research and Opportunities
K.G. Srinivasa (CBP Government Engineering College, India) Ganesh Chandra Deka (M. S. Ramaiah Institute of Technology, India) and Krishnaraj P.M. (M. S. Ramaiah Institute of Technology, India)
Engineering Science Reference • copyright 2018 • 189pp • H/C (ISBN: 9781522537076) • US $190.00 (our price)

Design Parameters of Electrical Network Grounding Systems
Osama El-Sayed Gouda (Cairo University, Egypt)
Engineering Science Reference • copyright 2018 • 316pp • H/C (ISBN: 9781522538530) • US $235.00 (our price)

Design and Use of Virtualization Technology in Cloud Computing
Prashanta Kumar Das (Government Industrial Training Institute Dhansiri, India) and Ganesh Chandra Deka (Government of India, India)
Engineering Science Reference • copyright 2018 • 315pp • H/C (ISBN: 9781522527855) • US $235.00 (our price)

Smart Grid Test Bed Using OPNET and Power Line Communication
Jun-Ho Huh (Catholic University of Pusan, South Korea)
Engineering Science Reference • copyright 2018 • 425pp • H/C (ISBN: 9781522527763) • US $225.00 (our price)

Transport of Information-Carriers in Semiconductors and Nanodevices
Muhammad El-Saba (Ain-Shams University, Egypt)
Engineering Science Reference • copyright 2017 • 677pp • H/C (ISBN: 9781522523123) • US $225.00 (our price)

Accelerating the Discovery of New Dielectric Properties in Polymer Insulation
Boxue Du (Tianjin University, China)
Engineering Science Reference • copyright 2017 • 388pp • H/C (ISBN: 9781522523093) • US $210.00 (our price)

Handbook of Research on Nanoelectronic Sensor Modeling and Applications
Mohammad Taghi Ahmadi (Urmia University, Iran) Razali Ismail (Universiti Teknologi Malaysia, Malaysia) and Sohail Anwar (Penn State University, USA)
Engineering Science Reference • copyright 2017 • 579pp • H/C (ISBN: 9781522507369) • US $245.00 (our price)

IGI Global
DISSEMINATOR OF KNOWLEDGE

701 East Chocolate Avenue, Hershey, PA 17033, USA
Tel: 717-533-8845 x100 • Fax: 717-533-8661
E-Mail: cust@igi-global.com • www.igi-global.com

Editorial Advisory Board

Table of Contents

Detailed Table of Contents

Wilk's Lambda and bounded Mahalanobis distance, are applied. Classification with features determined by bounded Mahalanobis distance statistically performs better than Wilk's Lambda. This chapter suggests that classification of e-nose with feature selection is a good choice to limit the cost of experiments and maintain good classification performance.

With new developments and upcoming technologies, new sensing techniques are becoming available. Unfortunately, none of these techniques provides output interpreted the way human perception works. An inability to improve the effectiveness of these technologies limits their use in dedicated applications and increases their complexity. The growing adoption of this technology makes it critical to create a system capable of handling e-nose challenging issues such as noise, drift, imbalanced data, dynamic environment, and high uncertainties. Without appropriate pattern recognition methods that allow inferences to be derived based on patterns observed within these data sets, it will not be possible to improve the performance of current e-nose systems. In this chapter, e-nose drift issue is introduced and the available drift counteraction methods is discussed.

Fluorescence quenching is a promising technique for chemical sensing applications such as the detection of explosive trace vapors. Nitroaromatic explosives are one of the primary targets for this approach enabling ultra-low detection limits down to sub parts-per-billion in air. Many studies, however, focus on enhanced sensor sensitivity, whereas practical applications often require the identification and quantification of detected species. Electronic noses and efficient sensor systems are a promising solution to address this challenge. The authors review recent approaches and trends for explosive trace vapor detection and discuss theoretical concepts for fluorescence quenching as well as target analytes, sensor materials, and fabrication techniques. Statistical learning techniques such as principal component analysis and linear discriminant analysis, sensor systems, and camera-based read-out strategies are in the focus of the chapter. The authors conclude with recommendations and solutions for the elaborated challenges and with visions on future research directions.

Industrial odor concentration measurement in continuous mode is a challenging task using olfactometers, as it's expensive and requires human involvement for a prolonged time. This chapter presents the development of an indigenous metal oxide sensor-based electronic nose system for measurement of industrial odor in ou/m3. The results of electronic nose and field olfactometer were correlated using multilinear regression and partial least square regression techniques. The results showed satisfactory prediction by both the models, with RMSE (6.70, and 4.02), RAE (0.29 and 0.16), and NAE (0.89 and 0.96), respectively, for MLR and PLS. The results indicated better performance of PLS compared to MLR. The objective of the present work is to train and employ artificial olfaction system for continuous measurement of obnoxious emissions emitted from industries bypassing involvement of olfactometer.

Chapter 6
Electronic Nose Applications 104
Ping Wang, Zhejiang University, China
Fan Gao, Zhejiang University, China
Xusheng Zhang, Zhejiang University, China
Min Wang, Zhejiang University, China

In this chapter, the authors introduce three typical applications of electronic nose, including the diagnosis of lung cancer and gastrointestinal diseases and food freshness detection. Firstly, an electronic nose system intended for lung cancer screening employs semiconductor and electrochemical sensors to compose an array in a chamber to conduct breath analysis with pre-concentration and multiple desorption procedures. Secondly, a design of virtual electronic nose was presented to detect the concentration of hydrogen, methane, and carbon dioxide in breath for the diagnosis of small intestinal bacterial overgrowth. Finally, combined with the metal oxide semiconductor sensors and electrochemical sensors array, the small size electronic nose system is to identify the freshness and classify different kinds of food stored in the refrigerator.

Chapter 7
Electronic Noses in Food Analysis 132
Runu Banerjee Roy, Jadavpur University, India
Rajib Bandyopadhyay, Jadavpur University, India
Bipan Tudu, Jadavpur University, India
Nabarun Bhattacharyya, Centre for Development of Advanced Computing, India

Food quality evaluation is a tedious job as it can be sensed either by its flavor, taste, or appearance. Sensory evaluation for food quality determination is very complex and depends on biological sensor systems and is subject to high variability depending on taster's mood, physical state, etc. Moreover, as it is purely subjective, the observations may vary for different tasters. For food industries the instrumental means for quality monitoring has significant appreciations which are repeatable, accurate, and reliable. Electronic nose is an array of sensors that senses based on aroma of samples. For the last two decades, electronic nose has been applied in several domains of applications in food analysis, for example, food quality monitoring based on seasonal effect, ageing, geographical origin, fermentation, etc. This chapter aims to focus on different domains of application of electronic nose in food analysis.

Electronic nose, which is designed to perceive artificially the odour-active molecules in a sample headspace, has seen an increased use in the food industry as a rapid and reliable tool for quality assessment, classification, and authentication of several food items. The use of chemometrics and pattern recognition methods, together with gas sensors, emerged to be a very powerful analytical approach. In this chapter, an overview of the recent achievements in the field of electronic nose applications on animal-source food is given. Moreover, the authors deal with the recent research trends to overcome the actual sensor shortcomings, including sensor fusion techniques and their applications to evaluate animal-source foods and novel electronic nose systems.

This chapter contains information about application of the electronic nose type instruments to evaluation of ambient air odour quality connected with such sectors of human activity as municipal landfills or sewage treatment plants. The authors present potential sources of emission from these sectors, characterize the chemical compounds responsible for presence of unpleasant odours, describe the influence of those compounds on human health and related discomfort. Legal aspects pertaining to admissible odour concentration levels in selected countries are also presented. The chapter describes instrumental and sensory methods utilized for determination of odour concentration, odour intensity, or hedonic quality. The chemical sensors potentially and currently employed in the electronic nose type devices are also characterized. Moreover, the future prospects of application of the electronic nose instruments to evaluation of ambient air with respect to odour nuisance are discussed.

This chapter presents a proposal of the use of electronic noses in the monitoring of indoor air quality. The main focus is put on the detailed characteristics of today's indoor air quality control methods, the types of pollution in the air, and the development of electronic noses for air testing. Currently, scientists seek methodological and structural solutions that would enable real-time online indoor air control. It has been shown that using electronic noses in this situation is advantageous. In addition, potential uses of these devices are discussed, with particular focus on closed food processing spaces. The authors of the chapter argue that in the near future, the proposed solution could improve the quality of indoor air and thus the health of the users of the indoor environments, as well as the quality of the products prepared there.

In this chapter, the benefit of equipping the robot with odor sensors is investigated. The chapter addresses the types of tasks the mobile robots can accomplish with the help of olfactory sensing capabilities, the technical challenges in mobile robot olfaction, the status of mobile robot olfaction. The chapter also addresses simple and complex electronic olfaction sensors used in mobile robotics, the challenge of using chemical sensors, the use of many types of algorithms for robot olfaction, and the future research directions in the field of mobile robot olfaction.

Out of all the components of a mobile robot, its sensorial system is undoubtedly among the most critical ones when operating in real environments. Until now, these sensorial systems mostly relied on range sensors (laser scanner, sonar, active triangulation) and cameras. While electronic noses have barely been employed, they can provide a complementary sensory information, vital for some applications, as with humans. This chapter analyzes the motivation of providing a robot with gas-sensing capabilities and also reviews some of the hurdles that are preventing smell from achieving the importance of other sensing modalities in robotics. The achievements made so far are reviewed to illustrate the current status on the three main fields within robotics olfaction: the classification of volatile substances, the spatial estimation of the gas dispersion from sparse measurements, and the localization of the gas source within a known environment.

One of the approaches to mimic the remarkable abilities of the human olfactory system is by the design of computer-controlled sensor arrays that are capable of detecting and distinguishing a different range of smells and odors with consistent monitoring, referred to as electronic noses. This chapter introduces the opportunity of integrating smell sense in robots by the use of artificial neural networks. The study proposes a structure for integrating electronic noses in robots to add the capabilities of smell-related assignments, typically to recognize hazardous substances such as sampling the air and decide its actions based on this information. Utilizing the proposed algorithm allows experts in this field to be aware of gas leakage areas and thus reduce unexpected incidences. The effectiveness of the algorithm is demonstrated using real-world samples, and the performance is examined via quantitative metrics and analysis. The results show that the proposed algorithmic framework outperforms state-of-the-art methods with an error rate of only 0.0999%.

Preface

There have been numerous works for odor recognition utilizing diverse sensor clusters and pattern recognition techniques in last decades. In spite of the fact that an odor is normally recorded using language expression, it is excessively troublesome for laymen, making it impossible to relate real odor with that expression. The odor detecting framework ought to be reached out to new applications since its standard style where the yield design from different sensors with in partially covered specificity is perceived by neural network system or multivariate investigation.

Electronic nose and odor sensing devices have gotten impressive considerations in the field of Artificial Olfaction and sensor technology during the last two decades. This is mainly because of the revelation of many applications derived from research in different fields of applied services. The potential for future improvements of imaginative electronic nose applications is colossal as researchers in numerous fields of scientific investigation and industrial development become more aware of the capabilities of the electronic nose. The ability of the electronic nose to detect odorless chemical makes it ideal for use in a wide range of applications such as clinical diagnosis, agriculture and forestry and food spoilage detection.

Therefore, the primary objective of this proposed book is to create an integrated and coherent source of electronic nose which will add value to the current research by focusing on emerging new applications of the electronic nose, and which will contribute to a better understanding of the steps involved in designing electronic nose with the aids of neural networks and other related technologies. Hence, this book will be of interest to researchers, academics, students, and practitioners of electronic nose applications.

The prospective audience for this book will be researchers, academic and practitioners from the following fields of research: electronic nose designers, odor sensing detections, clinical diseases diagnosis, agriculture and forestry and many other related fields.

This book aims to increase electronic nose new applications awareness by providing a clear direction for using existing and newly designed odor sensing sensors with the aid of neural networks and other related algorithms in enhancing the accuracy of electronic nose design and implementations.

The book will also be of value to a wider audience of readers who are interested in understanding electronic nose, including information technology students and anyone from other disciplines that are interested in learning about the electronic nose and its design and applications.

The first chapter of the book describes machine odor detection and how it is developed into an important aspect of our lives by illustrating different applications of it. From detecting food spoilage to the diagnosis of diseases, this chapter explains how electronic noses are developed and tested in various fields and industries for specific purposes.

This work, "A Framework for an Artificial-Neural-Network-Based Electronic Nose," is machine learning based e-nose system that has been developed for detection of various types of odors for a general purpose. The system can be trained on any odor using various e-nose sensors of various types.

It uses Artificial Neural Network as its Machine Learning algorithm along with an OMX-GR semi-conductor gas sensor for collecting odor data.

The system was trained and tested with five different types of odors collected through a standard data collection method and then purified which in turn had a result varying from 93% to 100% accuracy.

In summary chapter one aims at developing an Odor sensing system using Electronic nose as a hardware and Artificial Neural Network as software for general odor-sensing and identifying different odors

The second chapter introduces the issue of multicollinearity in electronic nose (e-nose) classification analysis and investigates some analytical solutions to deal with such problem. Statistically, multicollinearity effect may harm the classification analysis in producing bias estimate of the parameters during the development phase of the statistical classification rule. A common approach to deal with such effect is the feature extraction. However, the criterion used in extracting the raw features based on variances may not be suitable to the ultimate goal of classification accuracy. Alternatively, feature selection method would be advisable as it chooses only valuable features. Two distance-based criteria in determining the right features namely Wilk's Lambda and unbounded Mahalanobis distance are applied. Classification with features determined by unbounded Mahalanobis distance statistically performs better than Wilk's Lambda. The chapter suggests that classification of e-nose with feature selection is a good choice to limit the cost of experiments and maintain good classification performance.

This chapter overviewed both ideas of features extraction and feature selection on e-nose, data suggesting for improvement in data processing and accuracy of results. The arrangement of discussion is as follow; brief previous works related to huge dimensionality handling in e-nose is explained. Next, detail of the methodology designed purposely to investigate the potential uses of feature selection versus the feature extraction. Then, results from the designed methodology are discussed and future works propose potential work that can be planned for enhancement in e-nose technology. The conclusion is given in the final section.

The third chapter of the book shows how with new developments and upcoming technologies, new sensing techniques are becoming available. Unfortunately, none of these techniques provides output interpreted the way human perception works. An inability to improve the effectiveness of these technologies limits their use in dedicated applications and increases their complexity. The growing adoption of this technology makes it critical to create a system capable of handling e-nose challenging issues such as noise, drift, imbalanced data, dynamic environment and high uncertainties. Without appropriate pattern recognition methods that allow inferences to be derived based on patterns observed within these data sets, it will not be possible to improve the performance of current e-nose systems. In this chapter, e-nose drift issue will be introduced and the available drift counteraction methods will be discussed.

This chapter describes how this research work is motivated by research questions that have a significant technological impact and practical relevance to air pollution monitoring, medical diagnostics, food spoilage detection and security applications. Unlike the advances made in speech recognition and machine vision, e-nose pattern recognition requires significant improvements. The metal-oxide sensors used in e-nose have sensitivity, selectivity and stability issues which impact their performance. The authors have conducted a critical review of pattern recognition and drift compensation method for e-nose sensor array. Our analysis shows some problems in this technology which requires further investigation. E-nose drift remains the major issue in e-nose pattern recognition systems. Few research studies have addressed drift problem in e-nose. However, most studies which addressed drift considered synthetic data sets or very small private data sets.

In summary, in this chapter, a study is presented which covers prior and current research in e-nose pattern recognition. In particular, the taxonomy of pattern recognition and drift correction for e-nose is presented. We summarised drift correction methods for e-nose into four main categories: 1) adaptive methods, 2) reference based, 3) adaptive component correction and 4) signal processing. The methods covered in the literature are Neural Net, Self Organising Maps, K-Nearest Neighbors and Support Vector Machine.

The fourth chapter of the book covers fluorescence quenching and shows how it is a promising technique for chemical sensing applications such as the detection of explosive trace vapors. Nitroaromatic explosives are one of the primary targets for this approach enabling ultra-low detection limits down to sub-parts-per-billion in the air. Many studies, however, focus on enhanced sensor sensitivity, whereas practical applications often require the identification and quantification of detected species. Electronic noses and efficient sensor systems are a promising solution to address this challenge. The authors review recent approaches and trends for explosive trace vapor detection and discuss theoretical concepts for fluorescence quenching as well as target analytes, sensor materials, and fabrication techniques. Statistical learning techniques such as principal component analysis and linear discriminant analysis, sensor systems and camera-based read-out strategies are in the focus of the chapter. The authors conclude with recommendations and solutions for the elaborated challenges and with visions on future research directions.

Furthermore, this chapter presents how the utilization techniques of sensor arrays have led to the development of optoelectronic noses that are able to discriminate explosive trace vapors using pattern recognition techniques. With the help of PCA, a first glance onto the discriminatory ability of a sensor array is possible with very low effort. Furthermore, simple techniques such as LDA showed great potential in classification tasks with low error rates related to the identification and quantification of explosive species. In terms of sensor array fabrication, printing proved to be a promising candidate towards the realization of low-cost and reproducible sensor arrays. At the same time, printing enables the realization of customized and highly integrated sensor arrays for miniaturized systems. Obviously, fluorescence quenching has contributed significant findings to the field of explosive detection which becomes more and more important all over the world.

The fifth chapter of the book highlights the usage of olfactometers and describes how industrial odor concentration measurement in continuous mode is challenging task using olfactometers, as it's expensive and requires human involvement for a prolonged time. This paper presents the development of an indigenous metal oxide sensor-based electronic nose system for measurement of industrial odor in ou/m^3. The results of electronic nose and field olfactometer were correlated using multilinear regression and partial least squares regression techniques. The results showed satisfactory prediction by both the models, with RMSE (6.70, & 4.02), RAE (0.29 & 0.16) and NAE (0.89 & 0.96) respectively for MLR and PLS. The results indicated better performance of PLS compared to MLR. The objective of the present work is to train and employ artificial olfaction system for continuous measurement of obnoxious emissions emitted from industries bypassing involvement of olfactometer

The chapter presents, an electronic nose system and its development to measure the odor concentration (ou/m^3) at the industrial site. A field olfactometer has been used to provide training to the developed electronic nose system. The main objective of the study was to provide an alternative to tradition olfactometry procedure and bypass it using machine olfaction system. The results obtained prove that PLS can effectively train the electronic nose system to measure the unknown industrial concentrations.

Though the field olfactometer used in this study has limited dilution ratio which could be overcome using a more advance olfactometer. However, the results obtain encourages to explore the electronic nose system further for continuous industrial odor concentration measurement and bypass the human involvement in the process.

The sixth chapter of the book introduces three typical applications of electronic nose, including the diagnosis of lung cancer and gastrointestinal diseases, and food freshness detection. Firstly, an electronic nose system intended for lung cancer screening employs semiconductor and electrochemical sensors to compose an array in a chamber to conduct breath analysis with pre-concentration and multiple desorption procedures. Secondly, a design of virtual electronic nose was presented to detect the concentration of hydrogen, methane and carbon dioxide in breath for the diagnosis of small intestinal bacterial overgrowth. Finally, combined with the metal oxide semiconductor sensors and electrochemical sensors array, the small size electronic nose system is to identify the freshness and classify different kind of food stored in the refrigerator.

This chapter describes electronic nose as an instrument, which comprises a sampling system, an array of chemical gas sensors with differing selectivity, and a computer with an appropriate pattern-classification algorithm, capable of qualitative and/or quantitative analysis of simple or complex gases, vapors, or odors. Recently, lung cancer has the continuously greatest incidence and mortality among all cancers. Because lung cancer is almost asymptomatic in the early stages, it is always detected in advanced stage when the curative intervention is no longer possible. It is obvious that screening is the only systematic means of finding lung tumors at early stage. However, the present detection tools for lung cancer is not available to a general high-risk population. The breath test is a non-invasive and low-cost detecting method which has potential to be a screening tool. Additionally, exhaled breath air originates in the respiratory system, which makes it more specific to lung cancer than any other cancers. The present research is devoted to exploring lung cancer markers in exhaled gas and condensate, developing detecting instruments and detecting technology which would meet the requirements of physical examination. It is expected that the early diagnosis of lung cancer will be realized after developing a detection technology which can be applied conveniently in regular medical checkup.

The seventh chapter of the book shows how food quality evaluation is a tedious job as it can be sensed either by its flavor, taste, and appearance. Sensory evaluation for food quality determination is very complex which depends on biological sensor systems and subject to high variability depending on taster's mood, physical states etc. Moreover, as it is purely subjective, the observations may vary for different tasters. Nowadays for food industries, the instrumental means for quality monitoring has got significant appreciations which are repeatable, accurate and much reliable. The electronic nose is an array of sensors which senses based on aroma of samples. For last two decades electronic nose is applied in several of applications in food analysis, viz, food quality monitoring based on seasonal effect, aging, geographical origin, fermentation etc. This chapter aims to focus on different domain of applications of electronic nose in food analysis.

The chapter presents aroma as a major important characteristic for food quality but taste is another important feature. When we perceive food, we actually recognize the flavor of food which is amalgamation of smell as well as taste. In human physiology smell and taste are influenced by each other. If due to cold, nose gets blocked that is the smell sense becomes weak, and then we cannot feel the taste of food properly. Smell and taste sense have cross-correlation effects. Towards the direction of overall quality of food, combined analysis of electronic nose and electronic tongue has shown quite improved result than single sensor system. In the area of food analysis combination of electronic sensor responses have opened a new research direction.

The eighth chapter of the book handles electronic nose, which is designed to perceive artificially the odor-active molecules in a sample' headspace, have seen an increased use in the food industry as a rapid and reliable tool for quality assessment, classification, and authentication of several food items. The use of chemometrics and pattern recognition methods, together with gas sensors, emerged to be a very powerful analytical approach. In this chapter, an overview of the recent achievements in the field of electronic nose applications on animal source food is given. Moreover, the authors will deal with the recent research trends to overcome the actual sensors shortcomings, including sensor fusion techniques and their applications to evaluate animal source foods, and novel electronic nose systems.

This chapter describes how E-nose emerged as a rapid, precise, and accurate approach for the determination of volatile compounds, with minimal or no sample preparation and reagent consumption, is also suitable for real-time monitoring of several processes. Anyway, despite the promising results obtained so far, there is still room for improvement. The fusion of E-nose data with signals provided by other instrumental sources, as well as novel gas sensing technologies, is being studied nowadays. Therefore, keep exploring this area could result in successfully applications towards the next years.

The ninth chapter of the book contains the information about application of the electronic nose type instruments to evaluation of ambient air odor quality connected with such sectors of human activity as municipal landfills or sewage treatment plants. The authors presented potential sources of emission from these sectors, characterized the chemical compounds responsible for presence of unpleasant odors, described the influence of those compounds on human health and related discomfort. Legal aspects pertaining to admissible odor concentration levels in selected countries were also presented. The chapter describes instrumental and sensory methods utilized for determination of odor concentration, odor intensity or hedonic quality. The chemical sensors potentially and currently employed in the electronic nose type devices were also characterized. Moreover, the future prospects of application of the electronic nose instruments to evaluation of ambient air with respect to odor nuisance were discussed.

The tenth chapter of the book presents a proposal of the use of electronic noses in monitoring indoor air quality. The main focus is put on the detailed characteristics of today's indoor air quality control methods, the types of pollution in the air, and the development of electronic noses for air testing. Currently, scientists seek methodological and structural solutions that would enable real-time online indoor air control. It has been shown that using electronic noses in this situation is significantly advantageous. In addition, potential uses of these devices are discussed, with particular focus on closed food processing spaces. The authors of the chapter argue that in the near future, the proposed solution could improve the quality of indoor air, and thus the health of the users of the indoor environments, as well as the quality of the products prepared there.

The chapter concludes that use of electronic nose for indoor air measurements seems to be more desirable also due to the fact that sensors used in these devices are not selective. It is thus possible to identify individual chemicals present in indoor air, or multicomponent gas mixtures, if the research is carried out in accordance with the holistic concept of measurement. This is the research approach that is more useful when different chemical processes occur in a closed space, often in a similar, but not identical way. A standard example may be the use of electronic noses to control air quality in food processing facilities (restaurant kitchens, large food processing facilities), and in particular in indoor environments where the frying process is carried out, as it may cause the production of carcinogenic chemicals. The use of electronic noses to control indoor air quality is a promising solution that can be used for indoor air monitoring, also because their chemical sensors are more suitable for emission rather than emission measurements.

The eleventh chapter of the book investigates the benefit of equipping the robot with odor sensors. The chapter addresses the types of tasks the mobile robots can accomplish with the help of olfactory sensing capabilities, the technical challenges in mobile robot olfaction, the current status of mobile robot olfaction. The chapter also addresses simple and complex Electronic Olfaction sensors used in Mobile Robotics, the challenge of using chemical sensors, the use of many types of algorithms for robot olfaction and the future research directions in the field of mobile robot olfaction.

In this chapter, many topics were addressed like the benefit of equipping the robot with odor sensors, the types of tasks the mobile robots can accomplish with the help of olfactory sensing capabilities, the technical challenges in mobile robot olfaction, the current status of mobile robot olfaction. The chapter also addressed simple and complex Electronic Olfaction sensors used in Mobile Robotics, the challenge of gas sensing and the use of many types of algorithms for robot olfaction.

The twelfth chapter of the book shows how out of all the components of a mobile robot, its sensorial system is undoubtedly among the most critical ones when operating in real environments. Till now, these sensorial systems mostly rely on range sensors (laser scanner, sonar, active triangulation) and cameras, while electronic noses have barely been employed despite they can provide a complimentary sensory information, vital for some applications, as it happens for humans. This chapter analyzes the motivation of providing a robot with gas-sensing capabilities and also reviews some of the hurdles that are preventing smell to achieve the importance of other sensing modalities in robotics. The achievements made so far are reviewed to illustrate the current status on the three main fields within robotics olfaction: the classification of volatile substances, the spatial estimation of the gas dispersion from sparse measurements, and the localization of the gas source within a known environment.

The thirteenth chapter of the book describes the design of computer-controlled sensor arrays that are capable of detecting and distinguishing different range of smells and odors with consistent monitoring, referred to as the electronic noses. This chapter introduces the opportunity of integrating smell sense in robots by the use of artificial neural networks. The study proposes a structure for integrating electronic noses in robots to add the capabilities of smell related assignments, typically to recognize hazardous substances such as sampling the air, and decide its actions based on this information. The classification of the different odors is accomplished by backpropagation neural network; in which, prior knowledge of the desired odor is fed into the system in order to calculate the error rate and minimize it through iterations. Utilizing the proposed algorithm allows experts in this field to be aware of gas leakage areas, and thus reduce unexpected incidences. The effectiveness of the algorithm is demonstrated by using real-world samples, and the performance is examined via quantitative metrics and analysis. The results show that the proposed algorithmic framework outperforms state-of-the-art methods with an error rate of only 0.0999%.

Chapter 1
A Framework for an Artificial-Neural-Network-Based Electronic Nose

Mudassir Ismail
University of Bahrain, Bahrain

Ahmed Abdul Majeed
University of Bahrain, Bahrain

Yousif Albastaki
University of Bahrain, Bahrain

ABSTRACT

Machine odor detection has developed into an important aspect of our lives with various applications of it. From detecting food spoilage to diagnosis of diseases, it has been developed and tested in various fields and industries for specific purposes. This project, artificial-neural-network-based electronic nose (ANNeNose), is a machine-learning-based e-nose system that has been developed for detection of various types of odors for a general purpose. The system can be trained on any odor using various e-nose sensor types. It uses artificial neural network as its machine learning algorithm along with an OMX-GR semiconductor gas sensor for collecting odor data. The system was trained and tested with five different types of odors collected through a standard data collection method and then purified, which in turn had a result varying from 93% to 100% accuracy.

BACKGROUND

Technology has boomed over the past few decades bringing us all kinds of comfort and accessibility. The screens that we had once that used to be black and white are now colorful and the methods of interacting with a computer using a mouse and keyboard has now been complimented by touch, and even through motion and air gestures.

DOI: 10.4018/978-1-5225-3862-2.ch001

These advances have had the researchers work on different ways and methods to add more ways to interact with the computer and somehow give it more 'senses'. This vision has led to development of various types of sensors through which computers can interact with users and the environment around them. Developments have been made in the fields of various types of interactions such as touch/pressure, measuring temperature and even giving computers a 'vision' through cameras and image recognition. These advances have led to various applications in many industries.

One of these 'senses' has been enabling the computer to 'smell' or detect odor. Imagine being on a video call with a friend over another continent and being able to smell what perfume they are wearing. Or having the ability to not only take pictures and keep them as memories but also the smell of different places and occasions. Or walking through a Virtual world with an ability to know what it smells like.

The efforts of researchers in this field led to creating of an odor-sensing electronic nose first introduced in 1982 (Lee, et al., 2012) which utilized a multi-sensor array of gas sensors combined to classify odors by the detection of different gases present. Since then until now over a period of over three decades, many advancements have been made in the hardware technology of electronic noses with more sensors being introduced and many being created for very specific purposes such as detection of leakage in natural gas factories, analyzing amount of carbon dioxide in an environment. Though less as compared to special purpose sensors, there has been developments in general purpose sensors consisting of a wider array of different types of sensors as well.

Given the wide range of needs of detecting 'odors' and electronic noses still being developed. Humans were utilized in food factories and many other industries to differentiate spoiled from unspoiled food and in other applications. This approach can be unreliable at times due to the limitations of the range of odors a human nose can detect. An alternate that is used are dogs, but the cost of their training is high and they have a short life span.

OBJECTIVES OF THE RESEARCH WORK

As humans, our sense of smell is very important, and we rely on it for various tasks and functions some of which are daily activities and others which can be more important. Despite the importance of this sense, our sense of smell is usually limited both in its capabilities and can be influenced by external factors such as flu, our surroundings and other factors. Our human sense of smell can also only detect a limited number of gases due to which must be facilitated by adding compounds to different gases for humans to be able to detect it (Lee, et al., 2012).

These limitations of human olfactory system make it difficult to rely on humans for the job of odor detection in Industries. Moreover, the odor detection of dangerous gases, even though possible by humans, may be fatal. An alternate approach is to train and utilize dog for odor sensing. This too has limitations as it is expensive to train dogs, and their life span is short and limited. These limitations have led to the development of electronic noses which try to mimic the human olfactory system. Electronic noses have proved their significance in various fields of health and industries and have been used as sensors for detection of food spoilage and in diagnosis of various diseases and much more (Gurney, 1997). Despite advances in the hardware of electronic noses, there hasn't been much attention paid to the software side of electronic noses.

This research aims to develop a general purpose Artificial Neural Network that can be used in various kinds of application from differentiating between markers, detecting food spoilage and diagnosing diseases.

As discussed in the previous section, the research and development on the software side of Odor sensing systems has been relatively less compared to its hardware. Moreover, these developments have been for very specific purposes and uses.

This chapter aims on developing an Odor sensing system using Electronic nose as a hardware and Artificial Neural Network as software for general odor sensing and identifying different odors as shown in Figure 1.

ARTIFICIAL INTELLIGENCE TECHNIQUES

Various Artificial Intelligence techniques have been developed and are used in different fields. For this research, the researchers analyzed three most relevant techniques which are as follows:

- **Artificial Neural Networks:** An ANN is made up of nodes in different layers. The layers are input layer, intermediate hidden layer(s) and the output layer as shown in figure 2. The links between nodes of adjacent layers have weights. The goal of training is to assign correct weights for these edges. These weights determine the output vector for a given input vector (Wilson & Baietto, 2009).

At the start, all the edge weights are randomly assigned. For every input in the training dataset, the ANN is executed. The output obtained is compared with the intended output that is already known, and if there is an error then the error is propagated back to the previous layer. The weights are adjusted according to the error. This process is repeated until the output error is equal to or below a predetermined threshold.

Once the algorithm terminates, we have a trained ANN. We consider that the ANN is ready to work with new inputs. This ANN can then be trained further by providing more input data and therefore making the results more accurate.

Figure 1. The odor sensing system

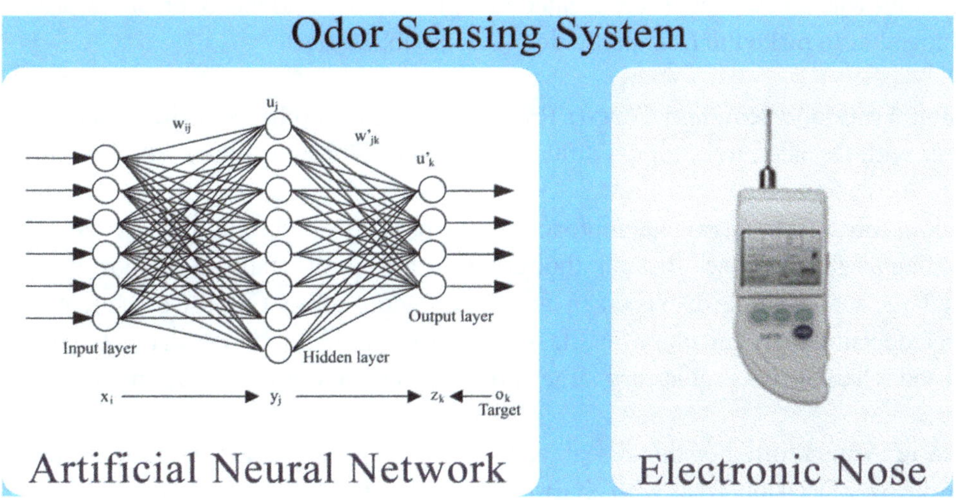

Figure 2. Architecture of an Artificial Neural Network

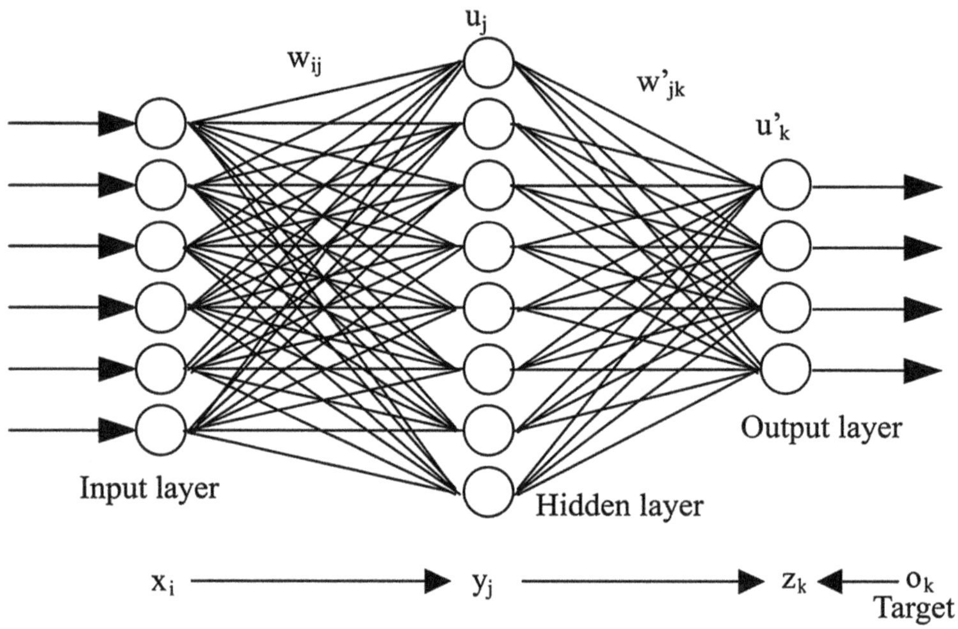

- **Decision Trees and XGBoost:** Classification or regression models in the form of a tree structure are built using a decision tree. It breaks down a dataset into smaller and smaller subsets while incrementally developing an associated decision tree at the same time. The result is a tree with decision nodes and leaf nodes. A decision node has two or more branches. Leaf node represents a classification or decision. The uppermost decision node in a tree which corresponds to the best predictor is called root node. Both categorical and numerical data can be handled by decision trees.

XGBoost stands for eXtreme Gradient Boosting. It is an implementation of gradient boosted decision trees designed for performance and speed. Its architecture is shown in Figure 3. It uses the technique of gradient boosting where new models are created that predict the residuals or errors of prior models and then added together to make the final prediction (Friedman, 2001).

- **Decision Trees and Random Forest:** Random Forest is another algorithm that is implemented together with Decision trees for classification problems, its architecture is shown in Figure 4.

The random forest is an aggregate approach that is used by combining many decorrelated random trees. It is an aggregate approach that can thought of as a form of nearest neighbor predictor. For this algorithm, a large set of sample data is taken, and then random subset of these data are generated. From these subsets, decision trees and classifier classes are created to classify the data. Random forest algorithms are used when the data set is very large and there is a possibility of missing data existing.

- **Choice of Algorithm:** For this research, the algorithm the authors choose is Artificial Neural Networks over Decision Trees with XGBoost and Decision Trees with Random Forest. The reason

Figure 3. Architecture of decision trees algorithm

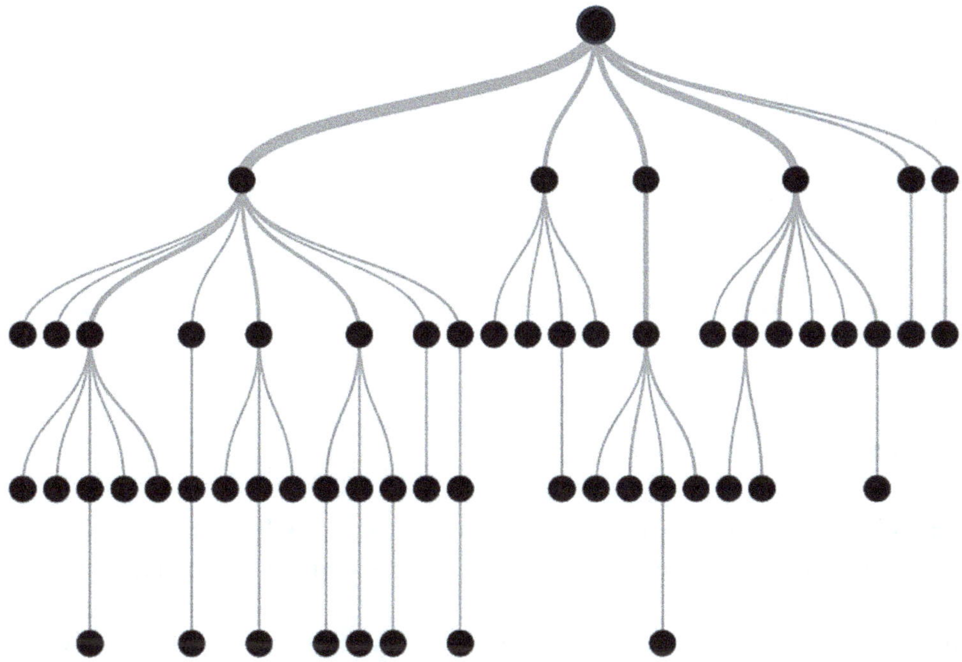

Figure 4. Architecture of decision trees and random forest algorithm

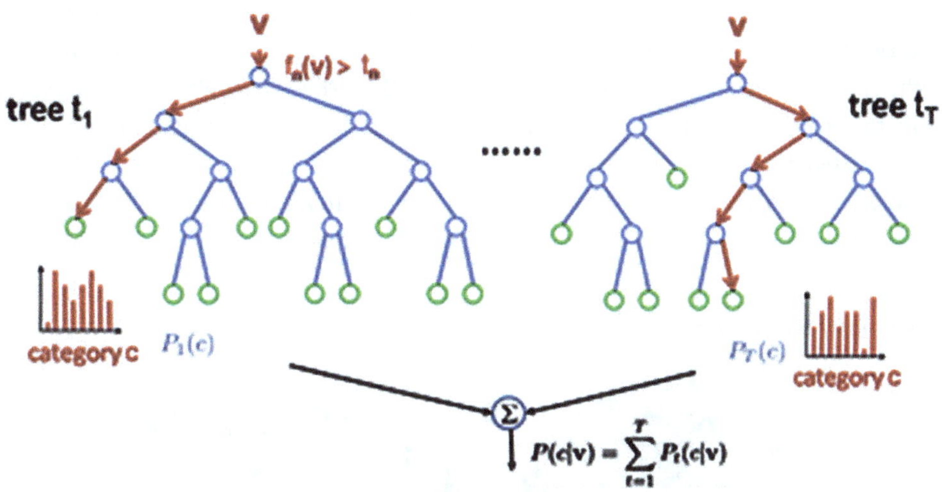

behind this are as follows. Firstly, Artificial Neural Networks (ANN) are more flexible in terms of their outputs, and can be used for Single or multiple outputs. Moreover, ANN provide the feature of 'Back propagation' due to which it requires minimal supervision and only needed to be provided with input data for it to train itself making it much simpler to train for large number of data and more flexible for various types of applications due to which giving the ability to use the same network in different fields. Finally, with more training, the 'weights' of the nodes of an ANN are adjusted automatically whenever trained to give more precise results, this feature can lead to simplifying our application, so it could be used by less tech-savvy people hence increasing its utility.

SCOPE OF THE WORK

The scope of the research, as shown in Figure 5, is to create a general odor sensing system able to differentiate between sampled odors using OMX-GR sensor as the hardware and Artificial Neural Network (ANN) as software.

The working of the hardware and the implementation details of the software are described in coming sections of this chapter.

RELATED WORKS

Many researches have been done on the electronic noses and their applications in various fields such as diagnosis, food industries, and as sensors. Some of these reviewed works are as follows.

- **Di Natale, and et al.** (2003) researched on the diagnosis of lung cancer using an electronic nose. The diagnosis system used a LibraNose electronic nose. The data collection was done in the premises of the hospital and samples of patients having diagnosed with lung cancer and those that had its symptoms were taken. The analysis of this data was done through partial least squares discriminant analysis. A result of 94% accuracy was obtained by the research in diagnosis of lung cancer.
- **Thaler, and et. al.** (2006) used a smell sensor for the diagnosis of sinnusitis. The smell sensor used by the research was a Cyranose 320. Patients with known sinnusitis were taken and samples were collected from them. Five samples each were collected from the patients with a rest-time of five minutes between each sample for the sensor to recover. For the analysis of this data, Support Vector Machine (SVM) algorithm was used. This resulted in an accuracy of 72% due to the limitations of the smell sensor.

Figure 5. The scope of the research

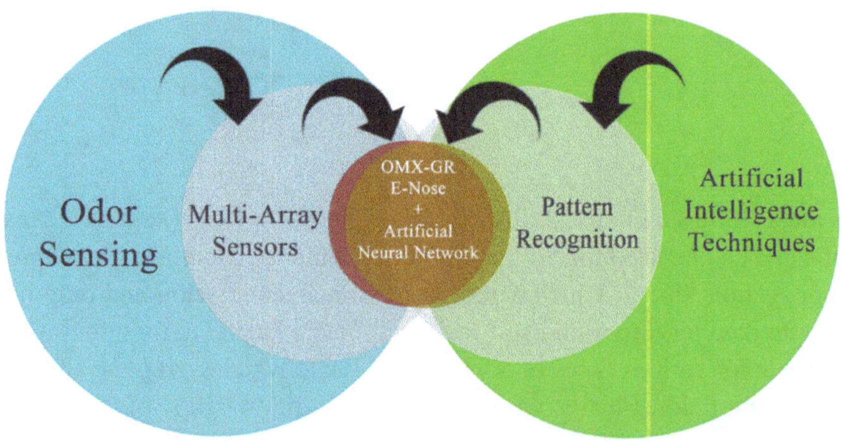

- **Mohamed, and et. al.** (2002) used a smell sensor for the diagnosis of diabetes. The researchers took urine samples from patients with known diabetes and healthy patients. These urine samples were then processed through a smell sensor and fed into an Artificial Neural Network, and also analyzed using Logistic regression and Principal Component analysis. This led to an accuracy of 92% when an Artificial Neural Network and Logistic regression was used and 88% when using Principal Component analysis.
- **Ampeuro, and et. al.** (2003) tested smell sensors in the applications of classifying dairy products, specifically, different kinds of cheese and milk. The research focused on different between different kinds of cheese, detecting molds of bacteria on cheese, classifying bacteria cultures in milk, aging of dairy products and detection of cheese that is gone bad. Various sensors were used to gather the data and the patterns were then analyzed to see results using Artificial Neural Networks. The results were mixed with the sensors unable to detect classifications such as defective cheese, different types of cheese and performing better on other classifications such as determining the age of a diary product.
- **Wilson, and et. al.** (2009) researched on various applications of electronic noses in different fields and industries. Per the research, there is intensive use of electronic nose sensors to determine the ripeness of fruits and vegetables, which has an accuracy of between 86% and 100%. Electronic noses have also been used to detect bacterial growth in dairy products and their shelf-lives and have done so accurately. Some electronic noses also have been specifically developed for detecting spoilages in different types of meats and their freshness having very precise results. The sweet/sourness of specific meat such as fishes could also be determined by using electronic noses for the detection of amount of lactic acid. The research also focuses on use of electronic noses in medicine and health where they have been used to detect traces of blood in urine samples and had an accuracy of 86.6% in doing so.
- **Pearce, and et. al.** (2006) in their research designed an electronic nose system that can be used to determine the quality of indoor-air. The system is composed of a hardware Metal Oxide Gas sensor and a fuzzy logic pattern recognition algorithm to determine the data from the sensor. These devices were low power and low cost and yielded high accuracy when tested for the detection of Nitrogen Oxide and Carbon dioxide gases in the indoor-air.
- **Luo, and et. al.** (2004) used a Cyranose 320 smell sensor with artificial neural network to detect the difference between different brands of cigarette. Four types of cigarettes from different brands were taken and data was gathered through the electronic nose. Three different artificial networks were trained on this data. These networks had similar characteristics except the input size, which varied between 32 or 35 inputs. After analysis, the network with the most inputs had better results with an accuracy varying between 80-100% for the given brands.
- **Gardener, and et. al.** (2000) in their research used a modified FOX 2000 smell sensor and CellFacts liquid analysis instrument to gather data to detect the presence of various bacteria such as Legionalla pneumphilia, Staphylococcus and Escherichia coli (E. coli). They gathered this data in strict lab conditions using various lab equipment to purify the sample before feeding it into the sensor. They analyzed the patterns of the collected data after gathering it and achieved 80.6% accuracy in detecting the presence of these bacteria. They also suggested that this result could've been improved on by using an artificial neural network.

- **D'Amico, and et. al.** (2010) used a custom made electronic nose to use for the detection of lung cancer. They took various breath samples from both healthy and cancer inflicted patients. Their samples mainly consisted of the last part of breath before a person ran out of breath to ensure that the air being collected was from the deeper part of the lungs. Moreover, the participants were required to fast for 9 hours before the samples were taken. The analysis of the data collected led a precision of 85.7% in detecting of lung cancer.
- **Hanson, and et. al.** (2005) developed a smell sensor with an array of 32 sensors for the detection of severity of pneumonia in Ventilator-associated pneumonia patients. They gathered samples from various patients in hospitals directly from their ventilators to ensure external factors did not affect the samples. The samples then were analyzed using Statistical analysis and each was given a 'Pneumonia score' based on the stage and the type of the bacteria present. The results were only partially convincing due to the limitations of the smell sensor used.

Summary of Review

Reviewing the literature in the field of electronic nose and its use in odor sensing shows that various smell sensors have been used for the different purposes alongside different methods of analysis which were mainly statistical in nature and not artificially intelligent. Moreover, the trend of using custom smell sensors is also seen, with many being developed for very specific purposes to detect specific odors only. An overview of this is given in the Table 1.

Table 1. An overview of different technology used in reviewed literature

Research Paper	Smell Sensor Used	Method/Algorithm for Odor Sensing	Accuracy
Lung cancer identification by the analysis of breath by means of an array of non-selective gas sensors. (Di Natale, et al., 2003)	LibraNose	Partial Least Squares Discriminant Analysis	94%
Use of an electronic nose to diagnose bacterial sinusitis. (Thaler & Hanson, 2006)	Cyranose 320	Support Vector Machine	76%
Predicting Type 2 diabetes using an electronic nose-based artificial neural network analysis. (Mohamed, et al., 2002)	Custom Smell Sensor	Artificial Neural Network	92%
Detecting quality of indoor-air using electronic nose. (Pearce, Schiffman, Nagle, & Gardner, 2006)	Custom Smell Sensor	Fuzzy Logic Pattern Recognition	90%-100%
Application of ANN with extracted parameters from an electronic nose in cigarette brand identification. (Luo, Hosseini, & Stewart, 2004)	Cyranose 320	Artificial Neural Network	80-100%
An electronic nose system to diagnose illness. (Gardner, Shin, & Hines, 2000)	Fox 2000	Statistical Analysis	80.6%
An investigation on electronic nose diagnosis of lung cancer. Lung Cancer. (D'Amico, et al., 2010)	Custom Smell Sensor	Partial Least Squares Discriminant Analysis	85.7%
Electronic nose prediction of a clinical pneumonia score: biosensors and microbes. (Hanson & Thaler, 2005)	Custom Smell Sensor	Statistical Analysis	<50%

Reviewing the aforementioned researches, this research uses a general-purpose sensor in form of OMX-GR and Artificial Neural Network for analysis due to its ability of being trained and being general purpose. The details of the working of the sensor and those of the Artificial Neural Network are given in coming sections.

BACKGROUND THEORY

In the previous section, related works was discussed. In this chapter, the functioning of electronic noses (e-noses), and an algorithm developed to mimic the biological neural network structure of the human brain known as artificial neural network will be discussed (Zhang, Patuwo, & Y. Hu, 1998).

Electronic-Nose

Electronic-nose or e-nose is a device intended to detect and classify odor. It's engineered to emulate the human olfactory system (Zhang, Patuwo, & Y. Hu, 1998). Interest among the scientists in the field of electronic-nose was recorded in the early 20th century when scientist Alexander Graham Bell famously said that "if you are ambitious to found a new science, measure a smell." (Bell, 1914). Nonetheless, developments in the field took place approximately 50 years later after Bell had put forth the idea of a new science. During the recent years, a lot of discoveries have been made in the field, yet there is a lot more to be learnt and explored.

- **Brief History of Electronic-Nose:** Electronic-Nose is a relatively new field, whose earliest developments can be dated back to 1961 when Moncrieff (1961) developed a mechanical nose to detect odors. Wilkens and Hatman (1964) built on the work of Moncrieff to create the first reported electronic nose in 1964. However, it was the work of Persaud and Dodd that introduced electronic nose as a system of chemical array sensor system in 1982 (Persaud & Dodd, 1982) and (Gardner & Barlett, 1994). It was also during the 1980's that the term 'electronic nose' started to appear, its specific usage can be dated to 1987 when it was used in the 8th International Congress of European Chemoreception Research Organization at the University of Warwick (Gardner J., 1987), (Lang, et al., 2007) and (Li, 2009).
- **The Term Electronic-Nose:** With the evolution of electronic-nose field, the term used for the electronic-nose device has also evolved. Initially, mechanical nose was used as the instrument was mechanical and currently, electronic-nose, artificial-nose and odor-sensing system are being interchangeably used for the same device (Al-Bastaki, 2009), (Moncrieff, 1961), (Gardner & Barlett, 1994), (Lang, et al., 2007) and (Li, 2009). Gardner and Barlett (Gardner & Barlett, 1994) defined electronic nose as an instrument, that is made up of an array of electronic chemical sensors having partial specificity and an appropriate pattern-recognition system which can recognize simple or complex odors. An array of electronic chemical sensors having partial specificity is a vital part of the definition as it implies that devices with only one sensor or the ability to identify only one compound or odor are not electronic-noses (Wilson & Baietto, 2009) and (Gardner & Barlett, 1994). Furthermore, Gardner and Barlett (1994) list the following components that are depicted in figure 6, as the basics of an electronic-nose:

- An odor delivery system that transfers the molecules of the source material to the sensor array system.
- A chamber where the sensors are located, usually having a fixed temperature and humidity, so that the absorption of the molecules is not affected.
- A transducer that converts chemical input to an electrical signal.
- A converter that converts electrical to digital signal.
- A pattern recognition system that takes the digital signal and predicts the odor.

The first four components generally form the smell sensor (Li, 2009) and (Kalita, Saikia, & Singh, 2015). Many smell sensors have already been developed and are going under constant improvements (Li, 2009). The last component can be either a microprocessor with an embedded algorithm or an application for pattern recognition. (Tang, et al., 2010) and (Wilson & Baietto, 2009). The smell sensor used in this research and the pattern recognition system are discussed in the following sections.

OMX-GR Handheld Odor Meter

An OMX-GR smell sensor as shown in the Figure 7 was used for odor sensing. It uses two semiconductor gas sensors to detect odors (Shinyei, n.a.). Strength and classification (ID) of the odor that are calculated using an original method of Shinyei technology are displayed on the screen through its LCD display (Shinyei, n.a.). There is no correlation between the indicated value of OMX-GR and the value of human sense of odor intensity. The ID is an integer with a range of 0 to 90 and the strength is also an integer with a range of 0 to 999 (Shinyei, n.a.). It has a different response rate to different compounds and liquids (Shinyei, n.a.) as follows:

- Ethanol 100ppm in 20 seconds.
- Methyl Mercaptan 0.16ppm in 50 seconds.
- Xylene 46ppm in 50 seconds.

A RS-232 cable was used to connect and read the strength and ID of odors from the sensor through the custom-made application.

Figure 6. Components of an electronic-nose

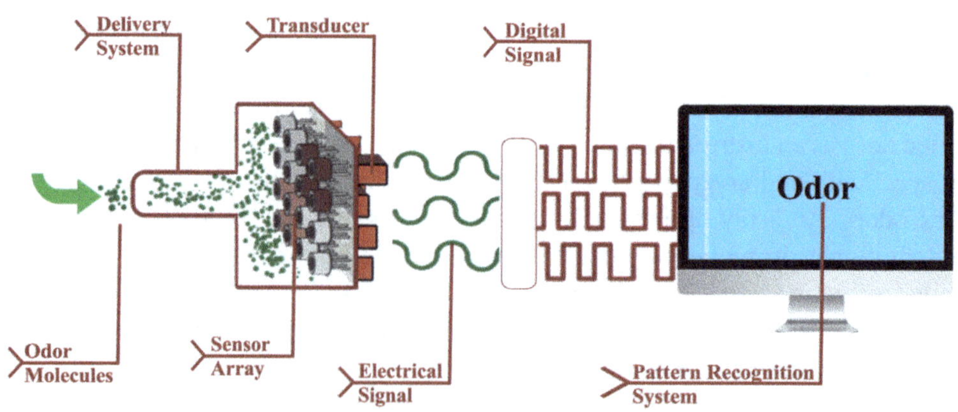

Figure 7. OMX-GR handheld odor meter

Pattern Recognition System

The final component of the electronic-nose from its definition is its Pattern Recognition System. The pattern recognition system uses computer algorithms to find regularities and key features in the output of the smell sensor to identify odor and to classify it into different categories (Bishop, 2006). For odor classification various algorithms have been used which include principal component analysis, cluster analysis, k-nearest neighbor, genetic algorithms, discriminant function analysis, canonical discriminant analysis, statistical pattern recognition, radial basis function and artificial neural networks (Nagle, Gutierrez-Osuna, & Schiffman, 1998), (Schaller, Bosset, & Escher, 1998), (Gardner & Bartlett, 1992), (Penza, Cassano, Tortorella, & Zaccaria, 2001), (Bicego, Tessari, Tecchiolli, & Bettinelli, 2002) and (Kim, et al., 2012). Among these the researchers will be using Artificial Neural Network (ANN) for the advantages mentioned in section Artificial Intelligence Techniques. A summary of ANN was given in Artificial Intelligence Techniques section, in the subsequent section ANN is discussed in detail.

Artificial Neural Network

Artificial neural network also known as neural network, has been an area of great interest ever since the discovery that the human brain works in a completely different way as compared to the conventional digital computer (Haykin, 2009). Moreover, the human brain does recognition computation much faster than the fastest computers in existence today (Haykin, 2009). On average, it can detect a familiar face among unfamiliar faces 10 times faster than a computer (Nagarajan & Stevens, 2008). Advantages like these have led to the efforts of simulating the brain electronically, thus it's important to understand the basic functioning of the human brain as described in the following section.

The Human Neural Network

The average human brain is composed of approximately 100 billion ($1x10^{11}$) tiny nerve cells called neurons (Herculano-Houzel, 2009), (Bartheld, Bahney, & Herculano-Houzel, 2016) and (Spitzer, 2006). These neurons have about 10 connections with each of the 1000 neurons that it is connected to, thus having 10, 000 connections individually and one quadrillion ($1x10^{15}$) connections collectively (Spitzer, 2006). The communication through the connections takes place through electrochemical signals (Ma, Forbes, Llano, Berger-Wolf, & Kenyon, 2016). Initially, the cell body of the neuron called a soma shown in Figure 8 collects and combines incoming information from other neurons, this information is passed as electrical signals to the axon hillock, which decides whether to fire an action potential or not based on a certain threshold. If an action potential is fired, then electrical signals are passed through the axon to the terminal endings which contain synapses. Synapses convert the incoming electrical signals to chemical signals. They then send signals to a dendrite of another neuron, these synaptic signals maybe inhibitory or excitatory i.e., they will make the next neuron less or more likely to fire their action potential (Madan & Dandina, 1994) and (Wan, 1993). It's due to the various sequences of the firing these neurons that humans can see, move, hear, smell, taste, touch, reason, think, dream, remember, plan and do all the things that we do through our brain (Solomon D, 2015). To make use of the neurons incredible abilities these neurons are electronically imitated as explained in artificial neural network model section.

ARTIFICIAL NEURAL NETWORK MODEL

After covering the human neural network in the previous section, in this section the various components of an artificial neural network will be covered. Starting off with the artificial equivalent of the human neurons that are represented as nodes or units as shown in figure 8 (Gurney, 1997).

Figure 8. Schematic diagram of a neuron and an ANN node

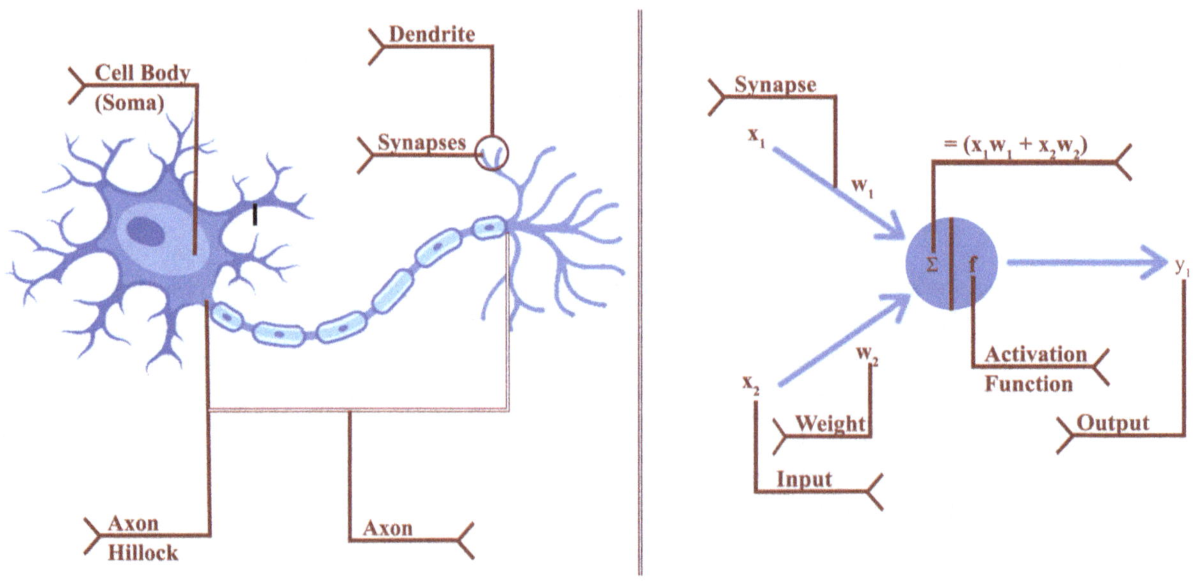

ANN Nodes

Like the neurons, the nodes also have synapses. However, the synapse of nodes functions a bit differently, each synapse is assigned a weight to it and the synapse takes the input from the nodes in the previous layer, and multiplies it with its weight. The product of all inputs with their respective weight is then summed together and passed to an activation function that works like the axon hillock of a human neuron and it is applied to obtain the output of the node.

Types of Networks

The inter-linking of the nodes through input, hidden and output layer results in an artificial neural network as displayed on the left side of Figure 9. Input and output layers can have any number of nodes however, both have only one layer. Hidden layer on the other hand, in addition to any number of nodes can also have any number of layers. When the number of layers in the hidden part of the network is greater than 1, then the network is known as a deep neural network as shown on the right side of Figure 9. Deep neural networks are used when transformations are required between layers i.e., in image processing the first hidden layer may be used to detect the edges of an image and the second layer may be used to extract the object from the obtained edges of the first layer.

Back to neural networks with a single layer, in the next subsection the activation function mentioned in section ANN nodes is explained in detail.

Activation Functions

Activation functions are sometimes called a transfer function or a squashing function. Its role is to limit the range of output (Haykin, 2011) and (Demuth, Beale, De Jess, & Hagan, 2014). Depending upon the chosen activation function the output range of a node is usually in the interval of [0,1] or [-1,1]. In neural networks, various activation functions are used. The choice of activation function comes down to the requirements of the network. Therefore, instead of covering majority of the activation functions, only the sigmoid activation function will be covered. But before covering the sigmoid function, let's first define the notations that will be used as illustrated in figure 10:

- $x_{ij:}$ The input of the jth neuron in the ith layer.
- $w_{ijk:}$ The weight coming from the kth neuron in the $(i-1)^{th}$ layer to the jth neuron in the ith layer.

Figure 9. An Artificial Neural Network and a Deep Neural Network

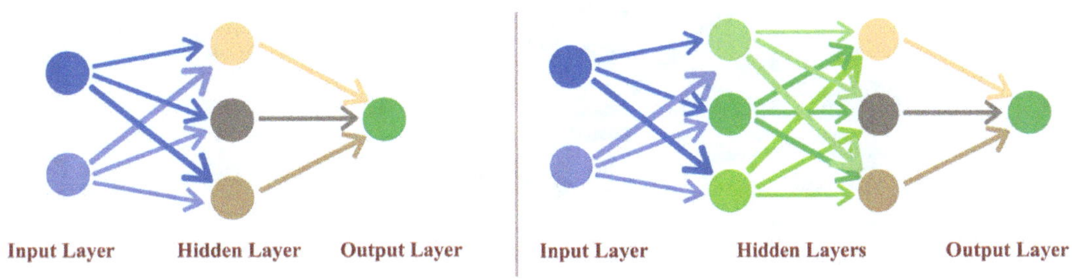

- y_{ij}: The output produced by the activation function for the jth neuron in the ith layer.
- z_{ij}: The activation value before the activation function is applied, which is equal to the sum of the product of all inputs with their respective weights coming into the jth neuron in the ith layer i.e., $\sum_k \left(x_{(i-1)k} * w_{ijk}\right)$.

Sigmoid Activation Function

The sigmoid activation function takes the input and applies the following formula to it:

$$y_{ij} = \frac{1}{1+e^{-(z_{ij})}}$$

These results are shown in Figure 11.

As visible in the graph 11, the sigmoid function takes input between minus and plus infinity and squashes it to a number between 0 and 1.

After covering the sigmoid activation function, in the next section the basics of feed forward algorithm will be covered.

Feed Forward Algorithm

The feed forward algorithm is used to carry the inputs obtained in the input layer to the hidden and then to the output layer to obtain an output from the network. The algorithm starts by assigning random numbers to the weights. It then takes in the inputs and does matrix multiplication of inputs with the weights coming into the 2nd layer. Next sigmoid activation function is applied to each element obtained

Figure 10. Illustration of notations used for sigmoid activation function

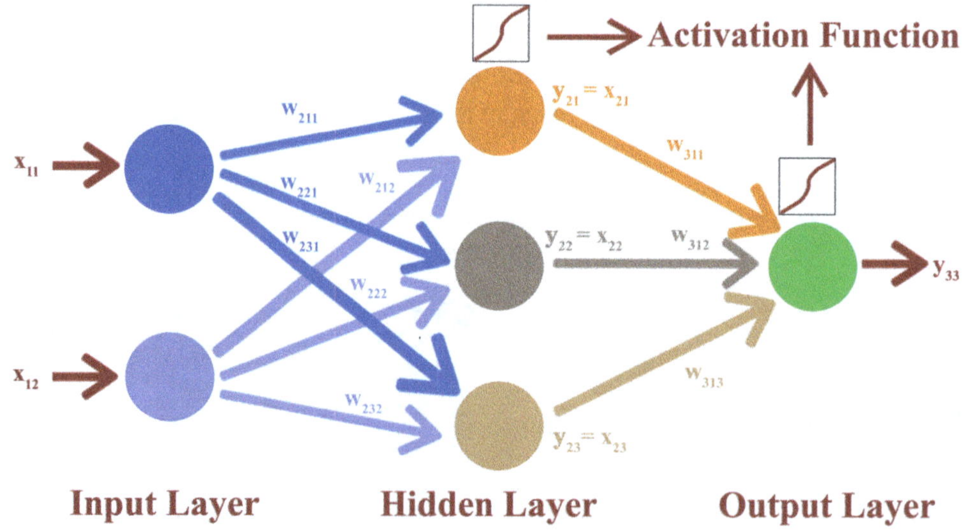

Figure 11. Graph of the Sigmoid Function
Plotted using desmos.com.

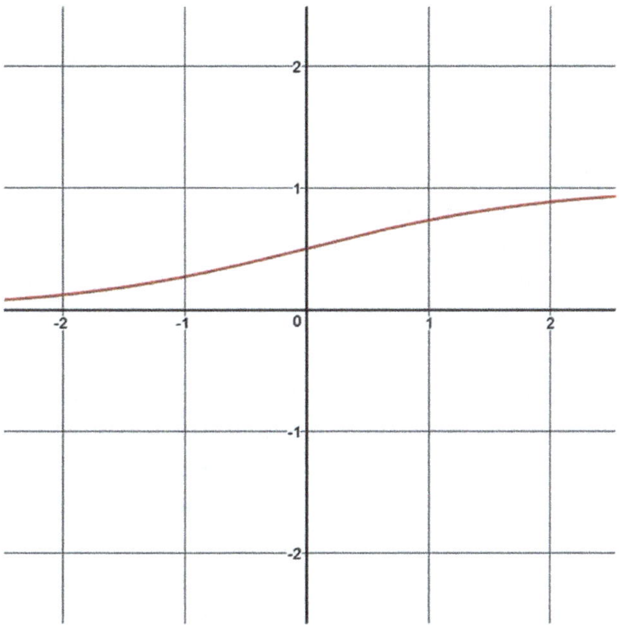

from the matrix multiplication, the new values represent the input to the next layer. This whole process of matrix multiplying the inputs with the weights and applying activation function is repeated till the algorithm reaches the output layer.

With the network now capable of producing an output, the training of the network that is done using the backpropagation algorithm is covered in the next subsection.

Backpropagation Algorithm

The backpropagation algorithm is a method used for training a multilayer neural network. Using the backpropagation algorithm, the weights in the network are modified to produce the expected outputs. First, feed forward algorithm described in previous section is used to produce an output. A cost function like the mean squared error i.e., $E = \sum\left(y - \hat{y}\right)^2$ is then used to calculate how far or close was the produced output ($\hat{y}$) to the expected one (y). The weights are then updated to minimize this cost function and the whole process is repeated till a satisfactory error rate is obtained.

DATA COLLECTION, ANALYSIS, AND RESULTS

After covering the background theory, data collection and its analysis will be discussed in this section. The discussion will start with data collection, in which the issues faced in data collection will be examined and solved. After which the collected data will be analyzed, processed and finally, the results will be discussed.

Data Collection

Odor data was collected using the OMX-GR sensor discussed in OMX-GR handheld odor meter section. However, many issues were faced in the collection of data as the sensor was more than a decade and half old, and couldn't be replaced with a new one due to limited budget of the research.

First, the authors realized that the sensor wasn't having a standard response time to the odors even when the odors were presented to the sensor from a uniform distance. This meant that the response time of odors couldn't be used as an input parameter to the neural network, which could've led to more accurate results. Response time of three odors is mentioned in the manual of the sensor (Shinyei, n.a.), from which it can be concluded that the sensor originally had a standard response time. Nonetheless, as the sensor had gotten old, its response time might have been affected.

Next, when different odors were read in quick succession, the sensor was giving completely different readings to the previously obtained readings of the odors. After researching, the authors came to know that different sensors have different recovery times in addition to response time (Wilson & Baietto, 2009) and (Bochenkov & Sergeev, 2010). However, unlike response time the recovery time wasn't mentioned on the sensor's manual. Therefore, to identify the recovery time, the researchers started by allowing the sensor a recovery time of 5 seconds and checked the consistency of the readings. The consistency of the readings didn't improve, so the recovery time was further increased, this continued till the recovery time was increased to 10 minutes and the readings' consistency hadn't improved, as sometimes the sensor wouldn't respond to an odor even after 10 minutes of recovery time. Therefore, the researchers concluded that the sensor no longer had a standard response or recovery time.

Then, it also came to our knowledge that environmental factors i.e., temperature and humidity affect readings of the sensors. This has also been confirmed in various researches (Wilson & Baietto, 2009), (Bochenkov & Sergeev, 2010) and (Powers, 2004). To reduce the effect of environmental factors pipes of varying sizes were used to expose odors directly to the sensor and restrict the environment's effect on them.

Even after taking all the previous mentioned steps to improve the readings, consistent readings weren't obtained. Therefore, the following purification method was used to obtain consistent readings.

1. Each substance's reading should be taken from a specific distance every time.
2. The readings should be taken through the pipe chambers.
3. The sensor should be given a recovery time of at least 10 minutes after every reading.
4. A reading is considered acceptable only if the reading's ID has a range difference of 3 for 60 seconds and this reading is consistent with previous readings of the same substance.

5 readings of various substances were taken, each for 60 seconds, however due to the sensor's limitation only 5 substances, namely, HiGeen Papaya hand sanitizer, Glue Stick, Obama perfume, Faber-Castell permanent marker and Pentel permanent marker gave acceptable readings for the purification method. The acceptable readings are discussed and analyzed in the following section.

Data Analysis

In data analysis, the readings of the 5 odors will be analyzed. A total of 300 readings were taken for each odor over 5 sessions of 60 readings each. Each reading has an ID as shown in Figure 12 and a strength value as shown in Figure 13.

Figure 12 shows the ID values of the odor readings. At the bottom of the figure the ID values of Obama perfume are represented by round dotted dashes. The ID values of Obama perfume range between 53 and 55. Above it we have the ID values of Glue Stick are shown that range between 57 and 59. In the middle the Pentel marker's ID values are shown that range between 77 and 79. At the top, the overlapping range values of HiGeen hand sanitizer and Faber-Castell marker are displayed. HiGeen's values range between 86 and 88, while Faber-Castell's values range between 87 and 89.

In Figure 12 strengths (the intensity) of the odors are plotted. Glue Stick's strength readings are shown at the bottom of the figure through round dotted dashes. Its readings gave a uniform strength reading that ranged between 16 and 19. Above the Glue Stick, we have the strength readings for Obama perfume that ranged between 70 and 82. The perfume's readings overlap a bit with the strength readings of Pentel marker that ranged between 80 and 143. In the middle, the strength readings of HiGeen that range between 241 and 352 are illustrated. Finally, at the top Faber-Castell marker's strength readings are shown that range between 779 and 999. The first two odors discussed i.e., Glue Stick and Obama perfume display uniform changes in strength. However, the strength range of the remaining odors vary widely especially the strength of Faber-Castell. This could be due to the reason that when these substances are exposed to air, their strength reduces as their odor disperses (Mukhtar, Ullman, Carey, & Lacey, 2004).

Figure 12. Odors IDs

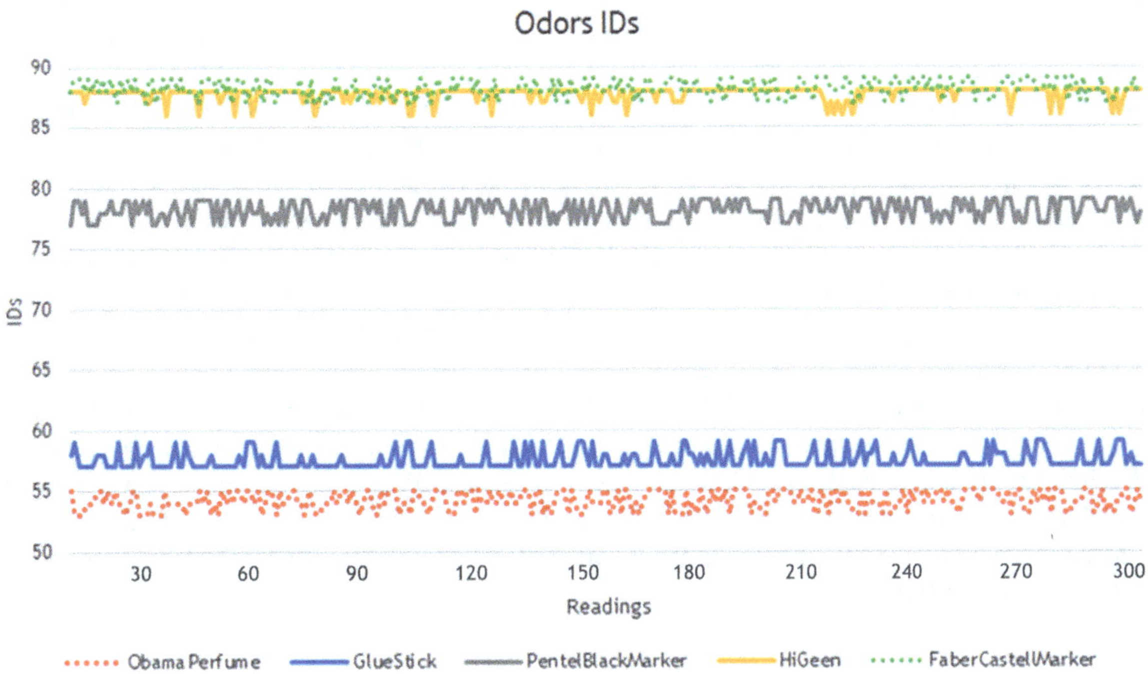

Figure 13. Odors strengths

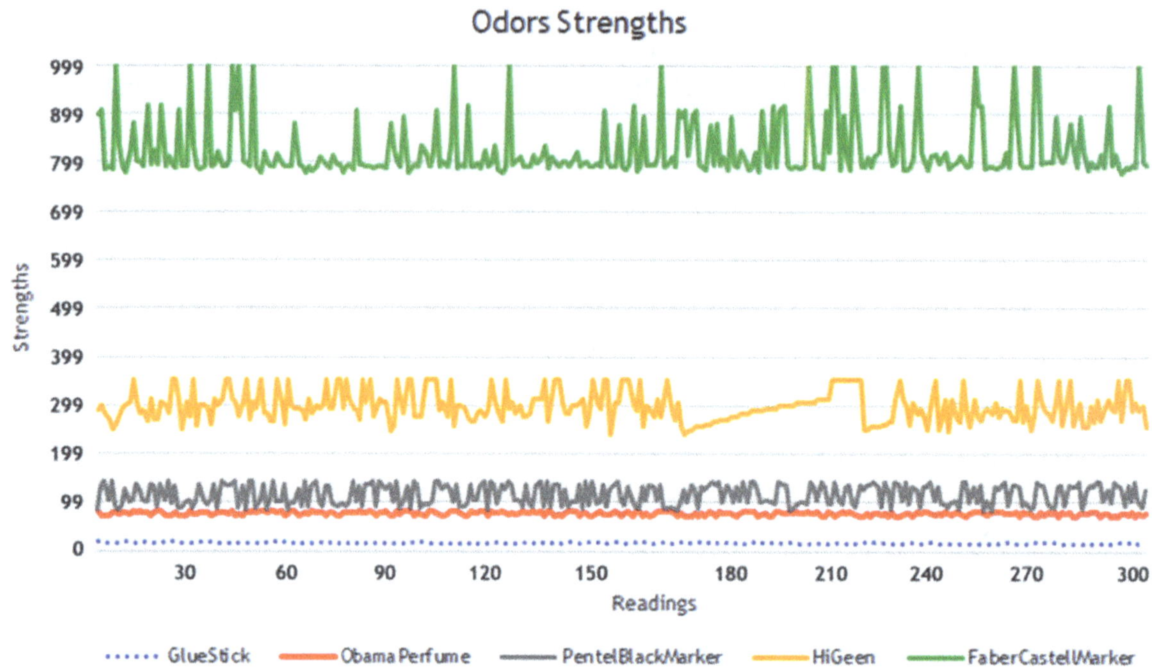

DATA PROCESSING AND RESULTS

For processing the data and identifying the odors using the neural network the readings were divided into a ratio of 20, 70 and 30 for validation, training and testing according to the findings of Data Division for Neural Networks (Shahin, Maier, & Jaksa, 2004). This means that 20% out of the total readings (1500) was selected randomly for validation set. Out of the remaining 1200 readings 70% were selected randomly for training the network and the remaining 360 readings (30%) were used for testing the trained network.

To train the neural network different architectures were used to evaluate which one would be the most appropriate. Training was done till an error rate of 0.1000 was obtained. The details of trainings using different architectures are given in Table 2, where the architecture of the neural network is represented as number of input layer: number of neurons in hidden layer: number of neurons in output layer.

From Table 2 we can conclude that for the data that was presented to the neural network the best architecture was 2:6:1. The training time decreased as the number of neurons in hidden layer was increased till the number of neurons in hidden layer was 7. Number of epochs decreased with the increase of number of neurons in the hidden layer. However, no significant change was seen in the validation rate, yet the 2:6:1 architecture acquired the best validation rate too. A simple 2:2:1 architecture was not suitable as there were overlapping data in both ID and Strength values, and therefore, the neural network took a lot of time to train itself to achieve a max error rate of 0.1000. It is to be noted that irrespective of which architecture is chosen for training, if it achieves the required max error rate while training, the testing and querying accuracy of the neural network will not be significantly affected by the network's architecture (Foody & Arora, 1997).

Table 2. Training results of different architectures

Architecture	Training Time	Epochs Completed	Validation Rate
2:2:1	1h 29m 14s	3,640,445	0.04680
2:3:1	0h 04m 45s	179,849	0.03773
2:4:1	0h 02m 19s	68,189	0.03816
2:5:1	0h 02m 07s	66,257	0.03841
2:6:1	0h 01m 39s	39,876	0.03752
2:7:1	0h 01m 52s	36,593	0.03791

After training the network with the purified data, testing was performed to test the performance of the network. All 360 testing samples were correctly classified by the neural network, which meant that a 100% accuracy rate was obtained as shown in Figure 14 generated by the program.

CONCLUSION AND FINDINGS

With advancements in technology and the never-ending goal of making life easier for humans, it is evident that odor sensing could lead to a better tomorrow. This paper discussed the various applications odor

Figure 14. Testing performance report

Network Testing Performance Report

Network Details

Name: ANNeNose
Type: ActivationNetwork
Layout: 6-1
Function: SigmoidFunction
Description:

Network Schema

Inputs: ID, Strength
Outputs: Output

* Underlined fields indicates a categorical (non numeric) field

Training Details

Training Entries: 840
Validation Entries: 300

Training Deviation: 0.0999998878999852

Testing Summary

Testing Entries: 360
Testing Items: 360

Hits: 360 100.00%
Correct: 360 100.00%
Incorrect: 0 0.00%

Errors: 0 0.00%

sensing could be used in and has been used for. Various researches have been done in this field using various other techniques to help develop this field of research. Most of research has been specifically focused on a single industry or field of application. Our paper focused on and developed a system using Artificial Neural Network with an E-Nose and laid the foundation for a general-purpose system which can be used for various applications and in various industries.

The smell sensor used in this research was an OMX-GR sensor which is a semiconductor based gas sensor. Along with the smell sensor as hardware, Artificial Neural Network was used as software which was built on top of Sinapse framework and the framework itself was heavily modified as well to fit the needs of the project. Due to the sensor being old, and its other various limitations. Many methods had to be tried for data collection. Adjustments had to be made to the distance between sample and the sensor while collecting readings, to its recovery times, and to other external factors to ensure the readings were precise. The data was also cleaned and went through a purification process as mentioned in detail in the previous sections.

After collecting data and its purification, it was fed to the ANN to train on and then the Artificial Neural Network was tested for accuracy. This process was repeated for all the samples that were taken. Backpropagation algorithm was used for the training and adjustment of weights to ensure the results were precise and accurate.

- **Findings:** The system developed in this research could differentiate between various types of odors with an accuracy of between 93-100%. It suggests that Artificial Neural Networks can be very efficiently utilized with smell sensors for the detection of various odors due to the large amount of data involved in the nature of the problem and the capability of ANN to learn and train further as more data is provided and hence increasing its precision.
- **Limitations:** There were various limitations that affected the research, mainly, those related to the smell sensor. The smell sensor used in this research is an OMX-GR sensor which was produced a decade and half ago. Due to this, data collection method had to be adapted per the sensor's limitation. Some of these limitations were related to connectivity issues as the Serial Port of the sensor uses legacy drivers which are rare to find. Therefore, necessary drivers had to be implemented within the system to make the sensor compatible. Furthermore, due to the lack of extensive documentation of the sensor, various experimentation and tweaks had to be made to the data collection methods to ensure the most accurate data was collected. Some of the data collection issues were related to the recovery and response time of the sensor.

Some software related limitations were the limitation of the framework used as well. This led to creation of various new functions to make the system perform better and faster and to make it much easier to use.

- **Future Implementations:** In the future, work could be done to test the system with various other sensors that are modern and precise to get readings of more odors with more parameters i.e., response time, to have a high accuracy with higher number of odors. Moreover, the ability to create a cloud-based datahub on which various researches could contribute with their data and samples to train the ANN could as well lead to further developments in the field of odor sensing. In terms of the developed software, it can be improved by combining the ANN with a Genetic algorithm to overcome getting stuck in local minima while training.

REFERENCES

Al-Bastaki, Y. (2009). An Artificial Neural Networks-Based on-Line Monitoring Odor Sensing System. *Journal of Computational Science*, *11*, 878–882.

Ampuero, S., & Bosset, J. O. (2003). The electronic nose applied to dairy products: A review. *Sensors and Actuators. B, Chemical*, *94*(1), 1–12. doi:10.1016/S0925-4005(03)00321-6

Bartheld, C., Bahney, J., & Herculano-Houzel, S. (2016). The search for true numbers of neurons and glial cells in the human brain: A review of 150 years of cell counting. *The Journal of Comparative Neurology*, *524*(18), 3865–3895. doi:10.1002/cne.24040 PMID:27187682

Bell, A. (1914). *Discovery and Invention*. Washington, DC: National Geographic Society.

Bicego, M., Tessari, G., Tecchiolli, G., & Bettinelli, M. (2002). A comparative analysis of basic pattern recognition techniques for the development of small size electronic nose. *Sensors and Actuators. B, Chemical*, *85*(1), 137–144. doi:10.1016/S0925-4005(02)00065-5

Bishop, C. (2006). *Pattern recognition and Machine Learning*. Springer.

Bochenkov, V., & Sergeev, G. (2010). Sensitivity, selectivity, and stability of gas-sensitive metal-oxide nanostructures. *Metal Oxide Nanostructures and Their Applications, 3*, 31-52.

D'Amico, A., Pennazza, G., Santonico, M., Martinelli, E., Roscioni, C., Galluccio, G., & Di Natale, C. (2010). An investigation on electronic nose diagnosis of lung cancer. *Lung Cancer (Amsterdam, Netherlands)*, *68*(2), 170–176. doi:10.1016/j.lungcan.2009.11.003 PMID:19959252

Demuth, H. B., Beale, M. H., De Jess, O., & Hagan, M. T. (2014). *Neural network design*. Martin Hagan.

Di Natale, C., Macagnano, A., Martinelli, E., Paolesse, R., D'Arcangelo, G., Roscioni, C., & D'Amico, A. (2003). Lung cancer identification by the analysis of breath by means of an array of non-selective gas sensors. *Biosensors & Bioelectronics*, *18*(10), 1209–1218. doi:10.1016/S0956-5663(03)00086-1 PMID:12835038

Foody, G., & Arora, M. (1997). An evaluation of some factors affecting the accuracy of classification by an artificial neural network. *International Journal of Remote Sensing*, *18*(4), 799–810. doi:10.1080/014311697218764

Friedman, J. (2001). Greedy function approximation: A gradient boosting machine. *Annals of Statistics*, *29*(5), 1189–1232. doi:10.1214/aos/1013203451

Gardner, J. (1987). Pattern recognition in the Warwick Electronic Nose. *8th International Congress of European Chemoreception Research Organization*.

Gardner, J., & Barlett, P. (1994). A Brief History of Electronic Noses. *Sensors and Actuators. B, Chemical*, *18*(1-3), 211–220. doi:10.1016/0925-4005(94)87085-3

Gardner, J., & Bartlett, P. (1992). Pattern recognition in odour sensing. *Sensors and sensory systems for an electronic nose*, 161-179.

Gardner, J. W., Shin, H. W., & Hines, E. L. (2000). An electronic nose system to diagnose illness. *Sensors and Actuators. B, Chemical, 70*(1), 19–24. doi:10.1016/S0925-4005(00)00548-7

Gurney, K. (1997). *An introduction to neural networks*. CRC Press. doi:10.4324/9780203451519

Hanson, C. W., & Thaler, E. R. (2005). Electronic nose prediction of a clinical pneumonia score: Biosensors and microbes. *The Journal of the American Society of Anesthesiologists, 102*(1), 63–68. PMID:15618788

Haykin, S. (2009). *Neural networks and learning machines*. Upper Saddle River, NJ: Pearson.

Haykin, S. (2011). *Neural Networks and Learning Machines* (3rd ed.). Pearson.

Herculano-Houzel, S. (2009). The human brain in numbers: A linearly scaled-up primate brain. *Frontiers in Human Neuroscience, 3*(31). PMID:19915731

Kalita, P., Saikia, M., & Singh, N. (2015). Electronic-Nose Technology and its Application -A Systematic Survey. *International Journal of Innovative Research in Electrical Electronics Instrumentation and Control Engineering, 3*(1), 123–128. doi:10.17148/IJIREEICE.2015.3126

Kim, E., Lee, S., Kim, J. H., Kim, C., Byun, Y. T., Kim, H. S., & Lee, T. (2012). Pattern recognition for selective odor detection with gas sensor arrays. *Sensors (Basel), 12*(12), 16262–16273. doi:10.3390/s121216262 PMID:23443378

Lang, H. P., Braun, T., Schmid, D., Hunziker, P., Jung, C., Hegner, M., & Gerber, C. (2007). An Artificial Nose Based on Microcantilever Array Sensors. *Journal of Physics: Conference Series, 61*, 663–667. doi:10.1088/1742-6596/61/1/133

Lee, S. H., Kwon, O. S., Song, H. S., Park, S. J., Sung, J. H., Jang, J., & Park, T. H. (2012). Mimicking the human smell sensing mechanism with an artificial nose platform. *Biomaterials, 33*(6), 1722–1729. doi:10.1016/j.biomaterials.2011.11.044 PMID:22153868

Li, S. (2009). Overview of Odor Detection Instrumentation and the Potential for Human Odor Detection in Air Matrices. *MITRE Innovation Program and US Government Nano-enabled Technology Initiative*, 5-43.

Luo, D., Hosseini, H. G., & Stewart, J. R. (2004). Application of ANN with extracted parameters from an electronic nose in cigarette brand identification. *Sensors and Actuators. B, Chemical, 99*(2), 253–257. doi:10.1016/j.snb.2003.11.022

Ma, C., Forbes, A., Llano, D., Berger-Wolf, T., & Kenyon, R. (2016). Swordplots: Exploring neuron behavior within dynamic communities of brain networks. *Electronic Imaging, 60*(1), 1–13.

Madan, M., & Dandina, H. (1994). *Neuro-Control Systems: Theory and Applications*. IEEE Press.

Mohamed, E. I., Linder, R., Perriello, G., Di Daniele, N., Pöppl, S. J., & De Lorenzo, A. (2002). Predicting Type 2 diabetes using an electronic nose-based artificial neural network analysis. *Diabetes, Nutrition & Metabolism, 15*(4), 215–221. PMID:12416658

Moncrieff, R. (1961). An instrument for measuring and classifying odours. *Journal of Applied Physiology, 16*, 742–749. PMID:13771984

Mukhtar, S., Ullman, J., Carey, J., & Lacey, R. (2004). A review of literature concerning odors, ammonia, and dust from broiler production facilities: 3. Land application, processing, and storage of broiler litter. *Journal of Applied Poultry Research*, *13*(3), 514–520. doi:10.1093/japr/13.3.514

Nagarajan, N., & Stevens, C. (2008). How does the speed of thought compare for brains and digital computers? *Current Biology*, *18*(17), 756–758. doi:10.1016/j.cub.2008.06.043 PMID:18786380

Nagle, H. T., Gutierrez-Osuna, R., & Schiffman, S. S. (1998). The How and Why of Electronic Noses. *IEEE Spectrum*, *35*(9), 22–24. doi:10.1109/6.715180

Pearce, T. C., Schiffman, S. S., Nagle, H. T., & Gardner, J. W. (2006). *Handbook of machine olfaction: electronic nose technology*. John Wiley & Sons.

Penza, M., Cassano, G., Tortorella, F., & Zaccaria, G. (2001). Classification of food, beverages and perfumes by WO 3 thin-film sensors array and pattern recognition techniques. *Sensors and Actuators. B, Chemical*, *73*(1), 76–87. doi:10.1016/S0925-4005(00)00687-0

Persaud, K., & Dodd, G. (1982). Analysis of discrimination mechanisms in the mammalian olfactory system using a model nose. *Nature*, *299*(5881), 352–355. doi:10.1038/299352a0 PMID:7110356

Powers, W. (2004). *The science of smell, part 3: odor detection and measurement*. Iowa State University Extension PM.

Schaller, E., Bosset, J. O., & Escher, F. (1998). 'Electronic Noses' and Their Application to Food. *Lebensmittel-Wissenschaft + Technologie*, *31*(4), 305–316. doi:10.1006/fstl.1998.0376

Shahin, M., Maier, H., & Jaksa, M. (2004). Data division for developing neural networks applied to geotechnical engineering. *Journal of Computing in Civil Engineering*, *18*(2), 105–114. doi:10.1061/(ASCE)0887-3801(2004)18:2(105)

Shinyei. (n.d.). *Handheld Odor Meter OMX-GR Operation Manual*. Kyomachi: Shinyei.

Solomon, D. A. (2015). Neuron the Memory Unit of the Brain. *IOSR Journal of Computer Engineering*, *17*(14), 48–61.

Spitzer, M. (2006). Brain research and learning over the life cycle. In Schooling for tomorrow: Personalising education (pp. 47-62). Academic Press. doi:10.1787/9789264036604-4-en

Tang, K. T., Chiu, S. W., Pan, C. H., Hsieh, H. Y., Liang, Y. S., & Liu, S. C. (2010). Development of a Portable Electronic Nose System for the Detection and Classification of Fruity Odors. *Sensors (Basel)*, *10*(10), 9179–9193. doi:10.3390/s101009179 PMID:22163403

Thaler, E. R., & Hanson, C. W. (2006). Use of an electronic nose to diagnose bacterial sinusitis. *American Journal of Rhinology*, *20*(2), 170–172. PMID:16686381

Wan, E. (1993). *Finite impulse response neural networks with applications in time series prediction* (Doctoral dissertation). Stanford University, Stanford, CA.

Wilkens, W., & Hatman, A. (1964). An electronic analog for olfactory processes. *Journal of Food Science*, *29*(3), 372–378. doi:10.1111/j.1365-2621.1964.tb01746.x PMID:14220555

Wilson, A., & Baietto, M. (2009). Applications and Advances in Electronic-Nose technologies. *Sensors (Basel)*, *9*(7), 5099–5148. doi:10.3390/s90705099 PMID:22346690

Zhang, G., Patuwo, E., & Hu, Y., M. (1998). Forecasting with artificial neural networks: The state of the art. *International Journal of Forecasting*, *14*, 35–60. doi:10.1016/S0169-2070(97)00044-7

Chapter 2
Artificial Odour Classification System

Nor Idayu Mahat
Universiti Utara Malaysia, Malaysia

Maz Jamilah Masnan
Universiti Malaysia Perlis, Malaysia

Ali Yeon Md Shakaff
Universiti Malaysia Perlis, Malaysia

Ammar Zakaria
Universiti Malaysia Perlis, Malaysia

Muhd Khairulzaman Abdul Kadir
Universiti Kuala Lumpur British Malaysian Institute, Malaysia

ABSTRACT

This chapter overviews the issue of multicollinearity in electronic nose (e-nose) classification and investigates some analytical solutions to deal with the problem. Multicollinearity effect may harm classification analysis from producing good parameters estimate during the construction of the classification rule. The common approach to deal with multicollinearity is feature extraction. However, the criterion used in extracting the raw features based on variances may not be appropriate for the ultimate goal of classification accuracy. Alternatively, feature selection method would be advisable as it chooses only valuable features. Two distance-based criteria in determining the right features for classification purposes, Wilk's Lambda and bounded Mahalanobis distance, are applied. Classification with features determined by bounded Mahalanobis distance statistically performs better than Wilk's Lambda. This chapter suggests that classification of e-nose with feature selection is a good choice to limit the cost of experiments and maintain good classification performance.

DOI: 10.4018/978-1-5225-3862-2.ch002

INTRODUCTION

The visibility of electronic nose (e-nose) in wide range of industrial applications such as food quality or authenticity (Yu et al., 2009; Falasconi et al., 2012; Gliszczyńska-Świglo & Chmielewski, 2016), medical diagnosis (Silvano Dragonieri et al., 2007; Yusuf et al., 2013), and indoor air quality (Abu H. et al., 2012) has been acknowledged by both academicians and practitioners. This electronic sensor has the capability of imitating the human sense of smell using sensor arrays and pattern recognition system, where e-nose assists for quick decision making and overcomes some drawbacks of dependency on olfactory experts or trained human panels (Masnan et al.; 2012). Despite of such advancement, the whole technology on e-nose is still emerging including the mathematical algorithms for classifying diligently smell/odour based on a systematic classification procedure. In details, classification procedure is a form of systematic judgment, often in a mathematical function that deals with a process of training a classification rule to familiarise with the behaviour of groups of instances (Mahat et al., 2015). The classification procedure generally is performed on three main steps; (i) construct an appropriate classification rule using a classified data, (ii) evaluate performance of the constructed rule to reflect its ability in classifying instances to their right groups, and finally (iii) use the constructed rule for allocating future instances with unknown group to one of the predetermined groups.

Human olfactory system includes nose and many parts of the brain. The ability to smell comes from specialised sensory cells, which termed as olfactory sensory neurons found in a small patch of tissue high inside the nose where these cells connect directly to the brain (Haddad et al., 2010). As printed in many biology textbooks, microscopic molecules released by substances will be detected by neurons. Then, these neurons will send messages to brain, which identifies the smell. Such nature procedure was then mimicked by engineers, which later devoted the electronic nose technology commonly known as e-nose. As a first step in producing a valid and reliable instrument, an e-nose needs to be trained with some qualified data sets, in attempt to build a database of reference. Then, the developed instrument may recognise future samples by comparing volatile compounds to those stored in the database. To achieve these, often quantitative method called supervised classification is chosen.

In supervised classification, having sets of raw e-nose data, one needs to construct the best possible classification rule such that the rule would enable one to determine the group of instances correctly. However, in common engineering setting for e-nose, sample data is obtained from an electronic setup consists of p features from g different groups. Unless the conducted experiments deal with adequate number of sample, small number of sample compared to the measured features ($n \leq p$) will lead to some mathematical problems. Among of these problems is singular covariance matrix that may hinder one to pursue with more complex analysis as the covariance matrix is a common statistics used in many statistical classification rules. Singularity may occur when most features in a data set are correlated among each other. Even, if the covariance matrix is free from singularity, dealing with many features in an analysis consumes great computational cost, where bias estimators could be produced (Masnan et al., 2015).

Several exercises have been executed in attempt to deal with multicollinearity in e-nose. Perhaps, the most common strategy among engineers is to implement feature extraction approach, i.e. principal component analysis (PCA) and partial least square (PLS), in an attempt to obtain much smaller size of data, as well as tackling the multicollinearity effects. Feature extraction offers ones to deal with much smaller data set by extracting some information of the original p features to k new latent features where

$k \leq p$. Feature extraction approach has an ability in removing the multicollinearity, but the obtained k latent features are subjective hence may lead ones to misinterpret the features. Besides, the choice on k latent features is subjected to the choice of criterion used, such as based on the largest variance explained, tolerate value of eigenvalue etc. which may not suit to the ultimate aim in performing classification, to seek for better features classification in future used (Mahat et al., 2007).

Alternative to feature extraction is feature selection where it chooses a set of m needed features from the original p features where $m \leq p$. In classification methodology, the choice of m needed features is often chosen such that the subset provides satisfactorily classification accuracy. The basic concept of this approach is to only consider features that are much informative to explain the differences between groups in the data. Although this technique is able to remove the multicollinearity in the data and let ones to deal with much smaller sample, its acceptance in e-nose community still minimum due to the challenge in performing the selecting features iteratively. Besides, a criterion for selecting informative features is often debated by many where one might prefers to use classification performance criteria, group separation criteria, or other criteria such as R^2 value, distribution based criteria etc. (Mahat et al., 2007).

This paper overviews both ideas of feature extraction and feature selection on e-nose data, suggesting for improvement in data processing and accuracy of results. The arrangement of discussion is as follow; the following section briefs previous works related to huge dimensionality handling in e-nose. Next section details out the methodology designed purposely to investigate the potential used of feature selection versus the feature extraction. Then, results from the designed methodology are discussed and future work proposes potential work that can be planned for enhancement in e-nose technology. Conclusion is given in the final section.

RELATED WORKS IN ODOUR CLASSIFICATION SYSTEM

Basic odour classification system includes three steps; capturing of sensor array, constructing of classification rule and decision making. However, some refinements are inserted in between of these steps when dealing multicollinearity e-nose data such as deleting highly correlated predictors prior to the classification process (Laura & Thomas, 2011) or reduce the number of predictors to a smaller set of uncorrelated transformed features (Hidayat et al., 2010; Jun *et al.*, 2012; Masnan et al., 2012). The common methods for the latter refinement are feature extraction and feature selection as depicted in Figure 1. Both feature extraction and feature selection have been proven worth to avoid the curse of multicollinearity, but the choice of these methods influence the performance of the constructed classification rule.

Feature Extraction

Feature extraction deals with large number of features by transforming all the raw features into new latent features using some available reduction methods such as principal component analysis (PCA) and partial least square (PLS). The strategy seeks a small number of new features that are able to explain as much information as possible of the original features. Then, the new features are used to construct a classification rule (Masnan et al., 2012).

Figure 1. Odour classification system with (a) feature extraction and (b) feature selection

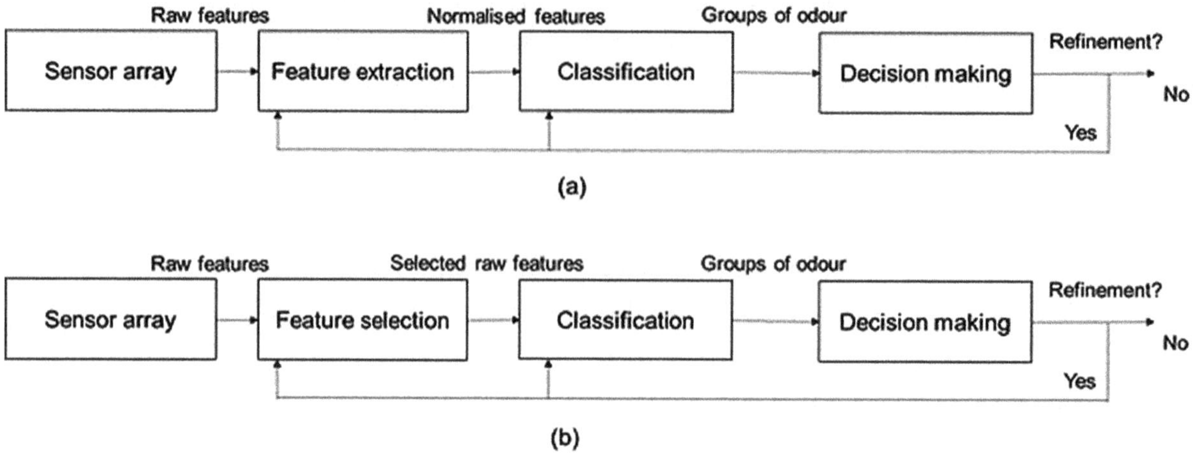

Principal Component Analysis (PCA)

Let $x \in R^p$ be the predictor features with covariance matrix Σ. The PCA summarises the p features as a single score, $z = \alpha^T x$, such that the variance of this score, $\text{var}(z)$, is maximised where α is a matrix of eigenvector of z. Various strategies can be used to maximise $\text{var}(z)$, but hold the constraint on α yield to definite values of α such that $\alpha_i^T \alpha_i = 1$ and $\alpha_i^T \alpha_j = 0$ where $i \neq j$ and $i, j = 1, 2, ..., p$. The new features are called components, z, possess a property such that the first component, z_1, explains most variance of the original features, followed by the second component, z_2, and so forth, $z_1 > z_2 > ... > z_p$. In practice, the first few k components, where $k \leq p$, will be used in the construction of a classification rule as the last $p - k$ components explain small amount variance of x which can be regards as noise. The choice of the *k* first components to retain can be finalised by several criteria which are:

- **Eigenvalues:** These values represent the variance of new components after the PCA has been executed. Often, eigenvalues which are greater than or equal to 1.0 will be used in the rule construction process, but Rencher (1998) warned that such rule of thumb may end up with many components.
- **Proportion of Total Variance Explained:** This indicator counts the proportion of the total of variance explained by the first *k* principal components over the overall variance in *p* principal components. Jollife (2002) proposed that retaining *k* components with total variance explained in a range of 70% - 90% would be good, but Jackson (1993) strongly argued such subjectivity of choice, especially when the components generated are important for much complex analysis.
- **Scree Plot:** This graphical display presents the association between eigenvalues with principal components in descending order. The choice of *k* components can be identified at the steep curve before the first point starts the flat line trend. This tool would be a good choice if the plotted scree plot having a steep curve, followed by a bend and a flat line. Otherwise, searching for *k* components would be troublesome (Dillon & Goldstein, 1984).

Partial Least Square (PLS)

PLS compromises between PCA and canonical correlation analysis in the data reduction process. It performs dimension reduction by considering the relation between groups, y, to the predictor features, $x \in R^p$. The strategy seeks for orthogonal x-components, $t_h = x\alpha_h$, and y-components, $u_h = y\beta_h$, that maximises the covariance between t_h and u_h (Barker, 2010) where α_h and β_h are score vectors of x and y respectively. Similar to PCA, PLS extracts orthogonally the original features to new latent features (called components). For classification purposes, only some k components will be used and the choice of k can be determined similar to criteria previously discussed in PCA. Both PCA and PLS are possible to be considered when one deals with greater number of features, especially when the correlation between features are major hurdles to the analyses (Dai et al., 2006). The different between PCA and PLS is that the former is based on the contribution of variance explained by only the predictor features, x, but the later considers the existence of groups, y, which provide greater insight about the extracted components suit with the aim of classification (Hoffmann et al., 2004). Both methods are competitive of each other and some recorded results can be viewed in Wentzell and Montoto (2003).

Feature Selection

Feature selection in classification problem chooses a subset of features from the overall measured features, and uses the subset to construct a classification rule. Apparently, using fewer features may be beneficial to avoid estimation problems, to improve classification performance and to save some costs of computation (Mahat et al., 2007). Some available criteria that are commonly used for distinguishing between useful and poor features in various field of studies are classification performance criteria (Wang et al., 2014) and group separation criteria (Mahat et al., 2007).

Classification Rule Performance Criteria

Selecting features based on classification rule performance seeks features that give the best performance, generally through maximum value of accuracy rate. Often, accuracy rate is a popular choice and easy to be interpreted, but might be misleading when the groups are imbalanced (Chawla, 2005). There are many other criteria that can be used to measure performance of the constructed classification rule such as false positive classification, false negative classification, area under the receiver operating characteristic (ROC) curve, and other aspects of the rule including imprecision and resemblance (Hand, 1997).

Separation of Groups Criteria

This criteria seeks a subset of features that appear to contribute the most relevant information for separating groups of instances. The criteria is preferable when the objective of the study is to understand the structure of groups (Baxter and Jackson, 2001) and would be an alternative to classification rule performance to avoid heavy computational tasks (Mahat et al., 2007). The distance between groups can be estimated using standard distance measures such as Euclidean distance, Mahalanobis distance, and much more.

Outside classification field of studies, there are other criteria that have been used for distinguishing between useful and poor features yet could be possible used in classification tasks. Model goodness-of-fit measure, e.g. AIC or BIC, would be a good option if one searches for a classification rule that is well suited to the data in hand (Daudin, 1986). Others may use correlation criterion, R^2, as it tells how good is the features in explaining the separation between groups (Rencher, 1993).

Classification Rule Based on Linear Discriminant Analysis

Linear discriminant analysis (LDA) can be considered as the most widely classification rule used in e-nose. The method was devised by Fisher in 1936 uses a linear combination of features, x, write as ($g(x)$), that consists p vector of the n objects to either π_1 or π_2. The work chooses for coefficients so that the ratio of the difference of the means of the linear combination in the two groups to its variance is maximised. Let the linear combination of x be written as $Y = w^T x$, μ_1 and μ_2 represent the means of Y in π_1 and π_2, respectively. Assuming both groups have the same covariance matrix, Σ, then the variance of Y is $w^T \Sigma w$ in both groups and w is chosen to maximize

$$\varphi(w) = \frac{(w^T \mu_1 - w^T \mu_2)}{w^T \Sigma w} \tag{1}$$

Whilst, the sample-based LDA is

$$D(w) = \frac{(w^T \overline{x}_1 - w^T \overline{x}_2)}{w^T S w} \tag{2}$$

where $\overline{x}_1$ and $\overline{x}_2$ are vectors of p features in π_1 and π_2 respectively, while S is a sample-based variance-covariance matrix. The best value of w is to choose the one that maximises $D(w)$ and this is given by $S^{-1}(\overline{x}_1 - \overline{x}_2)$.

METHODOLOGY

E-Nose Laboratory Materials

The experiment of e-nose was formed using the Cyranose320: the experiment on e-nose from Smith Detection^{TM} that consists of 32 non-selective sensors of different types of polymer matrix, blended with carbon black. This setup has been discovered satisfactorily in performing quantitative assessment of e-nose complex solutions (Hidayat et al., 2010; Markom et al. 2009; Masnan et al., 2015) and widely used at the Institute of Sustainable Agrotechnology, Universiti Malaysia Perlis. Figure 2 illustrates the experiment involved.

Figure 2. E-nose setup for headspace evaluation of sample

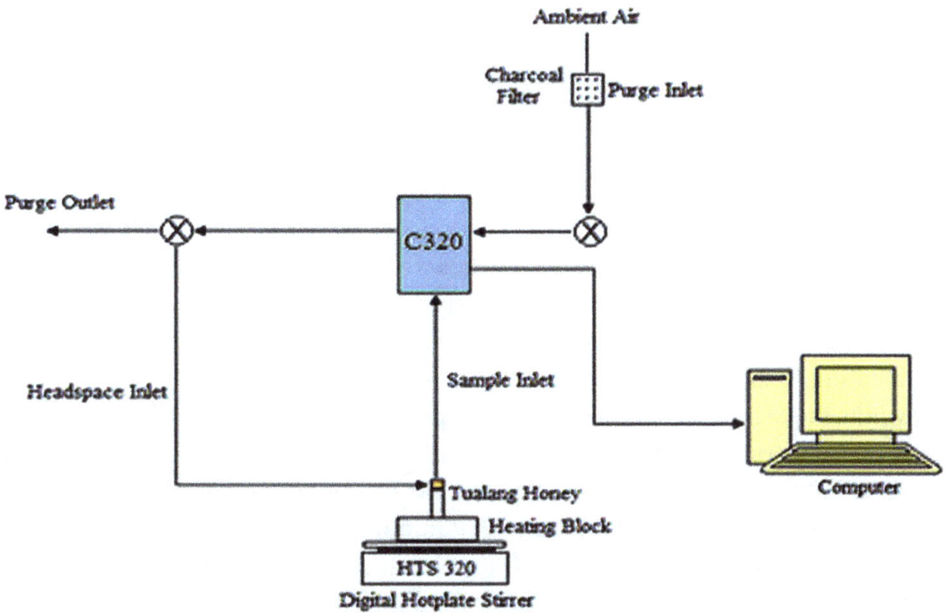

The sampling process was performed according to the substances of the setup experiment. There are three different data sets which are considered namely *rice*, *honey* and *foot infection*. As for *rice* and *honey* data sets, each sample was drawn from the bottle using 10ml syringe, was kept in a 13 x 100 mm test tube, and sealed with a silicone stopper. These steps were replicated ten times. Later, each sample was placed in a heater block and was heat up for 10 minutes to generate sufficient headspace volatiles before any measurement was taken. The temperature of sample was controlled at 50°C during the headspace collection (Masnan et al., 2012). The setting parameters in the conducted experiments were set as Table 1.

Meanwhile, the *foot infection* data set deals with the most common pathogens related to diabetic foot infections; *Escherichia coli*, *Staphylococcus aureus*, *Pseudomonas aeruginosa* and blank. The bacterial suspension solution obtained from clinical samples was adjusted at turbidity of standard MacFarland 1.0 ($3x10^8$cfu/mL) in NaCl solution which later subculture on blood agar plates. Next, the bacteria were incubated for 24 hours before analysed (Yusuf et al., 2013).

Table 1. E-nose parameter setting for rice and honey

	Cycle	Time (s)	Pump Speed
Sampling Setting	Baseline Purge	10	120 mL/min
	Sample Draw	40	120 mL/min
	Idle Time	3	-
	Air Intake Purge	40	120mL/min

Data Sets

The data sets used are the experimental outputs collected from several research studies held by Institute of Sustainable Agrotechnology, Universiti Malaysia Perlis. The data sets included in the discussion of this chapter are summarised in Table 2.

Investigation

Both feature extraction and feature selection approaches were investigated in dealing with multicollinearity e-nose data. Two common feature extraction approaches namely principal component analysis (PCA) and partial least square (PLS) were considered. The investigation set to remain at least 80% of the total variance explained in the data and no transformation of features was performed in the data reduction strategy. Meanwhile, the feature selection process was performed using two strategies. The first strategy is the traditional approach of stepwise selection where the choice of the features is determined by Wilk's Lambda value. The LDA was defined by starting with the feature that separates the groups most. Then, the LDA is extended by including further features that minimise the Wilk's Lambda value significantly.

Meanwhile, the second strategy is to include features that contribute to increase the distance between groups computed using the bounded Mahalanobis distance which earlier proposed by Masnan et al. (2015). The bounded Mahalanobis distance was obtained by first calculating the distance between g groups as follows

$$D_{ij}^2 = \left(\bar{X}_i - \bar{X}_j\right)^T S_{ij}^{-1} \left(\bar{X}_i - \bar{X}_j\right) \tag{3}$$

where $\bar{X}_i$ and $\bar{X}_j$ are the vector of means for groups i and j for $i = 1, 2, \ldots, g$, $j = i + 1, .., g - 1$ and $i \neq j$, and S_{ij}^{-1} is the inverse pooled covariance matrix for p features. In order to obtain bounded D_{ij}^2 of [0,1], each element in D_{ij}^2 was transformed using

$$D_{A(ij)}^2 = \frac{D_{ij}^2}{4 + D_{ij}^2} \tag{4}$$

Table 2. Data sets

Data	Description	Number of Groups	Size of Features	Total Number of Instances
Rice	The data consists 16 different type of rice, each with 275 number of samples, except for rice #6 with 365 samples.	16	32	4,490
Honey	The data comprises four different samples of *Tualang* honey, each with 117 samples. This data is a subset used by Masnan et al. (2012).	12	32	468
Foot infection	This data consists of four different type of bacteria as discussed by Yusuf et al. (2013). Sample of 250 instances were obtained for each type of bacteria.	4	32	1,000

where $D^2_{A(ij)}$ is a pairwise element of row and column in D^2_{ij}.

Once the bounded D^2_{ij} obtained, the features were ranked from the largest separation to the smallest separation. Then, the ranked features at the first quartile were used in the classification rule construction.

The investigation carried out used LDA as the classification rule due to its linear form of classification, which would be helpful to assist the investigation in comparing both feature extraction and feature selection approaches. The performance of each rule was measured based on accuracy rate that gives the proportion of instances being correctly classified by the classification rule. The whole experiment was conducted in a leave-one-out fashion with all the coding programs were executed in R-3.1.3 software. In leave-one-out setting, an instance was left out as a test instance whilst the remaining, called training instances, were used to extract or select features and train the classification rule. The constructed classification rule will estimate the group of the test instance and these processes were repeated until each instance in the data has been left out in turn. Figures 3 and 4 depict the analytical process with feature extraction and feature selection respectively.

RESULTS

Investigations on the three data sets were performed in three different settings. First, classification was performed by including all the raw features. Second, the construction of the LDA included the features

Figure 3. Analysis framework on e-nose with feature extraction

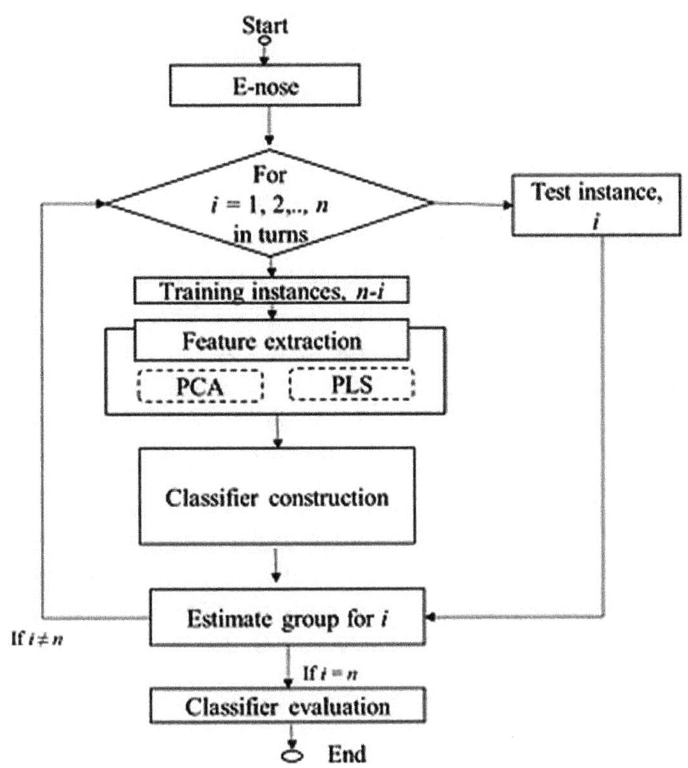

Figure 4. Analysis framework on e-nose with feature selection

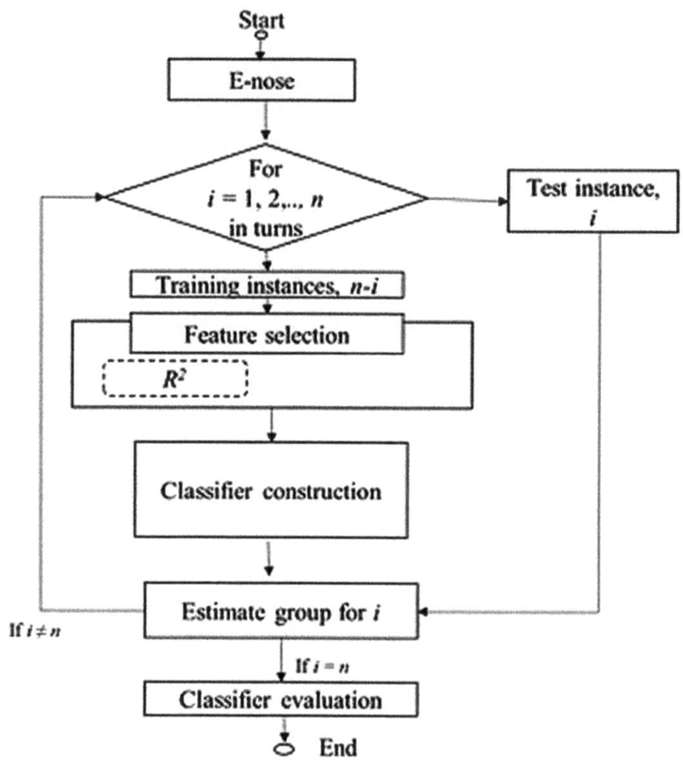

extracted using either PCA or PLS in the leave-one-out setting. Finally, the construction of the LDA included the features selected from the feature selection process either using the Wilks' Lambda criterion or the bounded Mahalanobis distance. As the whole classification studies were performed in the leave-one-out as depicted in Figures 3 and 4, therefore the selected features remained in the system and in each loop of the leave-one-out process chosen slightly different features.

Table 3 summarises the computation of leave-one-out accuracy rate for *rice*, *honey* and *foot infection* data sets in the three settings of classification. In general view, including all features in a classification rule leads to better accuracy than using some features. However, the *foot infection* data set provides evidences that using some features, either through feature extraction or feature selection, would perform as good as with all features included. This indicates that fewer features is adequate to provide decision for classification, which consume fewer cost rather than using all features. Also, the *honey* data set shows that LDA with bounded Mahalanobis distance used fewer feature, but recorded close accuracy rate to the full LDA.

Comparison on accuracy rate between feature extraction and feature selection reveals that performance of LDA is much better with the feature selection in all data sets, except equal performance in *foot infection* data. Meanwhile, the chosen of features based on the bounded Mahalanobis distance contribute to higher accuracy than the Wilk's Lambda in both *rice* and *honey* data sets.

Table 3. Performance of the classification rules on rice, honey and foot infection data sets

Classification Strategy	Leave-One-Out Accuracy Rate (%)		
	Rice	**Honey**	**Foot Infection**
All features included	99.6	91.1	100.0
Feature Extraction			
PCA	80.4	88.1	100.0
PLS	81.2	86.3	100.0
Feature Selection			
Wilk's Lambda	86.7	89.2	100.0
Bounded Mahalanobis distance	88.7	90.3	100.0

FUTURE RESEARCH DIRECTIONS

This study is a small part in the enhancement of e-nose technology. As the data are getting cheaper in digital era, more e-nose data sets are becoming available to be trained. In such a case, there is a challenge for researchers to improve related to e-nose product development using advanced and quick analytics. Feature extraction and feature selection are the current options to deal with large number of features, but there is a need to foresee much advanced analytics to deal with large amount of instances and features simultaneously.

CONCLUSION

E-nose data often exist with multicollinearity among the features which could harm the classification result. In order to avoid such problem, both feature extraction and feature selection provide solutions. The obtained results based on three real data sets shows that the accuracy of a classification rule is much better with fewer features. The evidence obtained show that LDA performs better when using selected features determined by feature selection rather than using latent features extracted from either PCA or PLS. Feature selection directly used the selected useful raw features in the classification while latent features may include poor raw features which harm the classification purposes. The used of separation between groups measures namely Wilk's Lambda and bounded Mahalanobis distance are promising though the latter gives better results than the former. The different is, bounded Mahalanobis distance considers the groups separation by including the covariance between groups, but Wilk's Lambda computes the separation of groups based on vector of means. Although various criteria are possible to be used for distinguishing between useful and poor features, it is advisable to consider a criterion that meet the purpose of performing the classification on e-nose.

ACKNOWLEDGMENT

Some part of the investigations discussed in this chapter was supported by Malaysian Ministry of Education, Fundamental Research Grant Scheme (FRGS, Ref: FRGS/1/2015/SG04/UUM/02/3).

REFERENCES

Barker M. (2010). *Partial least squares for discrimination*. Saarbrucken, Deutschland: LAP LAMBERT Academic Publishing.

Baxter, M. J., & Jackson, M. (2001). Variable selection artefact compositional studies. *Archaeometry*, *43*(2), 253–268. doi:10.1111/1475-4754.00017

Chawla, N. V. (2005). Data Mining for imbalanced datasets: An overview. In Data Mining and Knowledge Discovery Handbook. Boston: Springer. doi:10.1007/0-387-25465-X_40

Dai, J. J., Lieu, L., & Rocke, D. (2006). Dimension reduction for classification with gene expression microarray data. *Statistical Applications in Genetics and Molecular Biology*, *5*(1), 1–19. doi:10.2202/1544-6115.1147 PMID:16646870

Dillon, W. R., & Goldstein, M. (1984). *Multivariate analysis, methods and applications*. New York: John Wiley & Sons.

Dragonieri, S., Schot, R., Mertens, B. J. A., Le Cessie, S., Gauw, S. A., Spanevello, A., ... Sterk, P. J. (2007). An electronic nose in the discrimination of patients with asthma and controls. *The Journal of Allergy and Clinical Immunology*, *120*(4), 856–862. doi:10.1016/j.jaci.2007.05.043 PMID:17658592

Falasconi, M., Concina, I., Gobbi, E., Sberveglieri, V., Pulvirenti, A., & Sberveglieri, G. (2012). Electronic nose for microbiological quality control of food products. *International Journal of Electrochemistry*, *2012*, 1–12. doi:10.1155/2012/715763

Gliszczyńska-Świglo, A., & Chmielewski, J. (2016). Electronic nose as a tool for monitoring the authenticity of food - A review. *Food Analytical Methods*, *10*(6), 1800–1816. doi:10.1007/s12161-016-0739-4

Haddad, R., Medhanie, A., Roth, Y., Harel, D., & Sobel, N. (2010). Predicting odor pleasantness with an electronic nose. *PLoS Computational Biology*, *6*(4), e1000740. doi:10.1371/journal.pcbi.1000740 PMID:20418961

Hand, D. J. (1997). *Construction and assessment of classification rules*. Chichester, UK: John Wiley & Sons.

Hidayat, W., Shakaff, A. Y. M., Ahmad, M. N., & Adom, A. H. (2010). Classification of agarwood oil using an electronic nose. *Sensors (Basel)*, *10*(5), 4675–4685. doi:10.3390/s100504675 PMID:22399899

Hoffmann, K., Schulze, M. B., Schienkiewitz, A., Nothlings, U., & Boeing, H. (2004). Application of a new statistical method to derive dietary patterns in nutritional epidemiology. *American Journal of Epidemiology*, *159*(10), 935–944. doi:10.1093/aje/kwh134 PMID:15128605

Jackson, D. A. (1993). Stopping rules in components analysis: A comparison of heuristical and statistical approaches. *Ecology*, *74*(8), 2204–2214. doi:10.2307/1939574

Jollife, I. T. (2002). *Principal component analysis* (2nd ed.). New York: Springer.

Jun, F., Canqin, H., Jianguo, X., & Junbao, Z. (2012). Pattern classification using an olfactory model with PCA feature selection in electronic nose: Study and application. *Sensors (Basel)*, *12*(12), 2818–2830. doi:10.3390/s120302818 PMID:22736979

Laura, T., & Thomas, L. (2011). Classification with correlated features: Unreliability of feature ranking and solutions. *Bioinformatics (Oxford, England)*, *27*(14), 1986–1994. doi:10.1093/bioinformatics/btr300 PMID:21576180

Mahat, N. I., Krzanowski, W. J., & Hernandez, A. (2007). Variable selection in discriminant analysis based on the location model for mixed variables. *Advances in Data Analysis and Classification*, *1*(2), 105–122. doi:10.1007/s11634-007-0009-9

Mahat, N. I., Zakaria, A., & Shakaff, A. Y. M. (2015). Comparisons of parametric and non-parametric classification rules for e-nose and e-tongue. In AIP Conference Proceedings (vol. 1691(1)). doi:10.1063/1.4937017

Markom, M. A., Shakaff, A. Y. M., Adom, A. H., Ahmad, M. N., Hidayat, W., Abdullah, A. H., & Fikri, N. A. (2009). Intelligent electronic nose system for basal stem rot disease detection. *Journal of Computers and Electronics in Agriculture*, *66*(2), 140–146. doi:10.1016/j.compag.2009.01.006

Masnan, M. J., Zakaria, A., Shakaff, A. Y. M., Mahat, N. I., Hamid, H., Subari, N., & Saleh, J. M. (2012). Principal Component Analysis – A Realization of Classification Success in Multi Sensor Data Fusion. In P. Sanguansat (Ed.), *Principal Component Analysis/Book 2*. Rijeka, Croatia: InTech-Open Access Publisher.

Rencher, A. C. (1998). *Multivariate statistical inference and applications*. New York: Wiley.

Wang, X. R., Lizier, J. T., Nowotny, T., Berna, A. Z., Prokopenko, M., & Trowell, S. C. (2014). Feature selection for chemical sensor arrays using mutual information. *PLoS One*, *89840*. doi:10.1371/journal.pone.0089840 PMID:24595058

Wentzell, P. D., & Montoto, L. V. (2003). Comparison of principal components regression and partial least squares regression through generic simulations of complex mixtures. *Chemometrics and Intelligent Laboratory Systems*, *65*(2), 257–279. doi:10.1016/S0169-7439(02)00138-7

Yu, H., Wang, J., Xiao, H., & Liu, M. (2009). Quality grade identification of green tea using the eigenvalues of PCA based on E-nose signals. *Sensors and Actuators. B, Chemical*, *140*(2), 378–382. doi:10.1016/j.snb.2009.05.008

Yusuf, N., Omar, M. I., Zakaria, A., Abdullah, A. A., Kamarudin, L. M., Shakaff, A. Y. M., ... Yassin, M. S. (2013). Diagnosis of bacteria for diabetic foot infection using electronic nose technology. *IEEE Conference on Wireless Sensors 2013*. doi:10.1109/ICWISE.2013.6728791

Chapter 3
E-Nose Pattern Recognition and Drift Compensation Methods

Sanad Al Maskari
University of Queensland, Australia

Xue Li
University of Queensland, Australia

ABSTRACT

With new developments and upcoming technologies, new sensing techniques are becoming available. Unfortunately, none of these techniques provides output interpreted the way human perception works. An inability to improve the effectiveness of these technologies limits their use in dedicated applications and increases their complexity. The growing adoption of this technology makes it critical to create a system capable of handling e-nose challenging issues such as noise, drift, imbalanced data, dynamic environment, and high uncertainties. Without appropriate pattern recognition methods that allow inferences to be derived based on patterns observed within these data sets, it will not be possible to improve the performance of current e-nose systems. In this chapter, e-nose drift issue is introduced and the available drift counteraction methods is discussed.

INTRODUCTION

Big cities around the world are facing serious air quality and air pollution problems, which prompted governments to introduce more tightened emission and safety regulations. The study of gas identification and measurement using chemo-sensors known as electronic noses, is important in emission control. E-noses can provide an economical solution for air quality and air pollution monitoring. Despite their popular usage, they suffer from serious problems such as drift, noise, selectivity and their inability to provide true concentrations Vergara et al. (2012) ;Al-Maskari, Li, and Liu (2014). Electronic nose (e-nose) is a generic name for devices capable of identifying chemical analytes or measuring odour information. E-nose systems have been used in a wide range of applications including: food industry, agriculture, air quality and environment monitoring, odour monitoring in a poultry shed, medicine, water and waste water quality control Haugen et al (2000);Al Maskari, Kumar, and, Chiffings (2012). Despite their good

DOI: 10.4018/978-1-5225-3862-2.ch003

reputation, their practical values are affected by their poor stability, making them very vulnerable to drift and noise. These vulnerabilities can be resolved using periodic recalibration which in turn increases their cost significantly, making them overly complex. Compensation for the need of periodic recalibration and ability to postpone the recalibration period using machine learning will need to solve the aforementioned dynamic issues. In general, the performance of e-noses is effected by drift and noise. Sensor drift is the gradual and unpredictable variation of the chemo-sensory signal responses when exposed to the same analyte under identical conditions. Such variations could be caused by sensor poisoning, ageing and environmental changes such as humidity, temperature, pressure and system sampling non-specific adsorption. In the other hand, noise can be defined as any unwanted effect that obscures the measurement of a detected signal. Noise can be caused by system circuit faults, sensor poisoning, sensor ageing, and environmental effects. Because sensor drift is not deterministic, it is difficult, if not impossible, to distinguish it from noise and vice versa.

BACKGROUND

An extensive amount of research has been conducted in the past to develop and improve pattern recognition systems and correct drift for e-noses. Pattern recognition and machine learning have been employed to analyse data generated by e-noses in order to classify and quantify sensor responses Al-Maskari, B´elisle, et al. (2016b); Al-Maskari, Guo, and Zhao (2016); Al-Maskari, Saini, and Omar (2010). Despite the extensive research, the development of a robust and reliable pattern recognition system still remains a research issue for the chemical sensor community. Such limitations are attributed to the low sensitivity and stability of gas sensors as well as the multi-dimensionality and complexity of the surrounding environments. Current pattern recognition systems for e-noses still suffer from drift effects and noise which greatly affect gas sensor selectivity and sensitivity. Because of drift effect, sensors will not give exactly the same response even when they are exposed to the same gas with the same concentrations. Therefore, a method must be considered to correct gas sensor drift.

This section will discuss different pattern recognition and drift correction techniques found in the literature. Figure 1 provides a taxonomy of pattern recognition and drift methodologies applied to drift correction.

Signal Processing Methods

In signal pre-processing techniques, filtering strategies based on counteracting methods are intentionally used in order to filter out the drift-contaminated signal portions from the target signal. Generally, drift contamination will occur in a different frequency domain from the target signal and is typically associated with a lower frequency. Therefore, one possible way to filter out the drift is to transform the signal from the time domain to the frequency domain. After the transformation, the lower frequency components associated with drift can be removed carefully. The Discrete Wavelet Transform (DWT) method, which is based on feature extraction techniques, can provide a powerful tool to improve sensor responses by removing drift contamination from the low-frequency domain Huang (2004); hui, Jun-hua, and Zhong-ru (2003); Zuppa, Distante, Persaud, and Siciliano (2007). There are already some examples for the application of filtering techniques, such as the moving median filter and Fourier band-pass filter, which are applied to remove high-frequency fluctuations (such as noise and spikes) or low-frequency

Figure 1. Taxonomy of pattern recognition and drift correction techniques applied to E-Nose

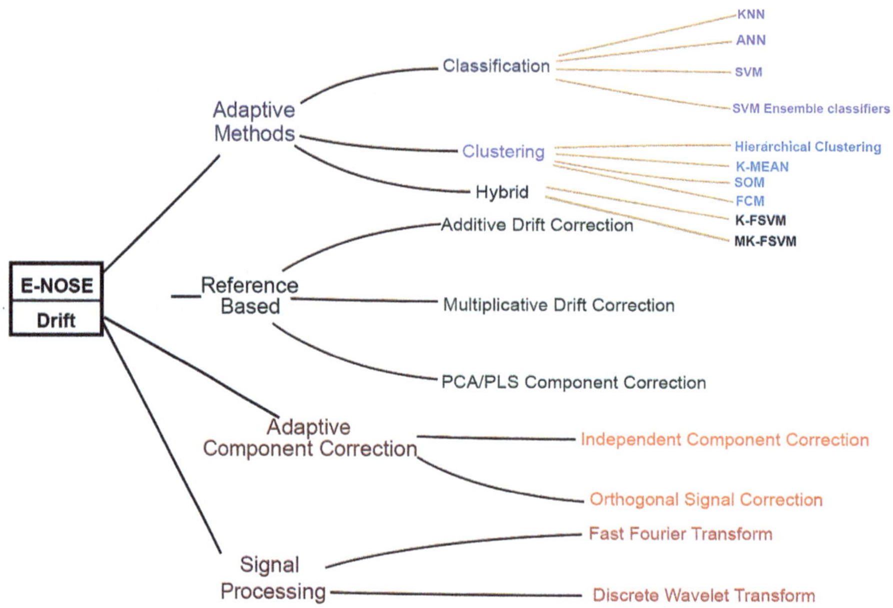

changes (such as drift) respectively. Compared with these filters mentioned above, the DWT technique applies iteratively high-pass and low-pass filters and provides a wide range of signal analysis with different resolutions. Using the DWT technique, it is very convenient to remove selected low-frequency components without deforming the main signal. Moreover, in the DWT technique, the signal is analysed at different frequency bands with different resolutions in order to provide a multi-resolution signal decomposition of sensor responses Mallat (1989); Zuppa et al. (2007). This feature is realised by applying shift and scaling operations to two prototype functions: the scaling function and the wavelet function. The scaling function is associated with a low-pass filter. The low-frequency components of the signal can be obtained from the convolution between the signal and scaling functions which produces approximation coefficients. The wavelet function, on the other hand, is associated with a high-pass filter, and high-frequency components can be obtained from the convolutions between the signal and wavelet functions. These components are called wavelet coefficients or detail coefficients.

The procedure for removing the drift-contaminated signal can be described as follows. First, wavelet decomposition and signal reconstruction are performed using the Mallat algorithm Mallat (1989). A sequence of increasingly smoothed and halved versions of the original signal is obtained by applying high-pass and low-pass filters iteratively on the vector of approximation coefficients. Then, define the pairs of low-pass and high-pass filters based on the chosen specific wavelet (e.g. Daubechies orthonormal functions). The associated approximation coefficients could have drift contamination. The decomposition level of the DWT can be fixed once the drifting trend is detected using the frequency domain analysis. Finally, discard the drift contaminated lowest frequency which is associated with approximation coefficients, and the drift contamination-free signal can be obtained by reconstructing the wavelet by computing the remaining coefficients. In Zuppa et al. (2007) the drift was generated artificially.

Reference Based Methods

To separate drift from sample changes, a stable sample called a reference gas is often measured. Different approaches to correct gas sensor data using a reference have been proposed. The drift effect can be estimated by measuring the change in sensor responses to one or more reference gas. The drift direction can be removed using a univariate method (sensor by sensor) or using a multivariate method to remove the directions responsible for the drift. One critical element in a reference-based method is using a good calibration reference sample. The sample should be:

- Chemically stable over time.
- Highly correlated with the real samples being measured. For example, when injected to the sensor it should give similar response as the real samples.
- The drift generated from the calibrated sample and real samples should be highly correlated.

The following section describes different methods based on periodic calibration.

Additive Based Methods

Fryder et al. Fryder, Holmberg, Winquist, and Lundstrom (1995) proposed a calibration-based method using a reference gas to correct sensor drift. In their experiment, they exposed 15 individual sensors to four different alcohols and water for 45 days. During that period, drift was observed and handled using the best reference gas from all calibration gases. The reference gas was selected by comparing the correction output of multiple calibration gases. Fryder et al. assumed that drift is additive, i.e. independent of signal level. The second assumption was that drift has a single direction. Thirdly, they assumed that different sensors drift in similar ways regardless of which gas they are exposed to. They reduced drift generated by the e-nose by subtracting the response to the reference gas from the sample response as shown in the following equation.

$$x'_{t,i} = x_{sample,t,i} - x_{reference,t,i} \tag{1}$$

where $x'_{t,i}$, $x_{sample,t,i}$, and $x_{reference,t,i}$ are the drift corrected sensor response, the uncorrected sensor response, and the response for the measured reference gas, respectively, all measured on sensor i at time t.

The experiment did not take into consideration signal poisoning, pollutants, and various gas type mixtures. Such elements can impact the results significantly. Furthermore, the drift was studied in a short period of 45 days, making the result hard to replicate in real-world situations.

Multiplicative Based Methods

The method introduced by Haugen et al Haugen, Tomic, and Kvaal (2000) is similar to additive base method, except that it tries to correct multiplicative drift effects found in fish and milk samples. The data set used in this study was based on fish and milk experiments. In the fish experiment, the salmon sample was kept in a controlled room where the temperature was set to $4 \pm 0.5^{0}C$ for more than 10 days. Only data from the first five days was used. For the milk experiment, the dataset contained two-day measurements for two types of milk with different qualities. The objective was to discriminate fresh pasteurised milk from slightly oxidised pasteurised milk. The method attempted to correct multiplicative drift effects generated by multiple sensors by regularly measuring calibration samples and real samples sequentially. A sequence of calibration samples and real samples was fed into the sensors and measured. The final complete sequence was then used to correct drift for all different sensors. Equation 2 was used to calculate the ratio between the responses at time t and at the initial time for each gas sensor. The compensation algorithm requires the real samples used between calibration samples to be kept constant during the measurement. Table 1 shows an example of correction sequences for the drift calibration algorithm.

The drift quotient is calculated by dividing the first calibration measurement by the second calibration sample and this will be repeated for all n calibration samples for all k sensors.

$$q_{1,n,k} = \frac{U_{cal,1k}}{U_{cal,nk}} \tag{2}$$

U_{cal} is the calibration measurement, n refers to the calibration sample and k refers to the sensor number. After calculating all quotients for all samples in k, a linear regression is used to predict the best correction factor as indicated in equation 1.3. Equation 3 shows the trendline function used to fit the data. The drift was eliminated by multiplying sensor response by the correction factor as shown in the above equation 4 refer to Figure 2.

$$f_{nk} = (ax_n - b)_k \tag{3}$$

Table 1. Sequance samples used for correction

n	Name	Sensor1	Sensor2	Sensor3	Sensork
1	calibration sample 1	Ucal,11	Ucal,12	Ucal,13	Ucal,1k
2	real sample 1	U21	U22	U23	U2k
3	real sample 2	U31	U32	U33	U3k
4	Real sample 4	U41	U42	U43	U4k
5	real sample 4	U51	U52	U53	U5k
6	calibration sample 5	Ucal,61	Ucal,62	Ucal,63	Ucal,6k
7	real sample 5	U71	U72	U73	U7k
8	real sample 6	U81	U82	U83	U8k
9	real sample 7	U91	U92	U93	U9k
10	real sample 8	U101	U102	U103	U10k
11	calibration sample 3	Ucal,111	Ucal,112	Ucal,113	Ucal,11k

Haugen et al. (2000).

Figure 2. Multiplicative drift: The drift is eliminated by multiplying fnk by sensor response.
Haugen et al. (2000).

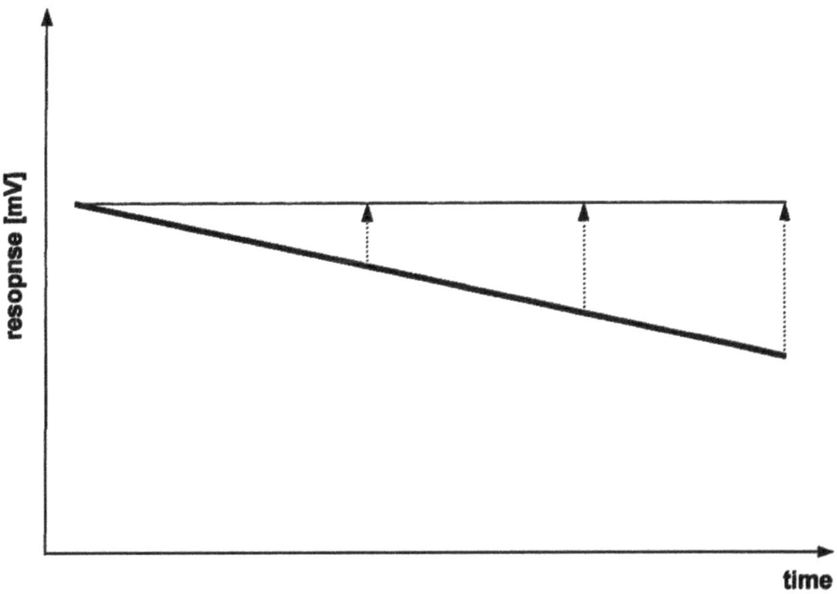

$$U_{corrected} = x_{tr} - f_{nk} \tag{4}$$

where fnk is the correction factor of sample n for sensor k, a is the slope of the trend line, b is the intercept, Xn is the sample number. To correct between different days an additive correction was used.

Component Correction

Various multivariate methods have been used to deal with sensor drift, such as partial least squares (PLS), principal components analysis (PCA), and canonical Correlation Analysis(CCA). In some studies PCA was used to find the directions of maximum variance of data x where is CCA identifies directions where two variables x, and y co-vary Gutierrez-Osuna (2000); Ziyatdinov et al. (2010). Multiplicative drift correction method (MDC) and CC are considered to be linear but the linearity assumptions are different in each method. In the MDC method, it is assumed that the relative change in each sensor is the same for the reference gas and all test cases. In the other hand, CC assumes that drift has a preferred direction, therefore, the direction is removed from the measurements to make the correction. This means that drift can only have one direction and will influence the sensor along a straight line. Tom et al. Artursson et al. (2000) used the CC method based on

PCA for drift handling. The sensor responses were analysed using PCA in order to detect the drift direction. PCA is a well-known dimensionality reduction method that can describe a high dimensional dataset by a few components. PCA can be used to project gas sensor data into a lower dimensional linear space (principal subspace) where the variance between data points in that subspace is maximised. The projection to the lower subspace can be used to identify significant drift. From the reference gas measurements, the projection matrix is generated using PCA. The drift direction is described by the first principal component p. It is assumed that samples will follow the same drift direction as the reference gas measurements. Therefore, the score matrix t is generated by projecting the samples into the p vector as shown in equation 5. The generated score matrix t contains the amount of drift for each sample, while the vector p comprises direction coefficients of the one-dimensional principal component. The drift is corrected by subtracting the samples from the drift component tp^T as described by equation 6.

$$t = x_p \tag{5}$$

$$x_{corrected} = x - tp^T \tag{6}$$

This method assumes a linear drift direction for sensor drift. The experiment was conducted based on measurements collected for over two months. The data was generated using an electronic nose by injecting four different gases (hydrogen, ammonia, ethanol, and ethane) using a gas-mixing system. The method attempts to correct drift using a reference gas without taking into consideration sensor ageing, poisoning, noise, and long-term drift. To avoid any information loss from other sensors, a suitable calibration sample must be used to provide a good drift correction for all sensors. The experiment was conducted in a very short period of time and didn't consider long term drift. Component correction methods assume that all chemical classes behave in the same way in the presence of drift, and this is not always the case (as it will be seen in the present study).

Component Correction by Training (CCT)

Unlike reference-based methods, CCT methods attempt to correct drift without the need to use a reference gas or calibration samples. These methods attempt to perform drift component correction using training data. A significant improvement in the classification rate can be achieved using these methods for only a fixed period of time. Furthermore, they can generate real responses that can be used for gas quantitative analysis.

Independent Component Correction

In PCA-based correction methods, such as CC Artursson et al. (2000), the principal components of the signal which is being computed are mutually uncorrelated. Since uncorrelated components don't necessarily imply that they are statistically independent, it cannot be concluded that relevant signals are completely separated by disturbances. Besides, a principal component usually carries information about both the signal and disturbances. A new approach has been proposed for filtering the drift from the sensor signal by deriving disturbances from the measurement environment Davide, Di Natale, and D'Amico (1994). In this study, a data matrix has been separated into a series of components, which are independent of each other, by applying Independent Component Analysis (ICA) Comon (1994) technique. In this method, the information carried by each component is independent and cannot be inferred from the others. The product of the probability of each unit determines the joint probability of independent quantities. When the ICA technique is applied to e-nose data, only those components which are correlated with sample features relevant to the application will be preserved. In ICA, the variance carried by each component cannot be determined, and consequently, it is not possible to select which components will be eliminated. This feature of ICA is different from, in fact it is contrary to, the Gaussian-based PCA and PLS models. Therefore, it is impossible to establish an order among the different components based on this information. In Di Natales research, the components were selected in a directed way. The independent components which are mostly correlated with the objective of the measurement were chosen and those more correlated with the disturbances were discarded. With respect to disposing drift effect, this method gives very good results due to external causes that can be monitored during the experiment, such as temperature and pressure. These external causes can be used as variables for selecting the components to be removed. However, selecting the correct components to be discarded may not be possible or difficult in some cases if the actual drift causes have not been fully identified or understood.

Orthogonal Signal Correction

A very interesting drift correction method based on Orthogonal Signal Correction (OSC) has been reported recently Padilla et al. (2010). The OSC signal processing technique was introduced by Wold et al. Wold, Antti, Lindgren, and hman (1998) and applied to spectra correction of NIR first. By using the OSC technique, a set of sensor array data X and concentration vector or class label vector C can be analysed, where X and C represent a set of independent and dependent variables, respectively. The main purpose of OSC is to remove the variance of X which is not correlated to any variables in C. It can be realised by constraining the deviation of non-relevant information in X, by which only information orthogonal to X can be removed. Using the OSC technique, it can be assured that as little as possible information is removed from the original data during the signal processing. Many different variants of the OSC algorithm have been reported in the literature. Padilla et al. applied the Wise implementation of the OSC algorithm. Despite the effectiveness of OSC, it still cannot completely correct the drift. One of the main drawbacks of OSC is that it requires a large dataset which contains a significant amount of drift. Without a large training data set, OSC will fail to identify the set of orthogonal components to be rejected. In real world scenarios, training data are collected over a short period of time. Furthermore, it can be very challenging to introduce new analytes to the recognition library because the removed components could be needed to effectively identify the new classes. Finally, in Wold et al. (1998) OSC performed better than unsupervised PCA-CC for the first 100 days. See Figure 3 for a comparison between OSC and PCA-CC.

E-NOSE PATTERN RECOGNITION AND DRIFT COMPENSATION BY ADAPTIVE METHODS

Adaptive pattern recognition methods take into consideration the drift effect by analysing pattern changes due to drift. The main advantage of these methods is to extend the recalibration period of the gas sensor system. The first attempt to use a pattern recognition system was made by Marco et al. in the early 90's. Since then, only a few methods have been proposed for drift counteraction and adaptive classifiers. Creating an effective and adaptive classifier to handle gas sensor arrays remains a challenging problem for the chemical sensor community.

Figure 3. Classification rates for corrected first test subset by PCA-CC and OSC methods and non-corrected data (No Corr) vs. number of samples in the calibration set
Padilla et al. (2010).

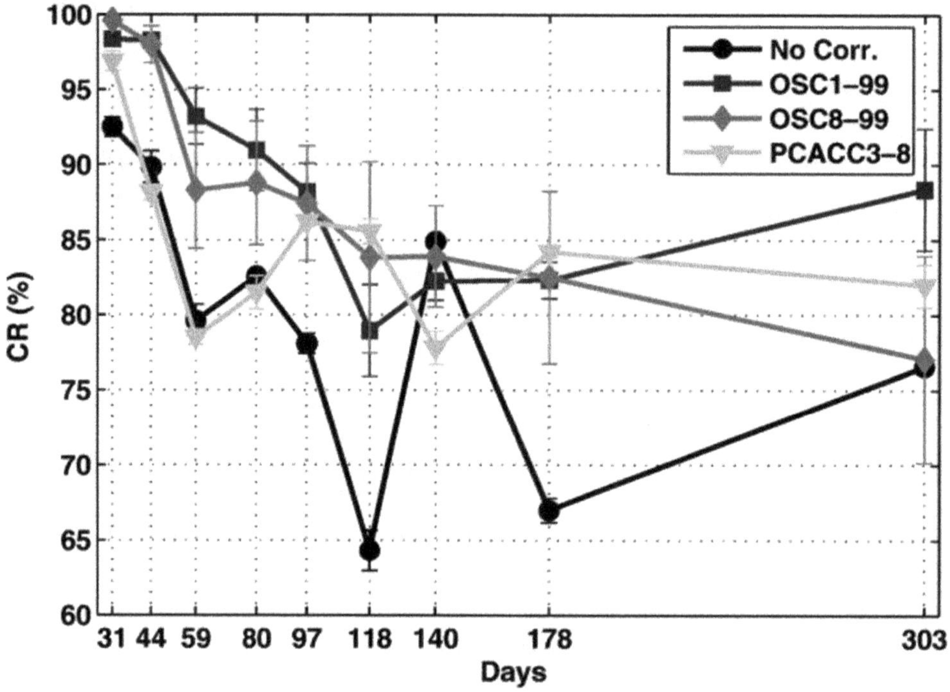

Drift Compensation Using Classification Methods

Neural Network

Neural networks are one of the most popular tools used for pattern recognition systems due to the following interesting features: 1) Unlike traditional model-based methods, artificial neural networks (ANNs) are data-driven self-adaptive methods in that they can adjust themselves to the data with only a few a priori assumptions about the models for a given problem. Thus they are able to learn the functional relationships from given examples even if the underlying relationship is unknown or undefined. 2) Neural networks can be used as universal function approximators. It has been shown in the literature that ANNs are capable of approximating any continuous function with arbitrary accuracy Al-Maskari, Guo, and Zhao (2016); Balkin and Ord (2000); Bishop (1995). Because the functional relationship between group membership and the attributes of an object is the key factor for any classification procedure, accurate identification of this underlying function is extremely important for the whole process. 3) Neural networks are a nonlinear data-driven approach Bishop (1995), a feature which makes them very flexible in modelling the complex relationships found in the real world, such as those presented by gas chemical sensors. They can be used to perform nonlinear modelling for given data without the need for a priori knowledge about the relationships between input and output variables. 4) The ability of neural networks to estimate the posterior probabilities can provide the basis for establishing a classification rule and performing statistical analysis Richard and Lippmann (1991).

Radial Basis Function (RBF)

A Radial Basis Function (RBF), changes the network building block from the perceptions based node, which is a non-linear function of a scalar product, to units. These units are activated based on the distance between the node input space and a prototype vector. RBF was first applied in interpolation and approximation theory field Bishop (1995); Broomhead and Lowe (1988), and also in the biological receptive area Moody and Darken (1988). The distance between the neurone centre and the input vector is calculated when each neurone is entered so that the neurone output is calculated by applying the basis function to this distance. The RBF output is computed based on the weighted sum of the neurone output and unity bias. A single RBF output can be described mathematically as follows:

$$f(x) = \sum_{j=1}^{m} w_j h_j(x) \tag{7}$$

where wj is a hidden output weight corresponding to the number of hidden neurones m in this linear model. hj is a nonlinear function of x. The main feature of these networks is the radial nature of the kernel used in the hidden neurones hj, which depend only on the distance between the centre cj and the input data x. Using kernels it is possible to build infinite basis functions. The basis function can be described as Gaussian,

$$h_j(x) = \exp\left(-\left(\sum_{k=1}^{n} \frac{\left(x_k - c_{jk}\right)^2}{r_{jk}^2}\right)\right) \tag{8}$$

where rj is the radius or width of the j-th hidden unit or basis function. The training is usually divided into two phases. In the first phase, the basis function parameters corresponding to the hidden units are determined using fast and unsupervised methods. The input space is transformed into feature space non-linear transformation or kernels. In the second phase, the final-layer weights are determined using the linear solutions, such as linear least square LS. Despite the availability of different ANN architectures, many researchers are focusing on searching for a new, improved ANN architecture which can perform better than the classical methods used in the gas sensor array. For example, in ? RBF is used to classify fresh edible oils using 346 samples generated from an e-nose. Also, RBF has been used to evaluate the quality of black tea Tudu et al. (2009). Compared to multilayer perceptron (MLP) and linear discrimination analysis, a dramatic improvement in PARC performance has been discovered by using Time-Delay Neural Network (TDNN) Zhang, Balaban, and Principe (2003). TDNN has been applied to recognise temporal patterns in time series data, such as phonemes in speech processing. The basic idea of TDNN is based on the idea that time series data provide additional information about the difference in time-related patterns among sensors responses. MLP and RBF methods have been quantitatively studied using VOC mixtures of acetone, hexanal, and 2-pentanone Taurino, Distante, Siciliano, and Vasanelli (2003) The classification performances of RBFs, Fuzzy-C Means, and SOM methods are compared on applications on different tea classes Dutta, Kashwan, Bhuyan, Hines, and Gardner (2003) . There are several pattern recognition and drift correction methods based on neural networks such as RBF, SOM

Marco, Pardo, Ortega, and Samitier (1997), MSOM Zuppa, Distante, Siciliano, and Persaud (2004), and Adaptive Resonance Theory (ART) Carpenter, Grossberg, and Reynolds (1991). The ART method has been applied to exploit different neural network architectures to enable the network to learn new information while preserving already learned patterns. The results obtained from ART for sensor gas classifications are very similar to the back propagation method (BP), but the training time is typically an order of magnitude faster than BP on small datasets Kim (2001). All proposed ANN-based drift correction methods exploit the fundamental neural network working principles. Commonly, a preliminary training phase is used in pattern recognition in order to train the neural network to identify the samples from the sensor array. In this phase, the similarities between different samples are learned using various learning rules Bishop (1995). The behaviour of each node in the network is dynamically changed based on the new incoming samples. Classification of input data can be achieved using supervised or unsupervised ANN methods. During the training phase, the network learns the characteristics of input patterns and the relationships between input and output variables in order to be able to classify new data points. It is very critical to select and maintain an appropriate learning rate in order to classify new data, and at the same time, compensate for drift. In order to avoid over-fitting of the model, the learning rate must be kept as low level. Although the first adaptive drift correction approach was applied using an artificial neural network (ANN), they have several drawbacks:

- Drift correction method is only applicable for slow and gradual drift process. Regardless of the time intervals between the exposures, discontinuity in response between consecutive exposures would immediately invalidate the classification model and would prevent adaptation.
- During normal operation of the system, it can be very complicated to select an optimal or an appropriate learning rate which may strongly affect the correction capability. Selecting the optimal learning rate still a research issue and no automatic method have been reported in the literature.
- Creating an adaptive model is a complex task and requires a high number of training samples.
- The neural network drift correction methods are mainly limited to gas classification applications. This makes it difficult for applying current ANN adaptive methods to obtain reliable gas concentration, especially when both classification and gas quantitative analysis are required.
- Tunning ANN is a very challenging task and finding a global optimum is not guaranteed.

KNN

k-Nearest Neighbour classifiers, also known as k-NN, is a classifying method based on the very simple idea of proximity. It is a non-parametric classifier model, which is characterized by the training data itself. In k-NN classification model, in order to classify incoming samples xt, first, the closest k samples are selected from the training set using some distance measurements.

Then xt point will be classified using the majority vote for k nearest points. As the amount of data approaches infinity, k-NN is guaranteed to converge with an error rate not worse than 2Pb (where Pb represents the Bayes error rate which is the lowest possible error rate for a given class of classifier) Cover and Hart (1967) . In most cases, K-NN is a very useful tool and performs very well when rapidly building wrappers for feature selection/extraction. K-NN has been and still being used widely in gas sensor applications Marcelloni (2001) Some literature discussed the drawbacks of k-NN and solutions for overcoming them. The generalisation capability of K-NN can be improved in some cases by replacing the Euclidean distance with Mahalanobis distance. Mahalanobis distance take into consideration data co-variance in the feature space which leads to elliptic decision boundaries in the 2D case compared to the circular boundary found in the Euclidean case.

$$D_{(u_i,\Sigma)}(x_i)^2 = (x_i - u)^T \sum^{-1}(x_i - u) \tag{9}$$

where u is the mean for some observations and Σ is the covariance matrix. Calculating the inverse of the covariance matrix Σ^{-1} can be problematic especially if the data samples is small in relation to the number of features involved. The Mahalanobis distance calculation (9) should be carefully handled in order to prevent problems during the computation of Σ^{-1}. Other extensions of K-NN, such as fuzzy classification of a Single Nearest Neighbor Fuzzy also have been reported Singh (1999). K-NN performance can be severely degraded in the presence of noise or irrelevant features. One challenging problem in K-NN is how to choose the optimal K that maximise the performance of K-N.

Drift Compensation Using Clustering Methods

A possible alternative approach to classification is a clustering approach which allows data to be clustered into n number of clusters based on their similarities. Clustering enables the grouping of different data points without the need for prior knowledge about the patterns or logical grouping. A good clustering method should reveal the true relationships between data points and allocate each data point to its correct group in order to reveal the inner structure of the data. Clustering methods can be useful for visualising a given data sample, revealing the data distribution, and can be used with other methods as an initialization step to enhance the performance.

Hierarchical Cluster Analysis

In the cluster analysis technique, usually a tree structure, also called a dendrogram, is used to represent the data. This tree structure can be constructed in two ways: agglomerative or divisive clustering. In agglomerative clustering, each sample is considered as a single cluster and iterated by merging clusters as a function of a similarity measurement until the whole data set is transformed into a single cluster, which is called the root. The divisive clustering is a top-down approach. The algorithm starts by placing all data points in a single cluster, then splits the cluster using a flat clustering algorithm based on similarities. This procedure is applied iteratively until each point is placed in its own cluster. Comparing these two approaches, although agglomerative clustering behaves

more logically with respect to computing, divisive clustering provides more interpretable results under some conditions Ripley (2007). Unlike agglomerative clustering (bottom-up), the top-down approach has the advantage of capturing the global distribution of the dataset, thus increasing its performance. Hierarchical Cluster Analysis has been used in the sensor array field for paper type identification Holmberg, Winquist, Lundstr¨om, Gardner, and Hines (1995). In this work, the clustering performance of sensors is considered as the quality indicator for sensor selection in a mixed array. Similar developments can be found elsewhere Aishima (1991); Gardner (1991); Shurmer, Gardner, and Corcoran (1990). The main drawback with hierarchical clustering is its sensitivities to outliers and noise. Data sets with noise or outliers can cause these methods to add new clusters or merge clusters. Secondly, the generation of unique clusters is not possible, and the user has to select the appropriate cluster. These methods have high complexity $o(n^3)$ though some optimal methods can achieve $o(n^2)$.

SOM

Self Organising Maps(SOM) was proposed by Marco et al Marco et al. (1997) to classify gases based on sensor responses. The objective of their work was to identify combustion gases using six tin oxide sensors. It was shown that the sensors exhibited nonlinear behaviours. With the use of conductance instead of resistance-based sensors, the nonlinearity still existed and was estimated to be 20% full scale of nonlinearity. To test SOM performance against long-term drift, 0-40% of artificial drift was introduced. The drift was modelled using $G(T) = G_0(1 \pm \alpha t)$, where a plus or minus sign was chosen randomly for every sensor. Marco et al attempted to classify different gases, including methane, carbon monoxide, carbon dioxide, hydrogen, and two binary mixtures: one of hydrogen and carbon monoxide and another of methane and carbon monoxide. Other researchers have used SOM to classify gases and handle sensor drift. In Di Natale, Macagnano, D'Amico, and Davide (1997), SOM was used to classify food according to its freshness. A fish data set was generated using an e-nose and the attempt was made to classify the freshness of the fish according to its storage time in a refrigerator.

The purpose of Kohonens SOM is to capture the topology and the probability distribution of some input data as shown in Figure 4. SOM is an unsupervised learning algorithm designed to produce a low-dimensional, discretized representation of the input space of the training samples. SOM consists of nodes or neurons which are associated with a weight vector. Neurones are arranged in a rectangular grid, and each neurone is represented by one rectangular cell. Each neurone has four logical immediate neighbours. Each neurone has a specific topological position (x, y coordinate in the grid) and contains a vector of weights w of the same dimension as the input vectors x. SOM tries to find a neurone with the closest weight vector w to the input vector x using Euclidean distance, as shown in equation 1.10. The neurone weights are initialized randomly and, in each step, the weight vector for each neurone and its neighbours gets updated using the learning rule presented in equation 11. When a new gas sensor response

Figure 4. SOM

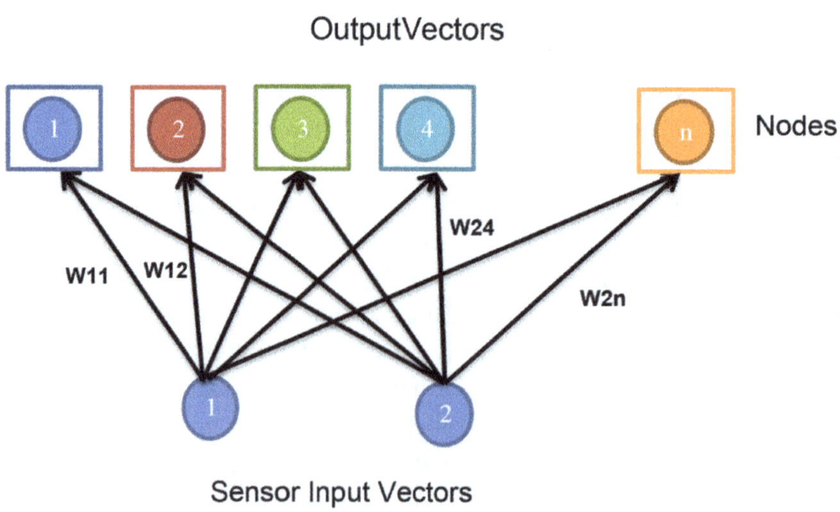

is presented to the network, the learning algorithm updates the neurones weights in order to group gases with similar properties together. Eventually, when SOM converges, all gases will be classified correctly. The topology visualisation characteristic of SOM has not been widely used in the sensory array field. Most of the literature applied SOM as a classification tool rather than as a visualisation tool. There is some evidence found in the literature demonstrating SOM has higher performance than MPL or RBFs.

$$Dist = \sqrt{\sum_{i=0}^{n} (v_i - w_i)^2} \tag{10}$$

$$w_{ji}^{new} = w_{ji}^{old} + \eta(t) \cdot \sigma(t) \cdot (x_i - w_{ji}^{old}) \tag{11}$$

$$\sigma(t) = \sigma_0 \exp\left(-\frac{t}{\lambda}\right), \ t = 1, 2, 3... \tag{12}$$

$$\eta(t) = \sigma_0 \exp\left(-\frac{dist^2}{2\sigma^2(t)}\right), \ t = 1, 2, 3... \tag{13}$$

where t represents the time step, $\eta(t)$ is the learning rate which decreases with time, and can be measured using different functions such as the exponential decay function shown in equation 12, $\sigma(t)$ is the neighbourhood area and this area shrinks over time. The neighbourhood can be calculated using the exponential decay function as shown in 12, dist is the distance between the neurone and the input vector, and can be calculated using the Euclidean distance. The method in Marco et al. (1997), only tries to handle a simulated drift imposed into sensor data without taking into consideration real drift scenarios. The result shows a rapid drop in classification accuracy as drift increases. The method needs to be validated against real drift, noise and simplicity, adaptivity, and non-linear dimensionality reduction. SOM maps high dimensional data into one or two dimensions which can cause important features to be lost or ignored. Due to the competitive learning nature of SOM, samples are usually surrounded by similar samples. However, similar samples could be placed in different cluster positions making them unrelated. SOM has poor tracking capabilities when the probability density of the data is changing. The competitive nature and the dynamic update of neurones of the algorithm make SOM knowledge history to be destroyed which result in unstable system. For instance, if one of the gas patterns is not measured for a long time, its neurones will be influenced by other neurones from other classes causing undesirable effects to the system. The final major problem with SOM is that they are very computationally expensive.

K-Mean

K-Mean is a non-hierarchical cluster analysis method, in which data is partitioned into a fixed number of components or clusters. During the initialization stage, data points are randomly assigned to clusters so that each cluster has roughly the same number of data points. Then, the distance from data point to each cluster centre is computed iteratively, and each observation is assigned to the cluster. Convergence is achieved when the assignment no longer changes. The initial partition selection, however, affects the final convergence significantly.

E-Nose Drift Counteraction by Ensemble Learning

Vergara et al. proposed classifier of ensembles to compensate for gas sensor drift. In their study, they collected data for e-nose over 36 months. The primary objective of their study was to develop a method to compensate for drift using an extensive dataset generated using metal-oxide gas sensors. The dataset consists of a six-gas/analyte classification problem dosed at various concentrations with the objective to discriminate between different analytes regardless of their concentration. SVM ensemble classifiers were trained using the current batch S_t at time t where $S_t = \{(x_1, y_1), \ldots, (x_{mt}, y_{mt})\}$ of size m_t. The final classifier is a weighted combination of all classifiers. Classifiers weights $\{\beta_1, \ldots, \beta_t\}$ can be estimated using prediction performance of classifiers at batch S_t. Prediction is made using a weighted majority voting technique:

$$h_{t+1}(x) = \arg \max_{m \in 1 \ldots M} \sum_{t: f_t(x) = y} \beta_t \tag{14}$$

Figure 5 shows the result acquired by Vergara et al. when training the classifier in three different settings. The classifiers were tested under 3 settings as described below:

- Setting 1 For every month, they trained a multi-class classifier with data from only the previous month and tested it on the current month.
- Setting 2 For every month, they trained an ensemble of multi-class classifier.
- Setting 3 Same as Setting 2 but with uniform weights on the individual (base) classifiers.

In Vergara et al. (2012), they assumed that drift is gradual and the noise was ignored. Ensemble learning (boosting) is particularly subject to over-fitting when there is significant noise in the training data. Furthermore, the proposed method is very prohibitive because it requires a large amount of labelled data from multiple batches. In real-time e-nose applications, it is not possible to acquire target labels due to the complexity and cost involved. Therefore, it is very critical to consider a semi-supervised adaptive system able to extract labels from unlabeled data.

Figure 5. Drift Compensation using Ensembl. Performance of classifiers under Setting 1 (black, dashed), Setting 2 (red, dotted) and Setting 3 (blue, dotted) described in the text. The black, continuous line corresponds to the setting where a classifiers was trained with batch 1 and tested on successive batches. *Vergara et al. (2012).*

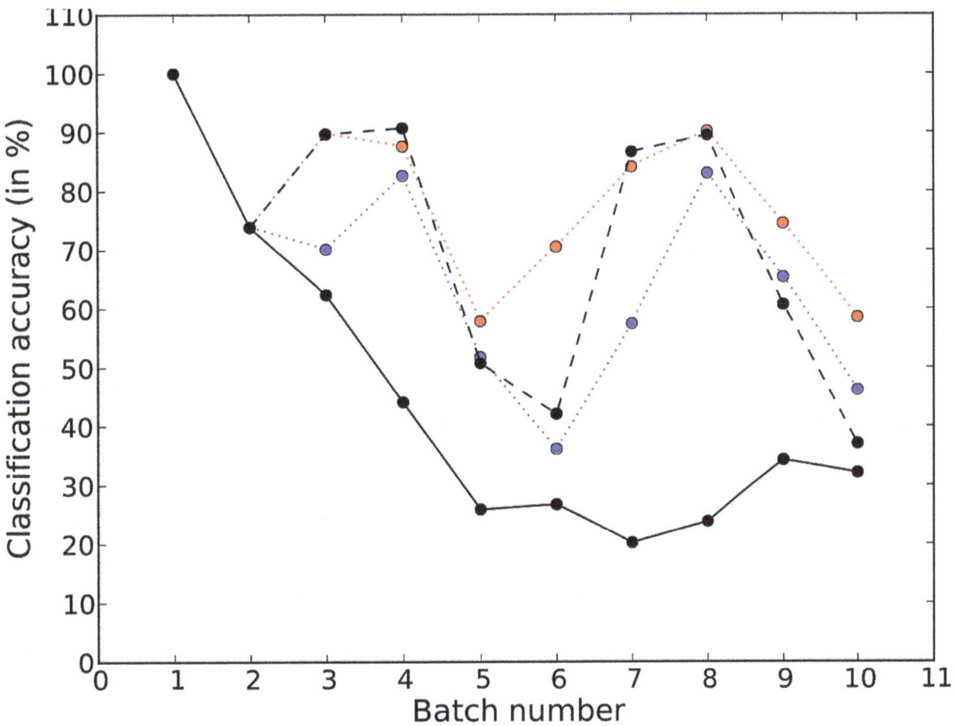

CONCLUSION

This work is motivated by research questions that have a significant technological impact and practical relevance to air pollution monitoring, medical diagnostics, food spoilage detection and security applications. Unlike the advances made in speech recognition and machine vision, e-nose pattern recognition requires significant improvements. The metal-oxide sensors used in e-nose have sensitivity, selectivity and stability issues which impact their performance. The authors have conducted a critical review of pattern recognition and drift compensation method for e-nose sensor array. Our analysis shows some problems in this technology which requires further investigation. e-nose drift remains the major issue in e-nose pattern recognition systems. Few research studies have addressed drift problem in e-nose. However, Most studies which addressed drift considered synthetic data sets or very small private data sets.

In this chapter, a study is presented which covers prior and current research in e-nose pattern recognition. In particular, a taxonomy of pattern recognition and drift correction for e-nose is presented. We summarised drift correction methods for e-nose into four main categories: i) Adaptive methods, ii) reference based, iii) adaptive component correction and iiii) signal processing. The methods covered in the literature are Neural Net, Self Organising Maps, K-Nearest Neighbors and Support Vector Machine.

REFERENCES

Aishima, T. (1991). Aroma discrimination by pattern recognition analysis of responses from semiconductor gas sensor array. *Journal of Agricultural and Food Chemistry*, *39*(4), 752–756. doi:10.1021/jf00004a027

Al-Maskari, S., B'elisle, E., Li, X., Le Digabel, S., Nawahda, A., & Zhong, J. (2016). Classification with quantification for air quality monitoring. Academic Press. doi:10.1007/978-3-319-31753-3_46

Al-Maskari, S., Guo, W., & Zhao, X. (2016). Biologically inspired pattern recognition for e-nose sensors. In *Advanced data mining and applications: 12th international conference* (pp. 142–155). Academic Press. doi:10.1007/978-3-319-49586-6_10

Al Maskari, S., Kumar, D., & Chiffings, T. (2012). Data mining for environment monitoring. In *International Conference on Software Technology and Engineering (ICSTE 2012)*. ASME Press. doi:10.1115/1.860151_ch87

Al-Maskari, S., Li, X., & Liu, Q. (2014). An effective approach to handling noise and drift in electronic noses. In *Databases theory and applications - 25th australasian database conference* (pp. 223–230). Academic Press. doi:10.1007/978-3-319-08608-8_21

Al-Maskari, S., Saini, D., & Omar, W. (2010). Cyber infrastructure and data quality for environmental pollution control - in Oman. *Proceedings of the 2010 DAMD International Conference on Data Analysis, Data Quality and Metada*. doi:10.5176/978-981-08-6308-1_D-038

Artursson, T., Ekl¨ov, T., & Lundstr¨om, I., Martensson, P., Sj¨ostr¨om, M., & Holmberg, M. (2000). Drift correction for gas sensors using multivariate methods. *Journal of Chemometrics*, *14*(5-6), 711–723. doi:10.1002/1099-128X(200009/12)14:5/6<711::AID-CEM607>3.0.CO;2-4

Balkin, S. D., & Ord, J. K. (2000). Automatic neural network modeling for univariate time series. *International Journal of Forecasting*, *16*(4), 509–515. doi:10.1016/S0169-2070(00)00072-8

Bishop, C. M. (1995). *Neural networks for pattern recognition*. New York: Oxford University Press, Inc.

Broomhead, D. S., & Lowe, D. (1988). *Radial basis functions, multi-variable functional interpola- tion and adaptive networks (Tech. Rep.)*. DTIC Document.

Carpenter, G. A., Grossberg, S., & Reynolds, J. H. (1991). Artmap: Supervised real-time learning and classification of nonstationary data by a self-organizing neural network. *Neural Networks, 4*(5), 565 - 588. Retrieved from http://www.sciencedirect.com/ science/article/pii/089360809190012T

Comon, P. (1994). Independent component analysis, a new concept? *Signal Processing*, *36*(3), 287–314. doi:10.1016/0165-1684(94)90029-9

Cover, T., & Hart, P. (1967). Nearest neighbor pattern classification. *Information Theory. IEEE Transactions on*, *13*(1), 21–27.

Davide, F. A., Di Natale, C., & D'Amico, A. (1994). Self-organizing multisensor systems for odour classification: Internal categorization, adaptation and drift rejection. *Sensors and Actuators. B, Chemical, 18*(1), 244–258. doi:10.1016/0925-4005(94)87090-X

Di Natale, C., Macagnano, A., D'Amico, A., & Davide, F. (1997). Electronic-nose modelling and data analysis using a self-organizing map. *Measurement Science & Technology, 8*(11), 1236–1243. doi:10.1088/0957-0233/8/11/004

Dutta, R., Kashwan, K., Bhuyan, M., Hines, E. L., & Gardner, J. (2003). Electronic nose based tea quality standardization. *Neural Networks, 16*(5), 847–853. doi:10.1016/S0893-6080(03)00092-3 PMID:12850043

Fayyad, U., & Uthurusamy, R. (1996). Data Mining and Knowledge Discovery in Databases (Introduction to the Special Section). *Communications of the ACM, 45*(8), 24–26. doi:10.1145/240455.240463

Fryder, M., Holmberg, M., Winquist, F., & Lundstrom, I. (1995, Jun). A calibration technique for an electronic nose. In *Solid-state sensors and actuators, 1995 and eurosensors ix. transducers '95. the 8th international conference on* (Vol. 1, p. 683-686). Academic Press. .717323 doi:10.1109/SENSOR.1995.717323

Gardner, J. W. (1991). Detection of vapours and odours from a multisensor array using pattern recognition part 1. principal component and cluster analysis. *Sensors and Actuators. B, Chemical, 4*(1), 109–115. doi:10.1016/0925-4005(91)80185-M

Gutierrez-Osuna, R. (2000). Drift reduction for metal-oxide sensor arrays using canonical correlation regression and partial least squares. In *Proceedings of the 7th international symposium on olfaction & electronic nose* (pp. 1–7). Academic Press.

Haugen, J.-E., Tomic, O., & Kvaal, K. (2000). A calibration method for handling the temporal drift of solid state gas-sensors. *Analytica Chimica Acta, 407*(12), 23–39. doi:10.1016/S0003-2670(99)00784-9

Holmberg, M., Winquist, F., Lundstr¨om, I., Gardner, J., & Hines, E. (1995). Identification of paper quality using a hybrid electronic nose. *Sensors and Actuators. B, Chemical, 27*(1), 246–249. doi:10.1016/0925-4005(94)01595-9

Huang, M.-C. (2004). Wave parameters and functions in wavelet analysis with filtering. *Ocean Engineering, 31*(7), 813 - 831. Retrieved from http://www.sciencedirect.com/ science/article/pii/S0029801803002129

Huang & Wang. (Eds.). (2016). *Advances in knowledge discovery and data mining: 20th pacific-asia conference*. Cham: Springer International Publishing. doi: 4610.1007/978-3-319-31753-3

Hui, D., Jun-hua, L., & Zhong-ru, S. (2003). Drift reduction of gas sensor by wavelet and principal component analysis. *Sensors and Actuators B: Chemical, 96*(12), 354 - 363. Retrieved from http://www.sciencedirect.com/science/article/pii/

Kim, D. (2001). Performance comparison of neural network models: Backpropagation vs. fuzzy artmap. *International Journal of Computer Mathematics, 78*(3), 365–382. doi:10.1080/00207160108805117

Mallat, S. G. (1989). Multiresolution approximations and wavelet orthonormal bases of 2. *Transactions of the American Mathematical Society*, *315*(1), 69–87.

Marcelloni, F. (2001). Recognition of olfactory signals based on supervised fuzzy c-means and knn algorithms. *Pattern Recognition Letters*, *22*(9), 1007–1019. doi:10.1016/S0167-8655(01)00040-X

Marco, S., Pardo, A., Ortega, A., & Samitier, J. (1997, May). Gas identification with tin oxide sensor array and self organizing maps: adaptive correction of sensor drifts. In Instrumentation and measurement technology conference, 1997. imtc/97. proceedings. sensing, processing, networking (Vol. 2, pp. 904-907). IEEE. doi:10.1109/IMTC.1997.610256

Moody, J., & Darken, C. (1988). *Learning with localized receptive fields*. Yale Univ., Department of Computer Science.

Padilla, M., Perera, A., Montoliu, I., Chaudry, A., Persaud, K., & Marco, S. (2010). Drift compensation of gas sensor array data by orthogonal signal correction. *Chemometrics and Intelligent Laboratory Systems, 100*(1), 28 - 35. Retrieved from http://www.sciencedirect .com/science/article/pii/S0169743909001877

Richard, M. D., & Lippmann, R. P. (1991). Neural network classifiers estimate bayesian a posteriori probabilities. *Neural Computation*, *3*(4), 461–483. doi:10.1162/neco.1991.3.4.461

Ripley, B. D. (2007). *Pattern recognition and neural networks*. Cambridge University Press.

Shurmer, H., Gardner, J., & Corcoran, P. (1990). Intelligent vapour discrimination using a composite 12-element sensor array. *Sensors and Actuators B: Chemical, 1*(16), 256 - 260. Retrieved from http://www.sciencedirect.com/science/article/pii/

Singh, S. (1999). A single nearest neighbor fuzzy approach for pattern recognition. *International Journal of Pattern Recognition and Artificial Intelligence*, *13*(01), 49–64. doi:10.1142/S0218001499000045

Taurino, A., Distante, C., Siciliano, P., & Vasanelli, L. (2003). Quantitative and qualitative analysis of vocs mixtures by means of a microsensors array and different evaluation methods. *Sensors and Actuators. B, Chemical*, *93*(1), 117–125. doi:10.1016/S0925-4005(03)00241-7

Tudu, B., Jana, A., Metla, A., Ghosh, D., Bhattacharyya, N., & Bandyopadhyay, R. (2009). Electronic nose for black tea quality evaluation by an incremental fRBFg network. *Sensors and Actuators. B, Chemical*, *138*(1), 90–95. doi:10.1016/j.snb.2009.02.025

Vergara, A., Vembu, S., Ayhan, T., Ryan, M. A., Homer, M. L., & Huerta, R. (2012). Chemical gas sensor drift compensation using classifier ensembles. *Sensors and Actuators. B, Chemical*, *166–167*(0), 320–329. doi:10.1016/j.snb.2012.01.074

Wold, S., Antti, H., Lindgren, F., & Hman, J. (1998). Orthogonal signal correction of near-infrared spectra. *Chemometrics and Intelligent Laboratory Systems, 44*(12), 175 - 185. Retrieved from http://www.sciencedirect.com/science/article/pii/

Zhang, H., Balaban, M., & Principe, J. C. (2003). Improving pattern recognition of electronic nose data with time-delay neural networks. *Sensors and Actuators B: Chemical, 96*(12), 385 - 389. Retrieved from http://www.sciencedirect.com/science/article/pii/S0925400503005744

Ziyatdinov, A., Marco, S., Chaudry, A., Persaud, K., Caminal, P., & Perera, A. (2010). Drift compensation of gas sensor array data by common principal component analysis. *Sensors and Actuators. B, Chemical, 146*(2), 460–465. doi:10.1016/j.snb.2009.11.034

Zuppa, M., Distante, C., Persaud, K. C., & Siciliano, P. (2007). Recovery of drifting sensor responses by means of fDWTg analysis. *Sensors and Actuators B: Chemical, 120*(2), 411 - 416. Retrieved from http://www.sciencedirect.com/science/article/pii/

Zuppa, M., Distante, C., Siciliano, P., & Persaud, K. C. (2004). Drift counteraction with multiple self-organising maps for an electronic nose. *Sensors and Actuators B: Chemical, 98*(23), 305 - 317. Retrieved from http://www.sciencedirect.com/science/article/pii/

Chapter 4
Fluorescence Quenching Sensor Arrays for the Discrimination of Nitroaromatic Vapors

Nico Bolse
Karlsruhe Institute of Technology (KIT), Germany

Anne Habermehl
Karlsruhe Institute of Technology (KIT), Germany

Carsten Eschenbaum
Karlsruhe Institute of Technology (KIT), Germany

Uli Lemmer
Karlsruhe Institute of Technology (KIT), Germany

ABSTRACT

Fluorescence quenching is a promising technique for chemical sensing applications such as the detection of explosive trace vapors. Nitroaromatic explosives are one of the primary targets for this approach enabling ultra-low detection limits down to sub parts-per-billion in air. Many studies, however, focus on enhanced sensor sensitivity, whereas practical applications often require the identification and quantification of detected species. Electronic noses and efficient sensor systems are a promising solution to address this challenge. The authors review recent approaches and trends for explosive trace vapor detection and discuss theoretical concepts for fluorescence quenching as well as target analytes, sensor materials, and fabrication techniques. Statistical learning techniques such as principal component analysis and linear discriminant analysis, sensor systems, and camera-based read-out strategies are in the focus of the chapter. The authors conclude with recommendations and solutions for the elaborated challenges and with visions on future research directions.

DOI: 10.4018/978-1-5225-3862-2.ch004

INTRODUCTION

The detection of trace vapors, especially those emitted from explosives, bombs and dynamites (EBDs) has become more and more important. Recent data from the National Consortium of the Study of Terrorism and Responses to Terrorism (2016) provide evidence that the worldwide number of terrorist attacks and those utilizing EBDs significantly increased as shown in Figure 1. According to a report from the Institute for Economics and Peace (2016), this rise is mainly related to increasing religious extremism, but also due to nationalism, separatism and politics. The report indicates that the increased extremism might be a direct result of the Syrian war starting in 2011 and of ISIS declaring caliphate in 2014. It is stated that private citizens and properties are targeted by most of the attacks which increased the number of deaths from terrorism from about 2.000 in the year 2000 up to more than 12.000 in the year 2015. Furthermore, the decline between 2014 and 2015 of roughly 10% is attributed to military actions and interventions in Iraq and Nigeria, even though both of which were among the countries with the highest number of deaths from terrorism in 2014 and 2015 as is explained in the report.

These trends demonstrate the urgent need for reliable explosives detection. However, the identification and quantification of detected targets remains challenging. Sniffer dogs, for example, can be trained to be highly sensitive on the one hand. On the other hand, their effectiveness critically depends on their constitution requiring frequent rests. Moreover, each dog needs a trainer and a handler resulting in a rather expensive approach. Other well-established detection approaches such as ion mobility spectrometers (IMS) are complex and expensive, too. Furthermore, IMS typically requires a radioactive ionization source which can cause regulatory difficulties. Therefore, active research is carried out on alternative technologies, each having its own advantages and drawbacks. A non-exhaustive comparison of various state-of-the-art technologies for explosive trace vapor sensing is given in Table 1.

Fluorescence quenching is just one of the competing detection technologies and has been recently commercialized with the Fido® device sold by FLIR for the US military. The ultra-sensitivity with

Figure 1. Worldwide number of terrorist attacks and those utilizing EBDs
Data retrieved from the National Consortium for the Study of Terrorism and Responses to Terrorism (2016). Used with permission.

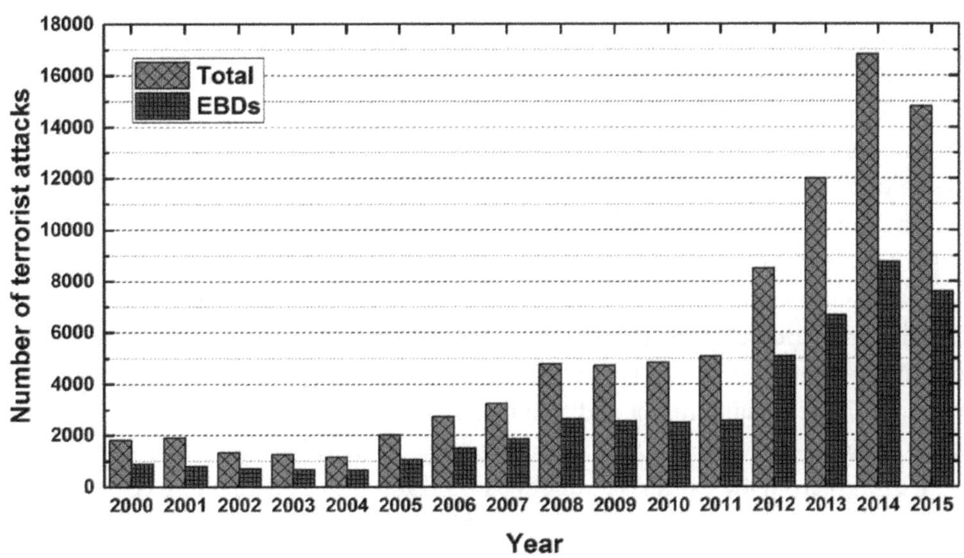

Table 1. Comparison of various detection technologies for explosive trace vapors; The table does not claim to be exhaustive.

Technology	Detection Principle	Advantages	Disadvantages
Sniffer dogs, bees and rats	Odor smelling	Highly sensitive, can be trained for specific purposes	Animal protection concerns, no quantification, limited operation time, expensive handling/training
Fluorescence quenching (FQ)	Charge and energy transfer based quenching	Highly sensitive, simple detection principle allowing highly-integrated and low-cost solutions	Non-selective, cross-sensitivities, stability and recovery issues such as photo-bleaching and irreversible quenching
Ion mobility spectrometry (IMS)	Molecular decomposition and separation in an electric field	Highly sensitive and selective, well-established approach	Rather complex systems, radioactive source, target needs to be evaporated
Gas chromatographic mass spectrometry (GC-MS)	Column-based separation combined with a mass spectrometer	Rapid throughput, high degree of confidence	Ultrahigh-pure gas supply required, long setup times when relocated, target needs to be evaporated
Infrared/terahertz spectroscopy	Fingerprint analysis of absorption spectra	Highly selective, low radiation dose	Low sensitivity, long optical paths require expensive mirrors
Laser-induced breakdown spectroscopy (LIBS)	Laser-pulsed microplasma analysis	Real-time standoff detection	Low selectivity, atmospheric interferences
Raman spectroscopy	Fingerprint analysis of Raman scattering	Real-time standoff detection	Low sensitivity
Cavity-ring-down-spectroscopy (CRDS)	Analysis of light pulse decay in resonant cavity	Discrimination of threats from complex backgrounds	Low selectivity, atmospheric interferences
Metal oxide gas sensors	Change in resistance	Well-developed technology, broad range of detectable targets, low-cost	Drift, cross-sensitivities, humidity- and flow-sensitive
Photoacoustic spectroscopy	Laser vibrometer analysis of pulse-excited target	Highly sensitive	Targets need to be evaporated, interferents in the infrared
Colorimetric sensors	Analysis of color shifts	Small, low-cost, simple	Long reaction times
Bio sensors	Antibody-antigen reactions	Highly specific	Slow response

Caygill, Davis, & Higson, 2012; Bielecki et al., 2012; Gares, Hufziger, Bykov, & Asher, 2016; Kumar & Murali., 2016.

detection limits on the parts-per-trillion (ppt) level render fluorescence quenching very attractive. The extreme sensitivity, however, comes with a lack of selectivity that can be attributed to the transduction based on charge or energy transfer mechanisms (Lakowicz, 2006). This lack is crucial for the identification and quantification of detected targets in the case of a single sensor. In contrast, the utilization of optical sensor arrays in combination with pattern recognition techniques, herein referred to as optoelectronic noses, help to bridge this gap. Such noses represent a powerful technique to detect explosive trace vapors by combining ultra-sensitivity and discriminatory ability. There are several groups working on different scientific approaches in which fluorescence quenching is utilized as the transduction mechanism in sensor arrays for the discrimination of explosive trace vapors. This includes but is not limited to multichannel detection and differentiation of explosives with a quantum dot array (Peveler, Roldan, Hollingsworth, Porter, & Parkin, 2016), a quantum dot and polymer cross-reactive array for chemical vapor detection (Bright, Nallon, Polcha, & Schnee, 2015) and a digitally printed optoelectronic nose for the selective trace detection of nitroaromatic explosive vapors using fluorescence quenching (Bolse et al., 2017a).

BACKGROUND

Fluorescence refers to the spontaneous emission of a photon in a fluorophore that has absorbed electromagnetic radiation (Lakowicz, 2006). The spontaneous emission is due to a radiative relaxation of an electron from the excited singlet state with the lowest energy (S_1) down to the ground state (S_0) as is indicated in the Jablonski diagram in Figure 2. It is a form of luminescence and can be differentiated from phosphorescence emitted from a spin-forbidden triplet state (T_1).

The theory of fluorescence quenching dates back to the work of Stern & Volmer (1919) on the decay time of fluorescence. They suggested a model to mathematically describe the quenching of fluorescence intensity and lifetime as a function of the target analyte concentration. In terms of fluorescence, the lifetime accounts for the time an electron spends in S_1 before it returns to S_0 by the emission of a photon. In their model, Stern & Volmer differentiated between dynamic and static quenching.

Dynamic Quenching

Dynamic quenching is based on collisional events between a fluorophore and a target analyte. These events can trigger the non-radiative deactivation of excited states if the target is able to diffuse to the fluorophore within the lifetime of the excited states. The mechanism is called collisional quenching, too, due to the kinetic character which renders the process reversible. For a continuous fluorescence excitation with a constant fluorescence over time, the concentration of excited states in the absence of a target is constant according to:

Figure 2. Jablonski diagram to illustrate optical transitions of electronic states

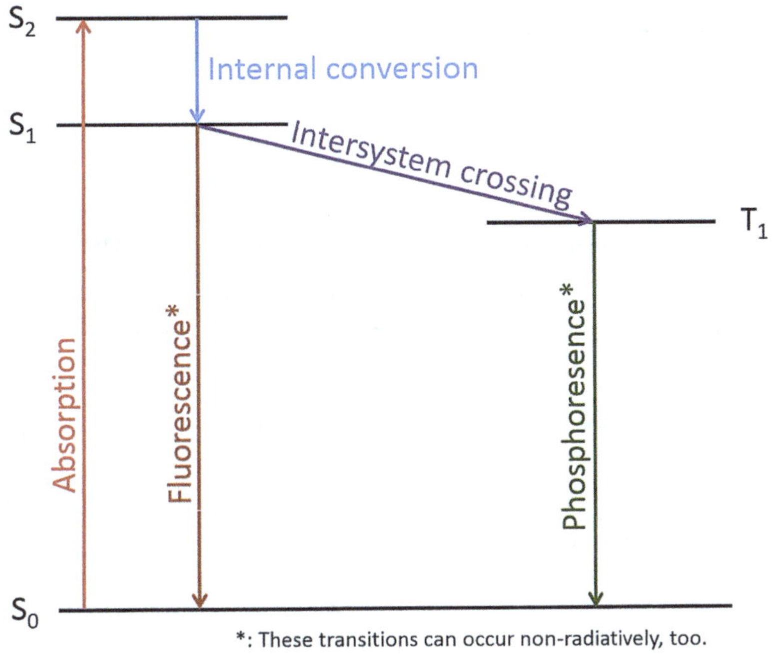

$$\frac{d\left[n^*\right]}{dt} = f\left(t\right) - \gamma_0 \left[n^*\right]_0 = 0, \tag{1}$$

where $\frac{d\left[n^*\right]}{dt}$ is the change of the concentration of excited states in the fluorophore over time, $f\left(t\right)$ is the time-dependent function of the fluorescence excitation, γ_0 is the sum of the non-radiative and the radiative decay rates of the fluorophore and $\left[n^*\right]_0$ is the initial concentration of excited states in the fluorophore. In the presence of a target, Stern and Volmer made the empirical assumption that the non-radiative decay rate is only affected by the contribution of the quenching such that:

$$\frac{d\left[n^*\right]}{dt} = f\left(t\right) - \left(\gamma_0 + k_q\left[Q\right]\right) \cdot \left[n^*\right] = 0, \tag{2}$$

where k_q is the bimolecular quenching constant and $\left[Q\right]$ is the target concentration. A rearrangement of (1) and (2) directly yields:

$$\frac{\left[n^*\right]_0}{\left[n^*\right]} = 1 + \frac{k_q}{\gamma_0}\left[Q\right] \tag{3}$$

The Stern-Volmer equation for dynamic quenching in its most common form can then be derived from the assumption that the fluorescence intensity is directly proportional to the concentration of excited electronic states in the fluorophore such that:

$$\frac{I_0}{I} = 1 + \frac{k_q}{\gamma_0}\left[Q\right] = 1 + K_D\left[Q\right], \tag{4}$$

where I_0 is the fluorescence intensity in the absence of the target, I is the fluorescence intensity in the presence of the target and K_D is the Stern-Volmer constant for dynamic quenching. From another perspective, the fluorescence intensity ratio can be expressed as a ratio of fluorescence lifetimes because the kinetic deactivation of excited states reduces the lifetime in the same manner. Mathematically, this is expressed by utilizing the reciprocal decay rates $\tau_0 = \gamma_0^{-1}$ and $\tau = \left(\gamma_0 + k_q\left[Q\right]\right)^{-1}$ such that:

$$\frac{\tau_0}{\tau} = 1 + k_q\tau_0\left[Q\right] = 1 + K_D\left[Q\right] = \frac{I_0}{I} \tag{5}$$

Static Quenching

On the other hand, the concept of static quenching is based on the assumption that a fluorophore and a target are able to form a non-fluorescent complex. This formation is often based on photochemical reactions that do not necessarily be reversible. The Stern-Volmer equation for static quenching can be derived from the association constant for the complex formation in the presence of the target, defined by:

$$K_S = \frac{\left[n^* - Q\right]}{\left[n^*\right]\left[Q\right]}, \tag{6}$$

where K_S is the Stern-Volmer constant for static quenching, $\left[n^* - Q\right]$ is the concentration of non-fluorescent complexes, $\left[n^*\right]$ is the remaining fluorophore concentration and $\left[Q\right]$ is the target concentration. Because the concentration of excited states is constant in the fluorophore due to constant fluorescence excitation, the total fluorophore concentration in the absence of the target is given by:

$$\left[n^*\right]_0 = \left[n^*\right] + \left[n^* - Q\right] \tag{7}$$

A substitution of (7) into (6) and a rearrangement lead to the Stern-Volmer equation for static quenching:

$$\frac{I_0}{I} = 1 + K_S\left[Q\right] \tag{8}$$

It is important to note that the fluorescence intensity ratio is not equal to the fluorescence lifetime ratio in the case of static quenching. Here, the observed fluorescence in a quenched state results from uncomplexed fluorophores that do not interact with the target. Thus, the decay rates are not affected such that:

$$\frac{I_0}{I} \neq \frac{\tau_0}{\tau} = 1 \tag{9}$$

Modified Stern-Volmer Quenching

In reality, measurements often deviate from the linear relationship between the quenching of fluorescence and the target concentration.

Combined Quenching

If both dynamic and static mechanisms are present, the observed fluorescence quenching would follow the multiplication of (4) and (8) such that:

$$\frac{I_0}{I} = \left(1 + K_D\left[Q\right]\right) \cdot \left(1 + K_S\left[Q\right]\right) \tag{10}$$

Sphere-of-Action Quenching

Another model, introduced by Perrin (1924), considers an exponential deviation that becomes significant at high concentrations. Such a deviation for a dynamic process can be attributed to a contribution by an apparent static quenching due to close contact between fluorophore and target at the instant of excitation. For this case, Frank & Wawilow (1931) introduced a sphere-of-action volume V in which a target is in direct contact to a fluorophore so that the probability for quenching is equal to one within the sphere-of-action. This leads to a modified form of the Stern-Volmer equation:

$$\frac{I_0}{I} = \left(1 + K_D\left[Q\right]\right) \cdot e^{VN_A\left[Q\right]}, \tag{11}$$

where N_A is the Avogadro constant. In this case, the Stern-Volmer plot will be nearly linear for low concentrations but it will follow an upward curvature for high concentrations.

Accessibility Quenching

A deviation for high concentrations in the downward direction can be modelled when considering the accessibility of fluorophores as suggested by Lehrer (1971). In this case, the fluorescence intensity in the absence of a target is given by:

$$I_0 = I_{0a} + I_{0b}, \tag{12}$$

where I_{0a} is the fluorescence intensity emitted by fluorophores that are accessible for a target and I_{0b} is the fluorescence intensity emitted by fluorophores that are inaccessible (e.g. buried) for the target. Here, only I_{0a} can be quenched such that the Stern-Volmer equation results in:

$$\frac{I_0}{I} = \frac{I_{0a} + I_{0b}}{\frac{I_{0a}}{1 + K_a\left[Q\right]} + I_{0b}}, \tag{13}$$

where K_a is the Stern-Volmer constant of the accessible fraction. In any case, a fraction of fluorescence remains independent of target concentration. Thus, the fluorescence quenching saturates for high concentrations.

The Stern-Volmer plots for the different models are indicated in Figure 3.

Figure 3. Stern-Volmer plots for different fluorescence quenching models: Dynamic, static, combined (dynamic and static quenching), exponential (sphere-of-action quenching) and saturating (accessibility quenching)

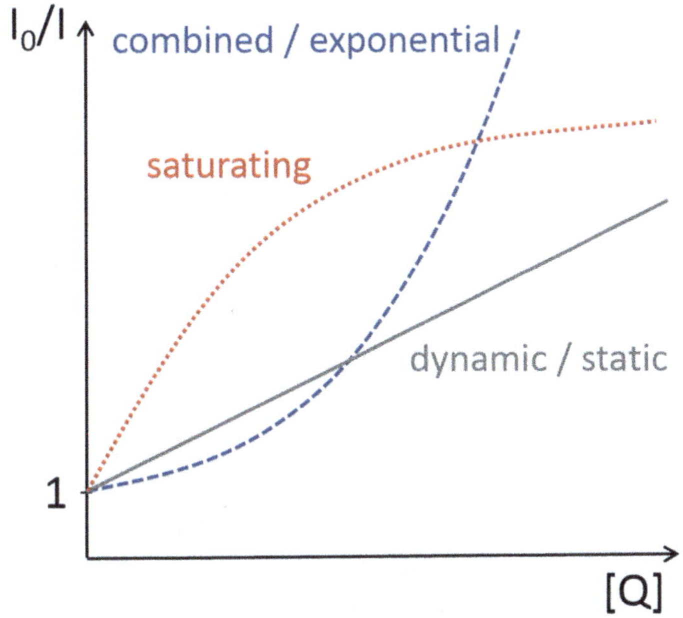

Deactivation Mechanisms

The non-radiative deactivation of the electronic states in a fluorophore due to the presence of a target can have many reasons. The most common processes include photo-induced electron transfer (PET), Förster resonance energy transfer (FRET) and Dexter energy transfer (DET) as mentioned by Sun, Wang, & Lei (2015).

Dexter Energy Transfer

DET is predominate for triplet transitions (phosphorescence) and is based on the exchange of electrons. For example, an excited electron in the fluorophore is transferred to the target while an electron in the ground state of the target is transferred to the fluorophore at the same time. However, the absorption spectrum of the fluorophore needs to overlap with the emission spectrum of the target and the wave functions of the electrons need to overlap requiring molecular distances on the lower Ångström level. The energy transfer rate reduces exponentially with increasing distance between fluorophore and target.

Förster Resonance Energy Transfer

FRET predominates for singlet transitions (fluorescence) and is based on dipole-dipole interactions between fluorophore and target. In this case, an electron in the lowest occupied molecular orbital (LUMO) of the fluorophore relaxes to the highest occupied molecular orbital (HOMO) while the released energy is transferred to the target such that an electron is excited from the HOMO to the LUMO of the target. The absorption spectrum of the fluorophore needs to overlap with the emission spectrum of the target for the resonance energy transfer to take place. Furthermore, the electronic oscillations of the fluorophore and the target should be oriented in parallel. The energy transfer rate is strongly dependent on the distance between fluorophore and target and scales with the inverse of the sixth-power distance. This requires short distances between fluorophore and target on the lower nanometer level.

Photo-Induced Electron Transfer

PET is often observed for the detection of explosives using conjugated polymers. The process is mostly based on an electron transfer from the fluorophore LUMO to the target LUMO as indicated in Figure 4(a). In this case, the target LUMO has to be lower in energy than the fluorophore LUMO in order to maintain a positive driving force between the two species. A different electron transfer can occur from the target HOMO to the fluorophore LUMO as shown in Figure 4(b). In the latter mechanism, the fluorophore acts as the acceptor while the target represents the electron donor.

Other Deactivation Mechanisms

There are other fluorescence quenching processes apart from PET, FRET and DET. These include the absorption of excitation radiation by the target if the absorption spectrum of the target and the emission spectrum of the excitation source overlap, absorption of fluorescence by the target if the emission spectrum of the fluorophore overlaps with the absorption spectrum of the target and reabsorption of the fluorescence by the fluorophore if the absorption and emission spectrum of the fluorophore overlap. However, these processes are not considered true quenching mechanisms in terms of Stern-Volmer theory and have to be corrected for in order to physically interpret Stern-Volmer constants. In contrast to fluorescence quenching processes, Andrew & Swager (2007), Germain & Knapp (2008) and others have utilized fluorescence enhancement for explosive sensing, too. The so-called "turn on" effect is based

Figure 4. PET mechanisms that quench the fluorescence. (a) The fluorophore is the donor and the target analyte is the acceptor. (b) The fluorophore is the acceptor and the target analyte is the donor.

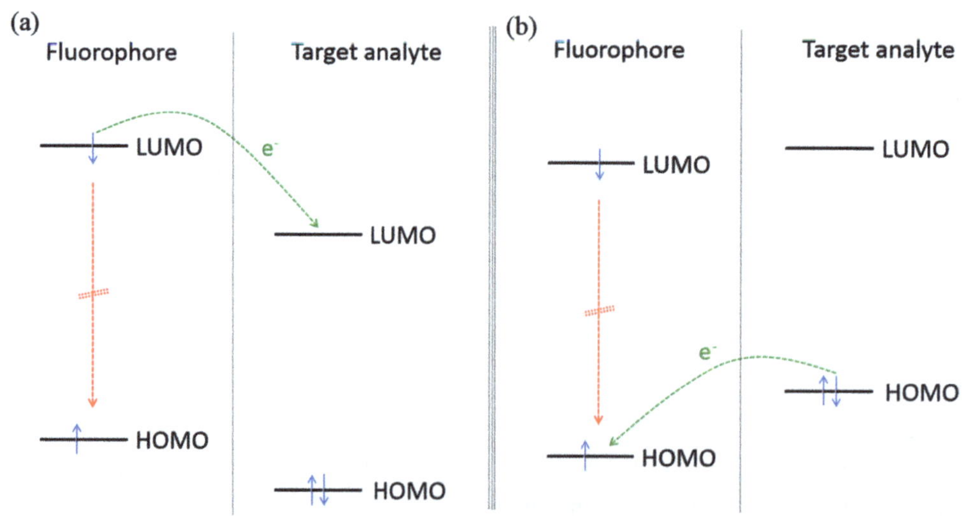

on additives forming non-fluorescent complexes with the fluorophore in the absence of target analytes. This initial step pre-quenches the fluorophore. Under certain conditions, the complexes can break up in the presence of a target analyte so that the fluorescence recovers from the pre-quenched state.

Sensor Materials

Within the last decades, scientists have focused on the development of highly-sensitive transducer materials for fluorescence quenching purposes. Organic polymers are just one class of materials that can be used for fluorescence quenching sensors (Toal & Trogler, 2006). In these polymers it is desirable to have long polymeric backbones that enable the delocalization of electrons over long distances. The long migration paths enhance the probability of target interactions at binding sites leading to an amplification of the quenching. Furthermore, rigid three-dimensional structures and twisted backbone galleries can offer interchain spacing to reduce self-quenching. These strategies helped to reach detection limits on the sub-ppb level. Instead of amplification, however, it was recently reported by Ali et al. (2015, 2016) that the diffusion and binding properties predominantly control the transduction. This means that the requirements for optimal transduction are still debated on.

Target Analytes

For fluorescence quenching sensors, the chemical class of nitroaromatic compounds is ideally suited as target analytes due to the beneficial electron withdrawing character of the nitro groups. Additionally, nitroaromatic residues remain detectable for a long period of time due to the stability of the benzene ring making such compounds resistant to oxidative degradation. Despite the high-explosive trinitrotoluene (TNT), many nitroaromatic compounds are not explosive themselves. However, they are often used in the manufacture of explosives and remain detectable in the final product. For example, a mixture of dinitrobenzene (DNB) with tetranitromethane and a mixture of nitrobenzene (NB) with nitric acid and water are highly explosive. The detection of DNB and NB is quite simple using appropriate fluorescent transducers but the detection of TNT is more challenging due to its low vapor pressure. In fact, the TNT concentration of a buried metal-encased landmine is estimated to be on the order of a few ppt which makes the detection very challenging. However, TNT is manufactured from the nitration of dinitrotoluene (DNT). The saturated vapor concentration of DNT is one to two orders of magnitude larger as compared to that of TNT which helps to indirectly detect TNT with reduced effort. In addition to their role in explosive manufacture, nitroaromatic compounds are a serious threat for human health and a critical environmental pollutant. A direct exposure to a few milligrams can already lead to carcinogenic, mutagenic and reprotoxic (CMR) effects. Thus, a reliable identification and quantification of such substances becomes even more important. The chemical structures of common nitroaromatic compounds and of two examples for organic sensor polymers are shown in Figure 5.

THE OPTOELECTRONIC NOSE

Usually, a fluorescent sensor consists of an ultraviolet (UV) excited thin film fluorophore whose fluorescence is monitored by a photodetector. An optoelectronic nose for the purpose of discriminating explosive trace vapors can be realized by the simultaneous fluorescence detection of multiple sensors

Figure 5. Chemical structures of common nitroaromatic compounds and of two exemplary sensor polymers

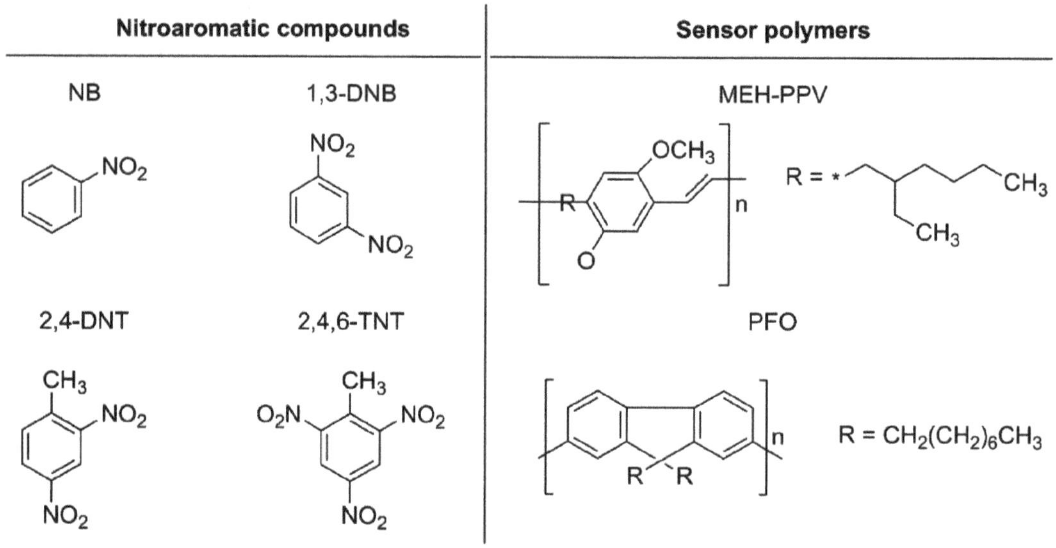

that respond differently to different target analytes in combination with pattern recognition techniques. In this section, the authors discuss the optoelectronic nose in terms of sensor array fabrication techniques, statistical discrimination techniques, sensor systems and signal readout for explosive detection underlined by theoretical and practical examples based on MATLAB and recent research.

Sensor Fabrication Techniques

In principle, one can imagine many techniques to be used for the fabrication of fluorescence quenching sensors and arrays. Apart from vacuum deposition techniques such as evaporation or sputtering, solution-processing is much more suited for low-cost applications. The most basic technique is to drop-cast the sensor elements onto conventional glass substrates. However, this process is accompanied by reproducibility issues due to coffee staining during drying leading to inhomogeneous layer thicknesses. Alternatively, spin-coating offers much more control in terms of homogeneity and thin film thickness. However, the substrates need to be fixed on a rotating stage in order to deposit a dissolved sensor material. Furthermore, spin-coating does not allow customized structures. These are critical drawbacks for large-scale production if multiple materials have to be deposited on the same substrate as it is the case for common sensor arrays. There are several other non-structuring techniques such as chemical bath deposition, spray-coating, dip-coating, blade-coating, Meyer bar coating, slot die coating and electrospinning. The large-scale fabrication of sensor arrays, however, requires a fast and reproducible deposition of patterned structures. Here, analog and digital printing techniques are ideally suited to fulfill these requirements. On the one hand, digital printing such as inkjet and aerosol jet printing enable highly customized sensor and array structures. On the other hand, analog printing such as gravure, screen and flexographic printing enable large-scale production with high throughputs. In the meantime, this does apply to inkjet printing, too. Figure 6 illustrates various solution-processing fabrication techniques.

Figure 6. Illustration of various solution-processing fabrication techniques

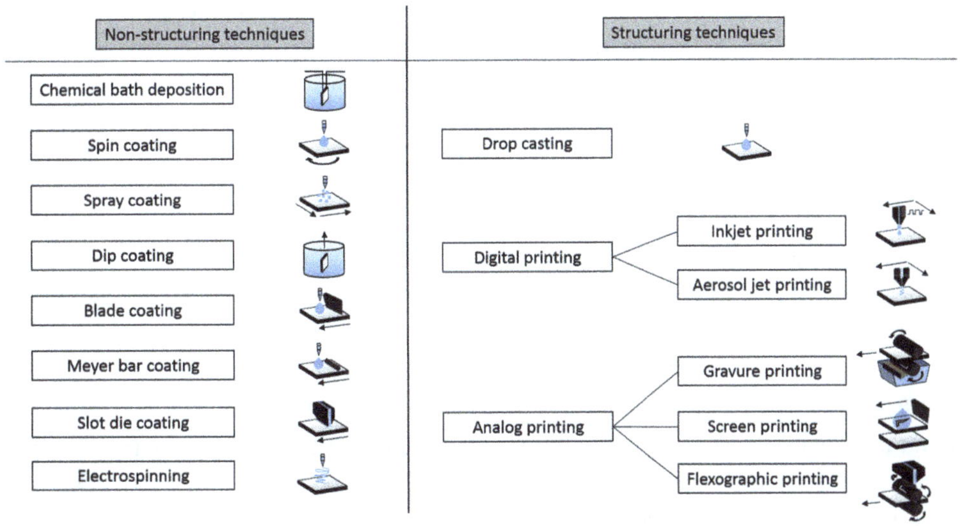

Statistical Discrimination Techniques

For many applications such as the detection of explosives and hazardous substances, the identification and the quantification of detected compounds is of high relevance. This challenge, i.e. the lack of selectivity, can be addressed with fluorescence quenching sensor arrays whose responses are interpreted with statistical discrimination techniques. In this section, the authors discuss two basic concepts based on a simple data set in order to illustrate the two techniques. To this end, the authors artificially created different normally distributed quenching efficiencies for the case of an array with four sensors. The artificial data corresponds to the sampling of 100 exposures to three different target analytes. It should be noted that this is a very basic example that does not necessarily require statistical analysis for target discrimination. However, a MATLAB implementation for such an experiment is given in the appendix. The provided source code can be easily adapted to analyze more complicated experiments.

Principal Component Analysis

One of the most basic methods to separate response signals from a sensor array is called principal component analysis (PCA). It is an eigenvector-based multivariate analysis in a very simple form (Pearson, 1901). This technique performs an orthogonal transformation of the sensor responses based on singular value decomposition or eigendecomposition in order to project the response signals onto so called linearly uncorrelated principal components. The algorithm for eigendecomposition is defined in such a way that the eigenvalues of the principal components are sorted in descending order so that the first principal component accounts for as much of the variability in the observations as possible. In scientific publications, this technique is often used to demonstrate the potential discrimination ability of sensor arrays. However, it is of less practical relevance for classification purposes due to its unsupervised nature. This means that the technique ignores pre-approved knowledge such as class labels. Nevertheless, PCA can be used to reduce the amount of data dimensionality with minimal loss of information, for example as input for other pattern recognition techniques. To this end, PCA performs dimensionality reduction by preserving as much of the variance for the whole data set as possible. Mathematically, this is carried out by computation of the eigenvector matrix M that diagonalizes the covariance matrix Σ_X of the data matrix X such that:

$$M^{-1}\Sigma_X M = D \tag{14}$$

Here, D is a diagonal matrix that contains the eigenvalues. X corresponds to the data matrix in which the columns represent the sensors in an array (features) and the rows represent the sensor responses for each measurement (observations) A rearrangement of the eigenvectors and eigenvalues in descending order allows to project the original data into principal component space, i.e. the space that is spun by the eigenvectors. Typically, most of the variance is contained in a low-dimensional space, such that higher dimensions can be neglected without significant loss of information. In this sense, the largest eigenvalues explain the most variance. For the input of the PCA, the sensor responses are typically given in terms of quenching efficiencies defined by:

$$QE = \frac{I_0 - I}{I_0}, \tag{15}$$

where I_0 is the fluorescence intensity in the absence of a target and I is the fluorescence intensity in the presence of the target. This means that the baseline in air has to be collected at first. After that, the quenching efficiencies can be calculated from the sensor array response. For this data, PCA performs the orthogonal transformation such that the quenching efficiencies can be projected into principal component space spun by the corresponding eigenvectors.

The authors performed PCA on the data set example, introduced in the beginning of this section, by utilizing the MATLAB code provided in the appendix. Table 2 shows the returned eigenvalues for each principal component and the amount of variance that can be explained by each of the components relative to the amount of variance that can be explained by all components.

Almost 96% of the overall variance in the data set can be explained by the first two principal components. This means that the data from the four-dimensional sensor array can be reduced to two dimensions while losing roughly 4% of the information that accounts for the variance. The projection of the quenching efficiencies onto the first two principal components is visualized in Figure 7.

Table 2. Eigenvalues and explained variance as a result of the PCA

Principal Component	Eigenvalue	Explained Variance (%)	Cumulative Explained Variance (%)
PC1	3.73	93.32	93.32
PC2	0.11	2.42	95.74
PC3	0.09	2.22	97.96
PC4	0.08	2.04	100.00

Figure 7. Projection of the data set example onto the two most important principal components; The numbers 1, 2 and 3 correspond to the three different target analytes.

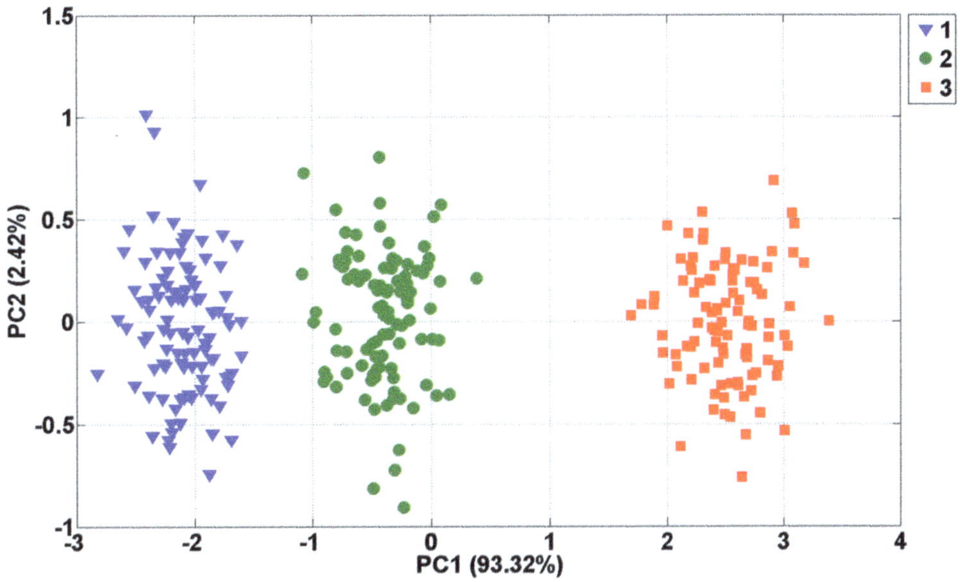

It can be seen that the projected data for each target analyte cluster into a specific group that is visually differentiable from the other groups. Visual inspection further shows that the first principal component is sufficient to discriminate between the three different groups because only very few information (2.42%) is contained along the second principal component axis. In conclusion, PCA can be used to visually inspect the discrimination ability of a sensor array for a given experimental data set.

Linear Discriminant Analysis

In order to tackle the challenge of classifying unknown observations from a sensor array, supervised statistical learning techniques such as linear discriminant analysis (LDA) are typically employed. For the application of explosive trace vapor detection, this includes the identification and quantification of a detected target analyte. Alternatively, LDA can be used for dimensionality reduction, too. In contrast to PCA, however, it preserves discriminatory information by taking class information into account (supervised). To this end, LDA finds linear transformations of the sensor array responses that minimize the variability for observations belonging to the same class while maximizing the variability for observations belonging to different classes (Fisher, 1936). Typically, a class is labeled by the target's name and its concentration in case of vapor sensing applications. LDA defines two matrices that describe the scatter within and between classes. The so-called within-class scatter matrix S_W is expressed as:

$$S_W = \sum_{k=1}^{K} \sum_{x \in k} \left(x - \mu_k\right)\left(x - \mu_k\right)^T, \tag{16}$$

with

$$\mu_k = \frac{1}{n_k}\sum_{x\in k} x \tag{17}$$

Here, K is the amount of classes, x is a vector containing the quenching efficiencies for a single observation, μ_k is the mean of the observations in class k and n_k is the amount of observations in class k. The scattering between classes is expressed by the between-class scatter matrix:

$$S_B = \sum_{k=1}^{K} n_k \left(\mu_k - \mu\right)\left(\mu_k - \mu\right)^T, \tag{18}$$

with

$$\mu = \frac{1}{n}\sum_{\forall x} x \tag{19}$$

Here, μ is the mean of all observations n.

Then, the objective function of the LDA solves the generalized eigenvalue problem by the eigendecomposition of the product $S_W^{-1}S_B$. Similar to PCA, the resulting eigenvalues are sorted in such a way that the first (largest) eigenvalue corresponds to the linear discriminant that contributes the most information in terms of minimizing within-class scattering while maximizing between class scattering. The amount of variance v explained by each of the d linear discriminants i can be computed from the eigenvalues λ_i according to:

$$v_i = \frac{\lambda_i}{\sum_{i=1}^{d} \lambda_i} \tag{20}$$

Alternatively, the Wilks' Lambda Λ_i is often calculated to test the significance of i linear discriminants according to:

$$\Lambda_i = \prod_{i}^{d} \frac{1}{1+\lambda_i} \tag{21}$$

It describes to proportion of the total variance in the discriminants not explained by differences among classes. Thus, values close to zero indicate a high discriminatory contribution.

For the same example as discussed for PCA, a linear discriminant model, also referred to as linear classifier, can be trained. Table 3 shows the eigenvalues and the explained variance for each of the linear discriminants using the same quenching efficiencies that were used to compute the PCA. The reader is referred to the appendix for a detailed MATLAB implementation.

Table 3. Eigenvalues, Wilks' Lambda and explained variance as a result of the LDA for the data set example in terms of dimensionality reduction

Linear Discriminant	Eigenvalue	Wilks' Lambda	Explained Variance (%)	Cumulative Explained Variance (%)
LD1	64.81	0.02	99.97	99.97
LD2	0.02	0.98	0.03	100.00
LD3	0.00	1.00	0.00	100.00
LD4	0.00	1.00	0.00	100.00

In comparison to the PCA example, the information on class separability can be explained by one to two linear discriminants without almost any loss (0 – 0.03%). Hence, LDA performs much better in separating the target analytes as compared to PCA for the given example. However, it should be noted that PCA can perform better as compared to LDA on other data sets (Martinez & Kak, 2001). Figure 8 visualizes the projection of the data onto the first two linear discriminants. To this end, LDA can be used to reduce the amount of data dimensionality similar to PCA.

In a second step, the trained linear classifier can be used to predict the class label (i.e. target analyte and concentration) for unknown observations. There are many different ways to predict unknown observations but often a Bayesian classifier is used to minimize the expected classification cost as performed by Mathworks (2017) according to:

Figure 8. Projection of the exemplary observations onto the two most important linear discriminants; The exemplary data is the same that was used for PCA. The numbers 1, 2 and 3 correspond to the class label for each of the three different target analytes.

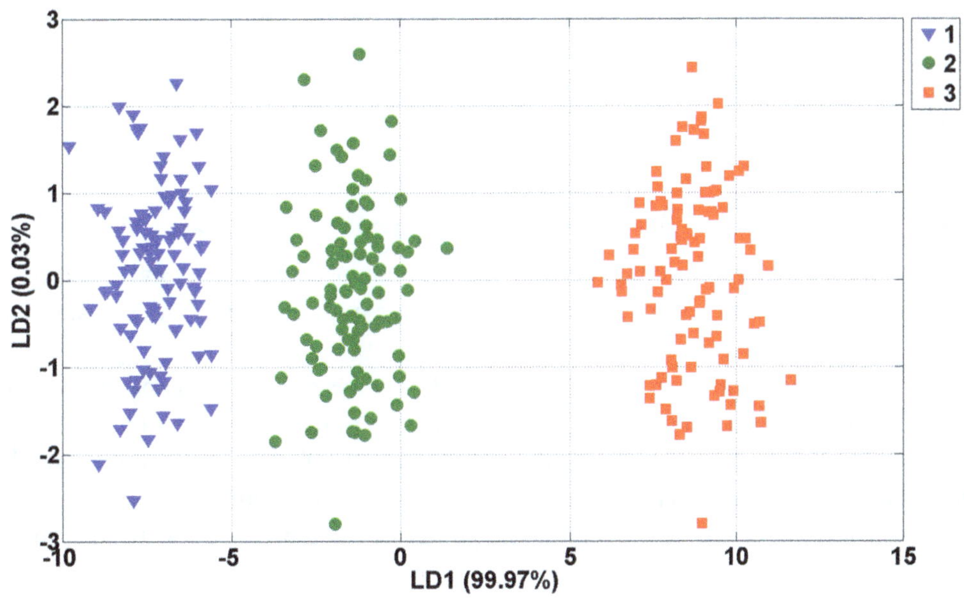

$$\hat{y} = \underset{y=1,\ldots,K}{\arg\min} \sum_{k=1}^{K} \hat{P}\left(k \mid x\right) C(y \mid k), \tag{22}$$

where $\hat{y}$ is the predicted class label, K is the number of classes, $\hat{P}\left(k \mid x\right)$ is the posterior probability of class k for observation x and $C(y \mid k)$ is the cost of classifying an observation as y when its true class is k. Here, the cost is defined to be equal to 0 for correct classification (diagonal of the matrix C) and equal to 1 for incorrect classification (remaining entries in the matrix C). However, a custom cost can be defined if specific classifications are more likely than others, for example. The posterior probability $\hat{P}\left(k \mid x\right)$ of class k for observation x is defined as the product of the prior probability $P\left(k\right)$ of class k and the multivariate normal density $P(x \mid k)$, normalized by $P\left(x\right)$ as the sum over k of that product:

$$\hat{P}\left(k \mid x\right) = \frac{P\left(x \mid k\right) P\left(k\right)}{P\left(x\right)} \tag{23}$$

In this example, the prior probability $P\left(k\right)$ of class k is the number of training observations for class k divided by the total number of training observations. The prior probabilities can be defined differently, of course. On the first glance, the minimization of the expected classification cost looks counter-intuitive. Bayesian classifiers as represented by $\hat{P}\left(k \mid x\right)$ are usually maximized because they represent the likelihood for an observation x belonging to class k. For each class, the multiplication with the cost matrix C, however, considers one class as the true class and computes the cost of classifying the observation into the remaining classes. In this case, it makes sense to minimize $\hat{P}\left(k \mid x\right) C(y \mid k)$. Most important for the prediction, however, is the multivariate normal density $P(x \mid k)$ with mean μ_k and covariance Σ_k for an observation x for class k according to:

$$P\left(x \mid k\right) = \frac{1}{\sqrt{2\pi \cdot \det(\Sigma_k)}} e^{-\frac{1}{2}\left(x-\mu_k\right)^T \Sigma_k^{-1}\left(x-\mu_k\right)} \tag{24}$$

The term $\left(x - \mu_k\right)^T \Sigma_k^{-1} \left(x - \mu_k\right)$ in the exponential function is the squared Mahalanobis distance which is a multidimensional measure for the distance between observation x and mean of class k. In the simplest case, the prediction returns the label for the class with the minimum Mahalanobis distance to the given test observation.

The prediction and projection of an unknown and randomly created artificial observation on the given exemplary training model is shown in Figure 9. The prediction returns class 2. Indeed, visual inspection of Figure 9 shows that it is the class with the minimal Mahalanobis distance to the unknown observation.

It is important to note, that LDA makes use of two theoretical assumptions for the prediction of unknown observations: normality of the data and equality of the covariance matrices. The validity of these assumptions can be tested in several ways. Two strategies are briefly described in this section.

Figure 9. Prediction and projection of the test observation onto the two most important linear discriminants; The numbers 1, 2 and 3 correspond to the class label for each of the three different target analytes.

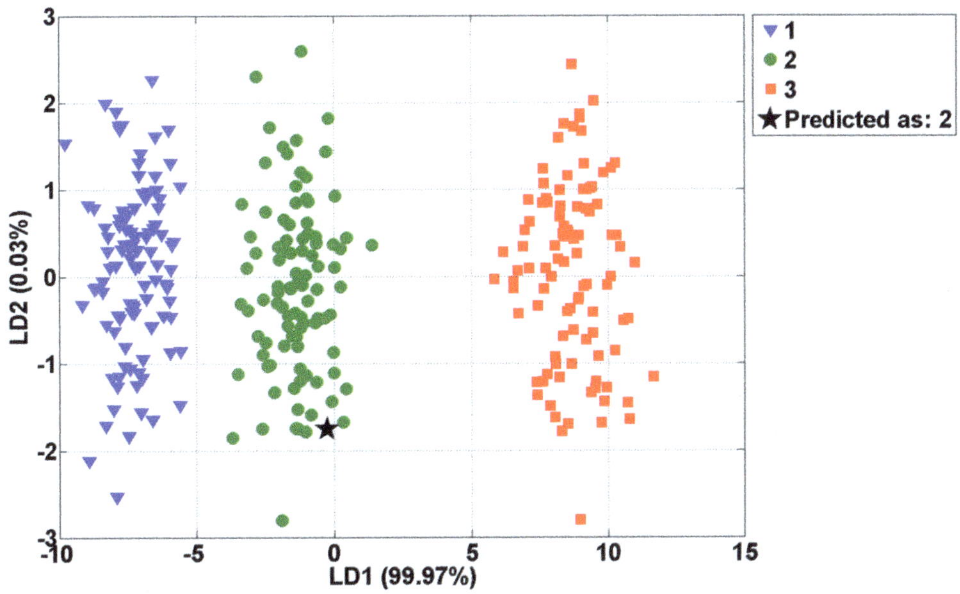

One way to test the normality of the observations is the inspection of the data in a χ^2 quantile quantile (QQ) graph that plots the observed quantiles (ascendingly sorted Mahalanobis distances in squared units) versus the expected quantiles (inverse of the χ^2 cumulative distribution) as explained by Blom (1958), for example. If the observations x are from a multivariate normal distribution with mean μ and covariance Σ, their squared Mahalanobis distances $\left(x-\mu\right)^T \Sigma^{-1}\left(x-\mu\right)$ have a χ^2 distribution with p degrees of freedom. Figure 10 shows the QQ plot for the given LDA example as discussed earlier in this section. The given data is in very good agreement to the bisecting line which confirms the normality of the data.

The equality of the covariances can be tested with the Bartlett test (Bartlett, 1937; Box, 1949). It checks the lack of uniformity of the covariance matrices across all classes according to the test statistics:

$$-2\log\left(V\right)=\left(n-K\right)\cdot\log\left(\det\left(\Sigma\right)\right)-\sum_{k=1}^{K}(n_k-1)\cdot\log\left(\det\left(\Sigma_k\right)\right), \tag{25}$$

with n observations in total, n_k observations per class k, K classes, the pooled-in covariance matrix Σ and the class-specific covariance matrices Σ_k. For comparable and large class sizes, this test statistic can be approximated by a χ^2 distribution with $\left(K-1\right)D\left(D+1\right)/2$ degrees of freedom, in which D is the amount of sensors in the array. In this case, the corresponding p-value can be calculated as:

$$p=1-f\left(-2\log\left(V\right),df\right), \tag{26}$$

Figure 10. QQ plot to visually inspect if the observations are normally distributed; The data points strongly agree to the bisecting line which means that the data is normally distributed.

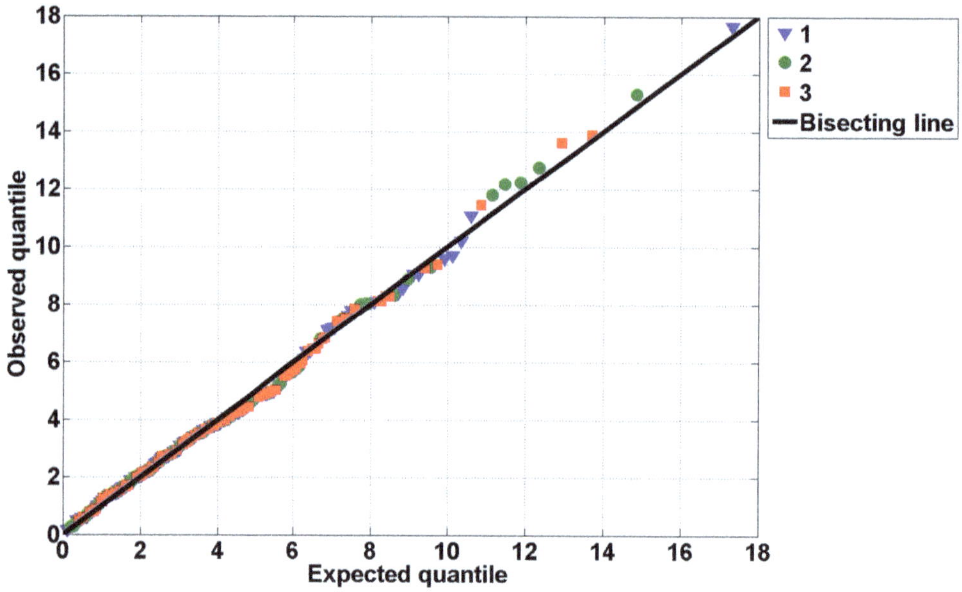

where $f\left(-2\log\left(V\right), df\right)$ is the χ^2 cumulative distribution function of the test statistics $-2\log\left(V\right)$ using the corresponding degrees of freedom specified in df. If p is greater than a given significance level α (typically $\alpha = 0.05$), the null hypothesis of equal covariance matrices is accepted. For the given LDA example, the logarithm of the determinant of the covariances, the degrees of freedom and the p-value are given in Table 4. The p-value is much greater than 0.05 which means that the hypothesis that all class covariances are equal against the alternative that at least two covariances are different is accepted. Thus, there is no evidence to believe that the covariances are different in the example.

If both normality of the data and equality of covariances are confirmed (as is the case in the example) a linear classifier is optimal for the data. In practice, however, one or both assumptions often do not hold. If this is the case, a quadratic classifier returned from quadratic discriminant analysis (QDA) might be better suited for a classification because QDA does not assume equal covariance across classes. However, the better suitability is not necessarily the case. It has to be investigated individually for each classification task. The assumption of normality is easily explained as it is a direct result of the utilization of a multivariate normal distribution in the prediction. However, the assumption of equal covariance is not that obvious. It is related to the objective function of a classifier which is quadratic for QDA and

Table 4. Logarithm of the determinant of the covariances, test statistic, degrees of freedom and p-value for the Bartlett test on the data set example

$\log\left(\det\left(\Sigma\right)\right)$	$\log\left(\det\left(\Sigma_1\right)\right)$	$\log\left(\det\left(\Sigma_2\right)\right)$	$\log\left(\det\left(\Sigma_3\right)\right)$	$-2\log\left(V\right)$	df	p
-9.73	-10.08	-9.76	-9.52	17.47	20	0.62

linear for LDA. Usually, the objective function δ for a class k is derived from the logarithm of the normalized Bayesian product $\hat{P}\left(k \mid x\right)$ according to:

$$\log\left(P\left(k\right)\right) - \frac{1}{2}\log\left(2\pi\right) - \frac{1}{2}\log\left(\det\left(\Sigma_k\right)\right) - \frac{1}{2}\left(x - \mu_k\right)^T \Sigma_k^{-1}\left(x - \mu_k\right) \tag{27}$$

Typically, all class-independent constants are discarded for the definition of objective function. Thus:

$$\delta_{k,QDA} = \log\left(P\left(k\right)\right) - \frac{1}{2}\log\left(\det\left(\Sigma_k\right)\right) - \frac{1}{2}x^T\Sigma_k^{-1}x - \frac{1}{2}\mu_k^T\Sigma_k^{-1}\mu_k + x^T\Sigma_k^{-1}\mu_k, \tag{28}$$

is quadratic in x. Here, both mean and covariance vary across classes. In LDA, the objective function becomes linear in x under the assumption $\Sigma = \Sigma_k$ according to:

$$\delta_{k,LDA} = \log\left(P\left(k\right)\right) - \frac{1}{2}\mu_k^T\Sigma^{-1}\mu_k + x^T\Sigma^{-1}\mu_k, \tag{29}$$

such that the model has the same covariance matrix for each class but class-varying means.

Validation Techniques

There are several strategies to check the validity of a trained discrimination model. A prominent method is the so called leave-one-out validation that investigates how stable the model reacts to out-of-sample observations. Leave-one-out validation is a K-fold cross validation technique with K equal to the number of observations n. In this case, the n observations are divided into K folds of equal size (here: $K = n$). For each observation, a linear classifier is trained using $K - 1$ folds while the remaining fold is used for prediction. The results of such a test are typically visualized in a confusion matrix that counts correct classifications (diagonal of the matrix) and misclassifications (remaining entries in the matrix) for each pair of classes. The summed confusion matrix for the leave-one-out-validation for the given example is shown in Table 5. The columns represent the predictions whereas the rows represent the actual classes known from training. Because all entries apart from the diagonal are equal to zero, the error rate for the given example is 0%. In addition to the general error rate, confusion matrices are used to calculate false-positive and false-negative rates, too.

Leave-one-out-validation is often considered as being too optimistic. Therefore, other validation techniques such as subset validation can be used additionally. This technique divides the n observations into two subsets. One subset is used to train the linear classifier while the other subset is used for predictions. The best validation technique, however, is to use known observations from repeated experiments.

Example: Optoelectronic Nose

Recently, the authors demonstrated the discrimination ability of a digitally printed optoelectronic nose for the selective trace detection of nitroaromatic explosive vapors (Bolse et al., 2017a). The authors

Table 5. Confusion matrix as a result of the leave-one-out validation for the data set example

Leave-One-Out Confusion Matrix		Predicted Class		
		Class 1	Class 2	Class 3
Actual Class	Class 1	100	0	0
	Class 2	0	100	0
	Class 3	0	0	100

reported on the average response of twelve sensor arrays that consist of six different organic polymers responding to nitroaromatic trace vapors through PET. The stability of the nose was increased by low power excitation to avoid photo-bleaching. The sensor polymers were printed onto conventional glass substrates using an aerosol jet printer and showed great sensitivity even though the polymers had been mainly used for other applications (Bocksrocker et al., 2012; Liu et al., 2012; Lindh, Sandström, Andersson, & Edman, 2016; Ullah et al., 2013). In the study, common nitroaromatic compounds such as NB, DNB and DNT were used as target vapors at concentrations between 1 – 10 ppb. In addition to the nitroaromatics, benzophenone (BP) was used as a control substance. The target analytes were delivered with a commercial vapor generator that utilized permeation tubes. These tubes contained the target analyte and were heated to a specific temperature. At constant temperature, the analyte's permeation rate is constant, too. Thus, a constant vapor concentration was delivered to the sensor arrays by the application of a constant air flow through the permeation oven. The authors performed LDA on the quenching efficiencies measured in their experiments as QDA did not lead to better results. An illustration of the aerosol jet printing process, the UV-excited sensor arrays and the experimental setup are depicted in Figure 11. It was shown that the detection limits for the nitroaromatic compounds are on the lower ppb level. As expected from the energetic configuration, BP was not detectable which supported the assumption that the transduction occurred through PET. The LDA model showed excellent stability to out-of-sample observations confirmed by leave-one-out validation with an error rate of only 1.1% and by subset validation with an error rate of only 0.8%.

Furthermore, Bolse et al. (2017b) investigated the within-batch reproducibility of the printed sensor arrays by LDA training and testing for each of the twelve arrays. Moreover, the samples were tilted relative to the camera to investigate the impact of the camera focus on the classification reliability. Sensor arrays within the same focus plane of the camera image showed comparable average error rates of only 4% as can be deduced from the underlined values in Table 6. Despite the results from Bartlett tests that there is no evidence to suggest that the variance in the quenching efficiencies is equal for all classes, QDA performed worse in terms of error rates for most of the cases.

In addition to the target- and concentration-specific LDA as shown in Figure 11, the quenching efficiencies for each observation can be normalized in order to retrieve a concentration-independent model. Here, each quenching efficiency of an observation needs to be normalized by:

$$x\_norm_{ij} = \frac{x_{ij}}{\sqrt{\sum_{j=1}^{D} x_{ij}^2}}, \tag{30}$$

Figure 11. (a) Illustration of the aerosol jet printing, (b) UV-excited sensor arrays, (c) experimental setup, (d) HOMO and LUMO level for sensor and target materials, (e) training model as computed from LDA. The different nitroaromatics NB, DNB and DNT were nicely separated at different concentrations. *a,b, d and e adapted from IOP (Bolse et al., 2017a), c adapted from SPIE (Bolse et al., 2017b).*

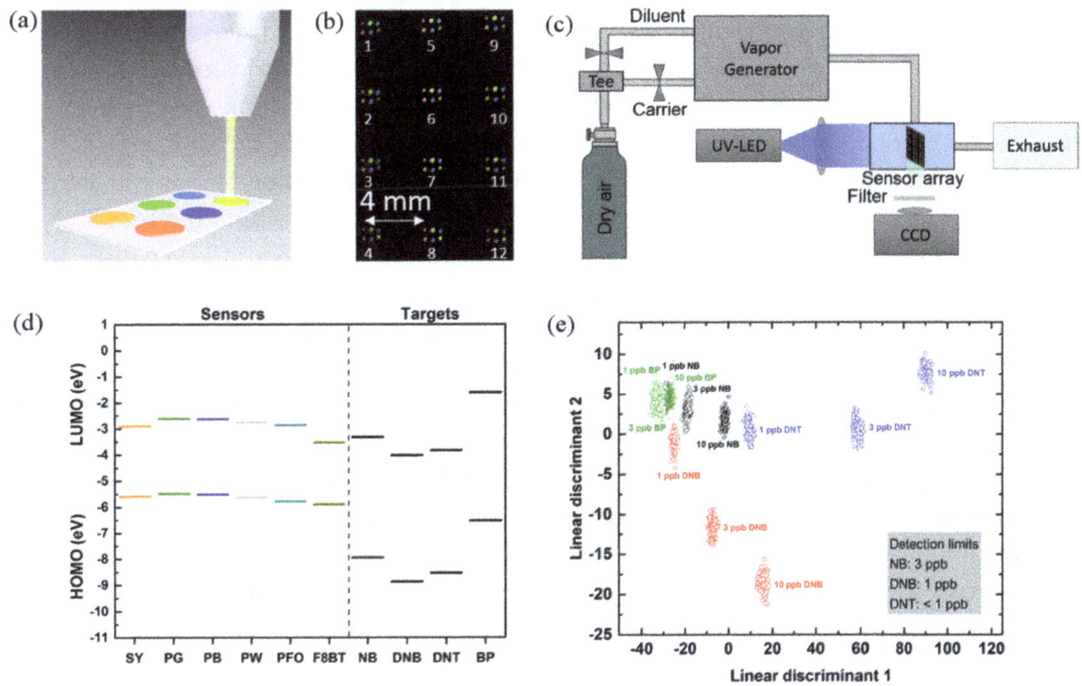

Table 6. Error rates for arrays within the same plane of the camera image when tested onto the LDA training model of the other arrays

Error Rates (%)		**Training**											
		A1	**A2**	**A3**	**A4**	**A5**	**A6**	**A7**	**A8**	**A9**	**A10**	**A11**	**A12**
Testing	**A1**	3	5	7	7								
	A2	3	2	2	3								
	A3	3	2	1	1								
	A4	8	6	6	5								
	A5					1	4	2	3				
	A6					2	0	5	6				
	A7					3	4	1	6				
	A8					3	16	11	2				
	A9									0	3	11	7
	A10									5	1	4	2
	A11									13	5	1	2
	A12									6	2	4	1

Compare Bolse et al., 2017b.

where x_norm_{ij} is the normalized quenching efficiency of sensor j for observation i, x_{ij} is the original quenching efficiency of sensor j for observation i and D is the amount of sensors in the array. Bolse, Schend, Habermehl, Eschenbaum, & Lemmer (2017c) recently discussed an example for concentration-dependent and concentration-independent discrimination of explosive-related nitroaromatic vapors using a microfibrous web array in contrast to the printed arrays as discussed earlier. Details on the study are shown in Figure 12. The polymer-coated microfibers were fabricated by drop-casting and the fluorescence was recorded with a Raspberry Pi based detector system discussed in the next section. Even though the thick polymer coating led to a rather slow sensor response, the authors demonstrated a clear separation of the target vapors (NB, DNB and DNT) at a concentration of 1 ppb in air after 60 s. In this case, the reported leave-one-out error rate dropped from 16.71% within the first 30 s down to 0.12% after 60 – 90 s. Furthermore, the LDA projection for each target vapor followed distinct trajectories for increasing concentration in three-dimensional discriminant space explaining 98.44% of the overall variance. A normalization of the LDA demonstrated the separation of the target vapors independent of the concentration. However, some projected data points scattered significantly. This was attributed to the normalization procedure described in equation (30) because observations with low quenching efficiencies received a much higher weight when normalized. This was crucial for the sensor array baseline (quenching efficiencies in normal air) which could not be included in the model. Hence, the utilization of concentration-independent models through response normalization is off less practical relevance due to the strong scattering of low quenching efficiency projections. Therefore, the authors suggested to make use of the direction and location of the target-specific trajectories in order to identify and quantify unknown observations.

Example: Sensor System

In the simplest case, a fluorescence quenching sensor system consists of just three components: an excitation source, an appropriate fluorescent transducer and a photodetector. Therefore, the technology offers great potential for low-cost solutions. Here, the authors describe a portable sensor system, mostly based on low-cost components (Bolse, Schend, Bell, Eschenbaum, & Lemmer, 2015). The system's measurement chamber includes an in- and outlet for gas and vapor sampling and is made out of PTFE-sealed aluminum. The chamber can be temperature-stabilized by using an external heating circuit. In addition, two optical windows provide access for optical excitation and fluorescence monitoring. The fluorescent responder, i.e. a sensor array, can be inserted into the chamber from the top. The sensor array is excited optically through one of the windows. To this end, the radiation emitted by a 370 nm UV-LED is collimated and bandpass filtered before it enters the chamber. Furthermore, a fraction of the emitted radiation is guided to a photodiode in order to monitor the excitation power of the UV-LED. A standard 8-bit CMOS-camera is used to monitor the sensor array fluorescence through the second optical window. A longpass filter absorbs the excitation radiation to prevent it from entering the camera. Furthermore, a temperature and humidity sensor is integrated in utmost proximity of the sensor array to monitor environmental conditions inside the chamber. Additional components such as a touch display unit, a Raspberry Pi, an extension board and a constant current source are integrated into the system's housing that can be vented with a small fan. The system is illustrated in Figure 13 including a sensor response signal from a DNT quenching experiment.

Figure 12. (a) SEM image, (b) image of the UV-excited microfibrous web array, (c) concentration-dependent LDA in three dimensions and (d) concentration-independent LDA in two dimensions
Compare Bolse et al., 2017c.

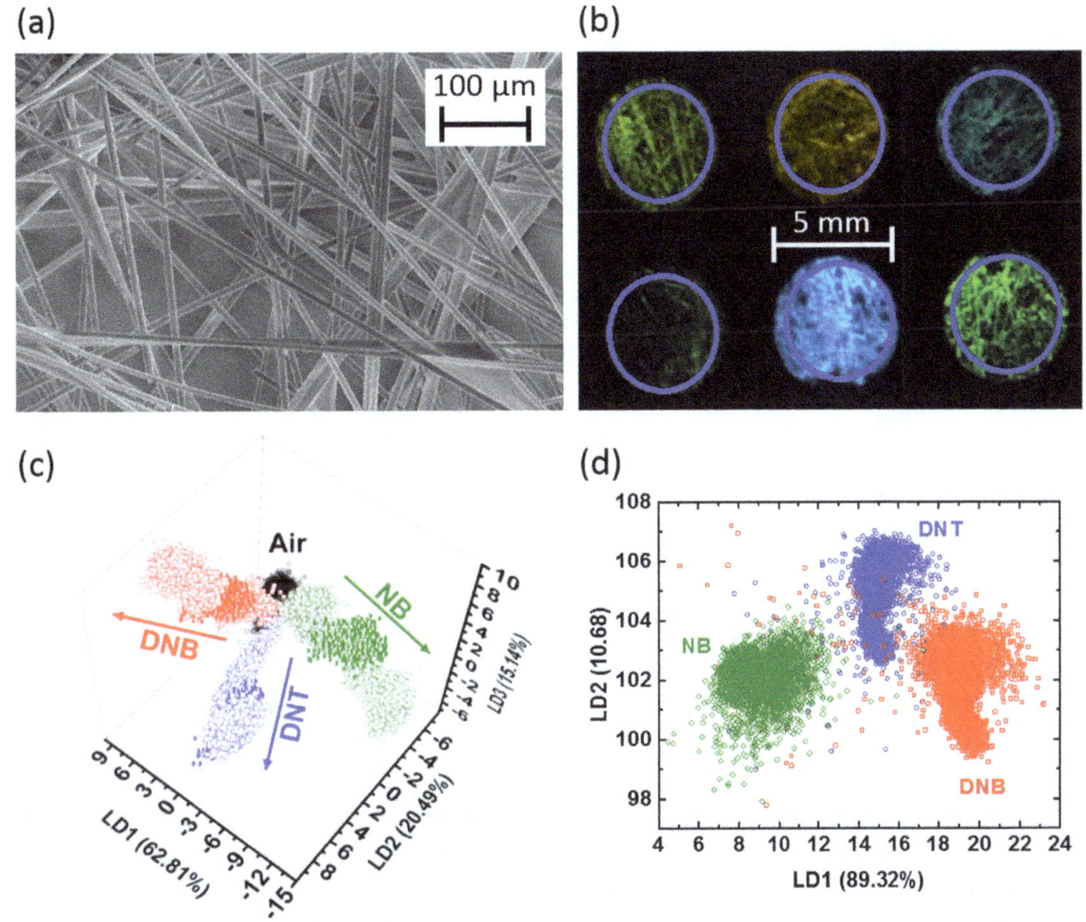

The system's measurement software was written with the Qt cross-platform application and user interface framework. Camera controls are accessible and regions of interest (ROIs) can be detected automatically. During measurements, the average intensity of each ROI is computed for each camera image which is acquired in a user-specified interval. The computed intensities are stored on the Raspberry Pi and shown on the user interface. Furthermore, temperature and humidity readings are taken to monitor environmental parameters which could affect the fluorescence of the sensors under test. The design of such sensor systems can be crucial to the system and sensor performance. A very important consideration is related to the vapor path through the detector unit. The sensitivity of the system can be strongly increased if the vapor is pushed through the sensor instead of bypassing it along the surface. Such a configuration can be easily realized for the case of micro- and nanofibrous sensors, but it is possible for thin film sensors, too. In comparison to a non-portable setup, in which the target vapor was not pushed through the sensor array but rather bypassed its surface, the response of the described system in terms of fluorescence quenching was about 20% faster for the same transducer under the same experimental

Figure 13. (a) Photograph and (b) sketch of the sensor system. (c) Sketch of the measurement chamber. (d) Detection of DNT vapor in nitrogen atmosphere
Bolse et al., 2015.

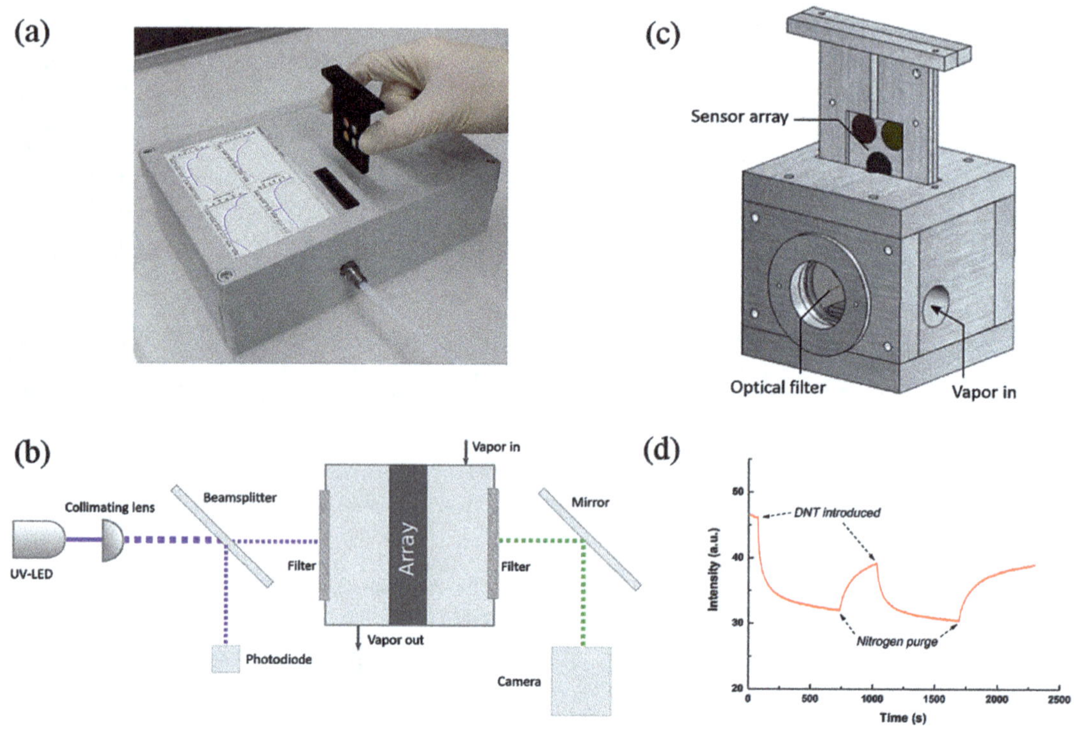

conditions as reported by the authors. Another important aspect is the illumination of the sensor array which needs to be homogeneous in order to avoid spatial intensity deviations. Therefore, the optical path has to be adjusted accurately and all components need to be mounted properly to withstand harsh operating conditions. Moreover, a stable UV-excitation is required in order to illuminate the sensor with a constant excitation power. However, most commercial LED drivers rely on an inverted pulse-width-modulated (PWM) signal that directly pulses the LED in a modulated fashion. This modulated pulsing can lead to disturbing flicker in the camera signal when utilizing rolling shutter based cameras such as CMOS-based cameras. Appropriate LED drivers have to be employed to avoid these artefacts. Nevertheless, a photodiode should always be used to monitor the excitation power during system operation and deviations should be corrected automatically or logged. In dependence on the application requirements, a temperature and humidity control might be required, too, as fluorescent sensors can be affected by these environmental conditions. For example, Bolse, Chouikh, Sellner, Eschenbaum, & Lemmer (2015a, 2015b) reported on a temperature-control for optical vapor sensors using an electrically heated thin film on the backside of the sensor substrate.

Most important in a fluorescence detection system, however, is the read-out of the camera because there are many potential obstacles when extracting fluorescence intensities from the camera images. For the read-out, suppose a sensor being quenched and an 8-bit RGB camera monitoring the fluorescence as visualized in the images in Figure 14.

Figure 14. (a) Overexposed, (b) appropriately exposed and (c) quenched image and histogram of an exemplary fluorescent sensor. The intensity scales between 0 – 255 because an 8-bit camera was used. A saturation corresponds to values of 255.

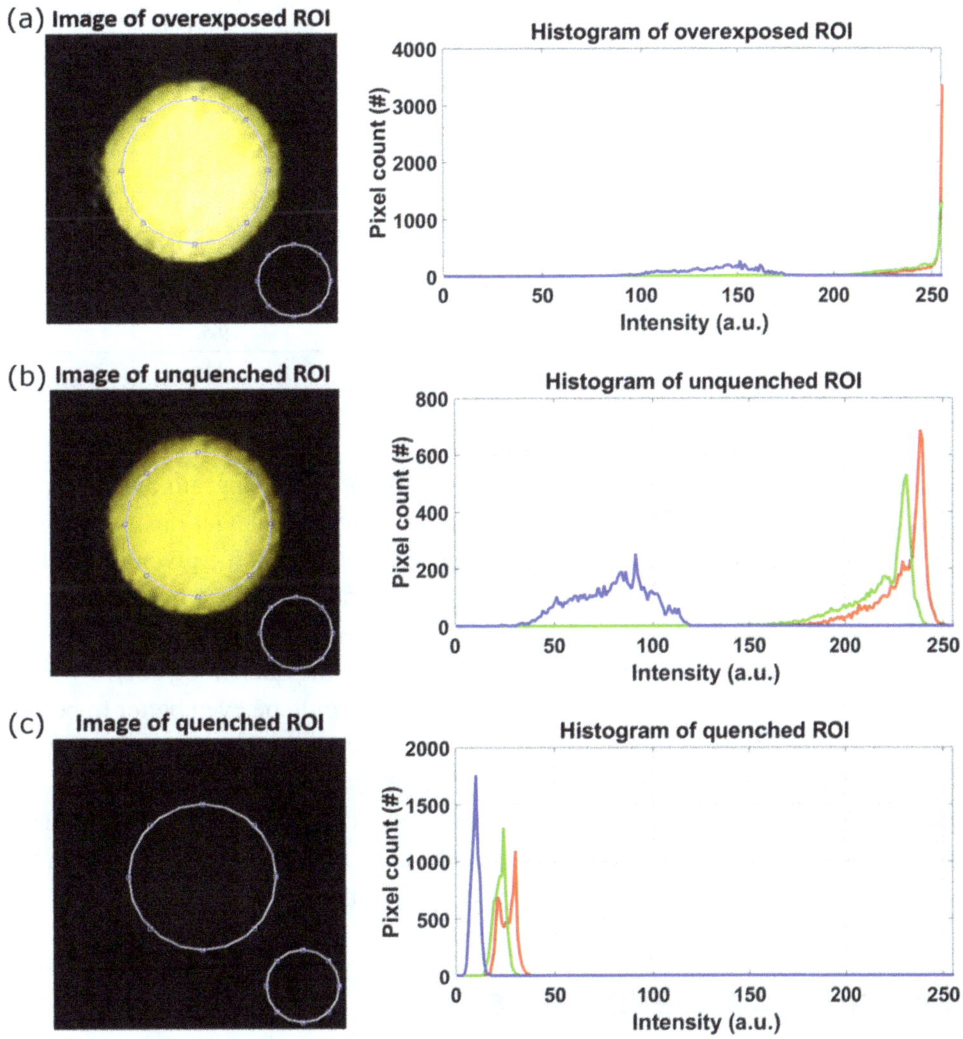

Here, an important mistake can occur due to overexposure leading to a saturation of one or more color channels in the RGB image. In many cases, this is not visible to the human eye by just looking at the camera image but rather becomes recognizable in the histogram of the image. In addition, there can be dark pixels within the ROI that have to be corrected for. This would be the case for pixels whose intensity is on the order of the background intensity. This is particularly important for fibrous sensors in which it is likely to have dark pixels between the fluorescent fibers. But there are even more potential obstacles. For example, they are related to grayscale conversion and background correction. The ratio of the fluorescence intensity in the absence (I_0) and in the presence of a target analyte (I) and the calculated errors for different read-out strategies and corrections are shown in Table 7 for the given example from Figure 14 (b) and (c).

Table 7. Fluorescence quenching and relative error for different camera read-out strategies and correction approaches

Strategy	Corrections	$\frac{I_0}{I}$ (a.u.)	Relative Error (%)
Histogram integration	Grayscale conversion, background correction	11.11	0.00
Histogram integration	Background correction	10.96	1.35
Histogram integration	Grayscale conversion	9.51	14.40
Histogram integration	None	9.17	17.46
Mean intensity	Grayscale conversion, background correction	11.11	0.00
Mean intensity	Background correction	10.96	1.35
Mean intensity	Grayscale conversion	9.51	14.40
Mean intensity	None	9.17	17.46

Of course, there is no difference between the integration of the histogram and the mean intensity because the first approach approximates the mean value. The relative deviations between the different corrections, however, are almost up to 20 percent in the extreme case of the given example, i.e. if there is no grayscale conversion and no background correction. Depending on the extent of quenching and the size of the ROI, the errors can become more severe. It is obvious that such errors – when not avoided - may have a significant impact on the physical interpretation of the quenching such when determining Stern-Volmer constants or when used in classification tasks. It would be even better to perform a superposition of the three color channels by taking into account the camera-specific channel sensitivity. The elaborated examples show that the development of efficient sensor systems is not trivial. In fact, great care has to be taken to avoid common obstacles and errors that could influence the system response and the pattern recognition reliability. However, it has also been shown that such systems can be built from low-cost components rendering fluorescence quenching even more attractive.

SOLUTIONS AND RECOMMENDATIONS

Today, fluorescence quenching is already being used commercially such as in the Fido® device as mentioned earlier and there are other players slowly showing up, for example Arborescent. However, the application of fluorescence quenching in commercial optoelectronic nose applications is still challenging. In practice, the training of pattern recognition techniques can be a critical drawback for industrial applications, because a lot of effort might have to be spent until a reliable discrimination model is established. Often such a model has to be retrained periodically to correct for sensor drift and other effects that influence the sensor array response. For fluorescence quenching sensor arrays, this includes time-consuming experiments to expose the sensor arrays to a large variety of target analytes because the discrimination model is limited to the target analytes included in the model, at least for PCA and LDA. In addition to a required long-term stability of electronic noses, reproducible fabrication is of utmost importance to ensure that only a subset of sensor arrays need to be trained to retrieve a reliable model. Ideally, such a model should be applicable to all the arrays produced. In fact, such issues have been crucial for the limited success of commercial applications in utilizing electronic noses for the detection of explosive trace vapors until today. Here, printing seems to be a reasonable solution for the fabrication of reproducible arrays as was shown by the authors. Furthermore, printing enables customizability and high throughput rates towards mass-production. The challenge of avoiding periodic retraining of pattern recognition techniques might be addressed with modern approaches in deep learning such as deep neural networks that can adapt to long-term effects such as drift and others. Those networks might be able to learn from each observation during operation such that the algorithm is performing self-taught and adaptive learning. The possibilities of integrating sensor arrays to a high extent and implementing resource-efficient pattern recognition techniques into compact systems, pave the way towards complex classification tasks with low-cost and mass-producible systems. Moreover, the combination of different detection technologies seems to be a very promising way, as is approached by Aryballe Technologies, for example.

FUTURE RESEARCH DIRECTIONS

The authors believe that printing techniques enable the realization of low-cost and mass-producible sensor arrays to detect explosives and other targets. Furthermore, deep learning techniques such as complex neural network approaches, offer the capabilities for much more efficient discrimination models that could help to reduce false-positive and false-negative rates beyond state-of-the-art limits. So far, organic polymers and small molecules have also been used to smell odors related to environmental contamination and human health issues. Possibly, not only arrays but complete sensor systems could be printed in near future. Currently, printed photodetectors reach comparable performances as commonly fabricated organic and inorganic detectors. With the strong efforts on printed organic light emitting diodes (OLEDs), one could soon imagine a printed UV-LED being used as excitation source. With this achievement, a complete sensor system could be deposited on a single substrate comprising a printed UV-LED, a printed fluorescent sensor array and a printed photodetector. This highly integrated configuration enables the low-cost production of complete sensor systems such as mobile or wearable sniffer systems. Imagine such a system being integrated into a smartphone that is able to raise early alarms if an explosive is close by or if the air quality is a risk to human health. If such a system is implemented into almost each smartphone, distributed sensing applications including early explosion warning and worldwide air quality monitoring could be realized. Presumably this will have a significant effect on our daily life.

CONCLUSION

It was shown that fluorescence quenching is a promising technique for explosive trace vapor detection. The ultra-low detection limits that can be reached with fluorescence quenching are one of the main advantages in comparison to other detection methodologies. With time, the theory of fluorescence quenching as introduced by Stern and Volmer has evolved into a thorough understanding of the quenching processes and mechanisms. Even though, not every piece of the puzzle has been resolved, experimental data can be reasonably interpreted with the help of state-of-the-art physical models. Furthermore, the utilization of sensor arrays has led to the development of optoelectronic noses that are able to discriminate explosive trace vapors using pattern recognition techniques. With the help of PCA, a first glance onto the discriminatory ability of a sensor array is possible with very low effort. Furthermore, simple techniques such as LDA showed great potential in classification tasks with low error rates related to the identification and quantification of explosive traces. In terms of sensor array fabrication, printing proved to be a promising candidate towards the realization of low-cost and reproducible sensor arrays. At the same time, printing enables the realization of customized and highly integrated sensor arrays for miniaturized systems. Obviously, fluorescence quenching has contributed significant findings to the field of explosive detection which becomes more and more important all over the world.

REFERENCES

Ali, M. A., Geng, Y., Cavaye, H., Burn, P. L., Gentle, I. R., Meredith, P., & Shaw, P. E. (2015). Molecular versus exciton diffusion in fluorescence-based explosive vapour sensors. *Chemical Communications*, *51*(98), 17406–17409. doi:10.1039/C5CC06367A PMID:26465590

Ali, M. A., Shoaee, S., Fan, S., Burn, P. L., Gentle, I. R., Meredith, P., & Shaw, P. E. (2016). Detection of explosive vapors: The roles of exciton and molecular diffusion in real-time sensing. *ChemPhysChem*, *17*(21), 3350–3353. doi:10.1002/cphc.201600767 PMID:27583839

Andrew, T. L., & Swager, T. M. (2007). A fluorescence turn-on mechanism to detect high explosives RDX and PETN. *Journal of the American Chemical Society*, *129*(23), 7254–7255. doi:10.1021/ja071911c PMID:17508754

Bartlett, M. S. (1937). Properties of sufficiency and statistical tests. *Proceedings of the Royal Society of London. Series A, Mathematical and Physical Sciences*, *160*(901), 268–282. doi:10.1098/rspa.1937.0109

Bielecki, Z., Janucki, J., Kawalec, A., Mikolajczyk, J., Palka, N., Pasternak, M., ... Wojtas, J. (2012). Sensors and systems for the detection of explosive devices – an overview. *Metrology and Measurement Systems*, *19*(1), 3–28. doi:10.2478/v10178-012-0001-3

Blom, G. (1958). *Statistical estimates and transformed beta variables*. New York, NY: John Wiley & Sons.

Bocksrocker, T., Preinfalk, J. B., Asche-Tauscher, J., Pargner, A., Eschenbaum, C., Maier-Flaig, F., & Lemmer, U. (2012). White organic light emitting diodes with enhanced internal and external outcoupling for ultra-efficient light extraction and Lambertian emission. *Optics Express*, *20*(56), A932–A940. doi:10.1364/OE.20.00A932

Bolse, N., Chouikh, G., Sellner, M., Eschenbaum, C., & Lemmer, U. (2015a). Temperature-control for optical vapor sensors based on photoluminescence quenching. In *Proceedings of OSA Advanced Photonics 2015, SeT2D.6*. Boston, MA: OSA. doi:10.1364/SENSORS.2015.SeT2D.6

Bolse, N., Chouikh, G., Sellner, M., Eschenbaum, C., & Lemmer, U. (2015b). Temperaturstabilisierung für Phosphoreszenz-Quenching-Sensoren zur Sauerstoffdetektion. In *Proceedings of MikroSystemTechnik Kongress 2015*. Karlsruhe, Germany: VDE Verlag.

Bolse, N., Eckstein, R., Schend, M., Habermehl, A., Eschenbaum, C., Hernandez-Sosa, G., & Lemmer, U. (2017a). A digitally printed optoelectronic nose for the selective trace detection of nitroaromatic explosive vapours using fluorescence quenching. *Flexible and Printed Electronics*, *2*(2), 024001. doi:10.1088/2058-8585/aa6601

Bolse, N., Eckstein, R., Schend, M., Habermehl, A., Hernandez-Sosa, G., Eschenbaum, C., & Lemmer, U. (2017b). Discrimination of trace nitroaromatics using linear discriminant analysis on aerosol jet printed fluorescent sensor arrays. *Proceedings of the Society for Photo-Instrumentation Engineers*, *10231*, 102310R. doi:10.1117/12.2265522

Bolse, N., Schend, M., Bell, L., Eschenbaum, C., & Lemmer, U. (2015). *An embedded system for the evaluation of ultra-sensitive optical sensor arrays based on photoluminescence quenching*. Paper presented at the 16th International Symposium on Olfaction and Electronic Noses (ISOEN 2015), Dijon, France.

Bolse, N., Schend, M., Habermehl, A., Eschenbaum, C., & Lemmer, U. (2017c). Transient response of a microfibrous web array for the discrimination of nitroaromatic explosive vapors. In *2017 ISOCS/IEEE International Symposium on Olfaction and Electronic Nose (ISOEN)*. Montreal, Canada: IEEE. doi:10.1109/ISOEN.2017.7968850

Box, G. E. P. (1949). A general distribution theory for a class of likelihood criteria. *Biometrika*, *36*(3/4), 317–346. doi:10.2307/2332671 PMID:15402070

Bright, C. J., Nallon, E. C., Polcha, M. P., & Schnee, M. P. (2015). Quantum dot and polymer composite cross-reactive array for chemical vapor detection. *Analytical Chemistry*, *87*(24), 12270–12275. doi:10.1021/acs.analchem.5b03559 PMID:26548712

Caygill, J. S., Davis, F., & Higson, S. P. J. (2012). Current trends in explosive detection techniques. *Talanta*, *88*(15), 14–29. doi:10.1016/j.talanta.2011.11.043 PMID:22265465

Fisher, R. A. (1936). The use of multiple measurements in taxonomic problems. *Annals of Eugenics*, *7*(2), 179–188. doi:10.1111/j.1469-1809.1936.tb02137.x

Frank, J. M., & Wawilow, S. J. (1931). Über die Wirkungssphäre der Auslöschungsvorgänge in den fluoreszierenden Flüssigkeiten. *Zeitschrift fur Physik*, *69*(1), 100–110. doi:10.1007/BF01391516

Gares, K. L., Hufziger, K. T., Bykov, S. V., & Asher, S. A. (2016). Review of explosive detection methodologies and the emergence of standoff deep UV resonance Raman. *Journal of Raman Spectroscopy: JRS*, *47*(1), 124–141. doi:10.1002/jrs.4868

Germain, M. E., & Knapp, M. J. (2008). Turn-on fluorescence detection of H_2O_2 and TATP. *Inorganic Chemistry Communications*, *47*(21), 9748–9750. doi:10.1021/ic801317x PMID:18828585

Institute of Economics and Peace. (2016). *Global Terrorism Index*. Retrieved April 10, 2017, from http://economicsandpeace.org/research/iep-indices-data/global-terrorism-index

Kumar, R. V. K., & Murali, G. (2016). A survey on the present state-of-the-art of explosives, detection methods and automatic explosive detection using wireless sensor network. *International Journal of Applied Engineering Research*, *11*(1), 504–510.

Lakowicz, J. R. (2006). Principles of fluorescence spectroscopy (3rd ed.). Baltimore, MD: Springer Science+Business Media. doi:10.1007/978-0-387-46312-4

Lehrer, S. S. (1971). Solute perturbation of protein fluorescence. Quenching of the tryptophyl fluorescence of model compounds and of lysozyme by iodide ion. *Biochemistry*, *10*(17), 3254–3263. doi:10.1021/bi00793a015 PMID:5119250

Lindh, E. M., Sandström, A., Andersson, M. R., & Edman, L. (2016). Luminescent line art by direct-write patterning. *Light, Science & Applications*, *5*, e16050. doi:10.1038/lsa.2016.50

Liu, X., Klinkhammer, S., Sudau, K., Mechau, N., Vannahme, C., Kaschke, J., & Lemmer, U. (2012). Ink-jet-printed organic semiconductor distributed feedback laser. *Applied Physics Express*, *5*(7), 072101. doi:10.1143/APEX.5.072101

Martinez, A. M., & Kak, A. C. (2001). PCA versus LDA. *IEEE Transactions on Pattern Analysis and Machine Intelligence*, *23*(2), 228–233. doi:10.1109/34.908974

Mathworks. (2017). Discriminant Analysis. In *MATLAB Documentation*. Retrieved April 3, 2017, from https://de.mathworks.com/help/stats/discriminant-analysis.html

National Consortium for the Study of Terrorism and Responses to Terrorism. (2016). *Global Terrorism Database*. Retrieved April 10, 2017, from http://www.start.umd.edu/gtd

Pearson, K. (1901). On lines and planes of closest fit to systems of points in space. *Philosophical Magazine*, *2*(11), 559–572.

Perrin, J. (1924). Loi de décroissance du pouvoir fluorescent en fonction de la concentration. *Comptes rendus de l'Academie des sciences, 178*, 1978-1980.

Peveler, W. J., Roldan, A., Hollingsworth, N., Porter, M. J., & Parkin, I. P. (2016). Multichannel detection and differentiation of explosives with a quantum dot array. *ACS Nano*, *10*(1), 1139–1146. doi:10.1021/acsnano.5b06433 PMID:26579950

Stern, O., & Volmer, M. (1919). Über die Abklingungszeit der Fluoreszenz. *Physikalische Zeitschrift*, *20*, 193–188.

Sun, X., Wang, Y., & Lei, Y. (2015). Fluorescence based explosive detection: From mechanisms to sensory materials. *Chemical Society Reviews*, *44*(22), 8019–8061. doi:10.1039/C5CS00496A PMID:26335504

Toal, S. J., & Trogler, W. C. (2006). Polymer sensors for nitroaromatic explosives detection. *Journal of Materials Chemistry*, *16*(28), 2871–2883. doi:10.1039/b517953j

Ullah, M., Tandy, K., Yambem, S. D., Aljada, M., Burn, P. L., Meredith, P., & Namdas, E. B. (2013). Simultaneous enhancement of brightness, efficiency, and switching in RGB organic light emitting transistors. *Advanced Materials*, *25*(43), 6213–6218. doi:10.1002/adma.201302649 PMID:23963863

ADDITIONAL READING

Askim, J. R., Mahmoudi, M., & Suslick, K. S. (2013). Optical sensor arrays for chemical sensing: The optoelectronic nose. *Chemical Society Reviews*, *42*(22), 8649–8682. doi:10.1039/c3cs60179j PMID:24091381

Förster, T. (1951). *Fluoreszenz organischer Verbindungen*. Göttingen, Germany: Vandenhoeck & Ruprecht.

Gardner, J. W., & Yinon, J. (Eds.). (2004). Electronic noses & sensors for the detection of explosives. Dordrecht, Germany: Springer Science+Business Media. doi:10.1007/1-4020-2319-7

Patel, H. K. (2014). *The electronic nose: artificial olfaction technology*. New Delhi, India: Springer. doi:10.1007/978-81-322-1548-6

Woodfin, R. L. (Ed.). (2007). *Trace chemical sensing of explosives*. Hoboken, NJ: John Wiley & Sons.

KEY TERMS AND DEFINITIONS

Bartlett Test: A null hypothesis test to check for the equality of covariances across classes.

Fluorescence Quenching: A decrease in fluorescence intensity (or lifetime) due to the non-radiative deactivation of excited states.

Leave-One-Out-Validation: K-fold cross validation technique in which K equals the total number of observations. It allows the investigation of the robustness of a statistical learning technique to out-of-sample observations.

Linear Discriminant Analysis (LDA): A supervised statistical learning technique that is able to create a linear classifier to separate and predict multiple classes of objects for pattern recognition purposes. It can be used for dimensionality reduction, too.

Mahalanobis Distance: A multidimensional distance measure of how much a projected data point differs from the mean of a class distribution.

Nitroaromatics: Organic compounds that contain one or more aromatic rings and one or more nitro functional groups (-NO_2).

Optoelectronic Nose: An electronic nose that utilizes optical sensors.

Principal Component Analysis (PCA): A simple eigenvector-based multivariate analysis that can be used for unsupervised statistical learning. Most often, it is used for dimensionality reduction of high-dimensional data sets.

Quantile Quantile (QQ) Analysis: A graphical technique to check whether a data set is normally distributed.

APPENDIX

This appendix contains MATLAB implementations for the statistical discrimination techniques discussed in the chapter. To this end, a sensor array with four sensors is supposed to sample 100 quenching efficiencies for each of three different target analytes.

PCA Implementation

A simple MATLAB implementation shows how the PCA is computed for an artificial data set based on eigenvalue decomposition:

```
%PCA input data
sensors = 4; %amount of sensors
observationsPerClass = 100; %amount of measurements per class
classes = 3; %amount of classes
observations = observationsPerClass * classes; %amount of observations
inputMatrix = zeros(observations,sensors); %data matrix
inputLabels = cell(observations,1); %label vector
for ii=1:classes
inputMatrix((ii-1)*observationsPerClass+1:ii*observationsPerClass,:) = randn(o
bservationsPerClass,sensors)+ii^2; %generate artificial observations (quench-
ing efficiencies) for each class
inputLabels((ii-1)*observationsPerClass+1:ii*observationsPerClass,1) =
{num2str(ii)}; %label each class
end
inputMatrix = zscore(inputMatrix); %center and scale sensor data
%PCA execution
[coeff, score, latent, tsquared, variance] = pca(inputMatrix,'algorithm','eig'
);
%PCA visualization
h1 = figure; hold on;
colors = 'bgrymc';
for ii=1:classes
plot(score((ii-1)*observationsPerClass+1:ii*observationsPerClass,1),score((ii-
1)*observationsPerClass+1:ii*observationsPerClass,2),'x','Color',colors(ii));
%plot projections along the two most important dimensions
end
xlabel(['PC1 (',num2str(variance(1)),'%)']); %explained variance for PC1
ylabel(['PC2 (',num2str(variance(2)),'%)']); %explained variance for PC2
legend(unique(inputLabels)); %plot legend
```

LDA Implementation

The LDA can be implemented in MATLAB with a similar effort. However, the following code includes prediction, leave-one-out-validation, QQ analysis and a Bartlett test, too.

```
%LDA input data
sensors = 4; %amount of sensors
observationsPerClass = 100; %amount of measurements per class
classes = 3; %amount of classes
observations = observationsPerClass * classes; %amount of observations
inputMatrix = zeros(observations,sensors); %data matrix
inputLabels = cell(observations,1); %label vector
for ii=1:classes
inputMatrix((ii-1)*observationsPerClass+1:ii*observationsPerClass,:) = randn(o
bservationsPerClass,sensors)+ii^2; %generate artificial observations (quench-
ing efficiencies) for each class
inputLabels((ii-1)*observationsPerClass+1:ii*observationsPerClass,1) =
{num2str(ii)}; %label each class
end
inputMatrix = zscore(inputMatrix); %center and scale sensor data
%LDA training and data projection
discriminationModel = fitcdiscr(inputMatrix,inputLabels,'DiscrimType','line
ar');
[P,E] = eig(discriminationModel.BetweenSigma,discriminationModel.Sigma); %P
contains the eigenvectors for projection, E contains the eigenvalues
[E_sorted, index_sorted] = sort(diag(E),'descend'); %sorted eigenvalues
P_sorted = zeros(length(inputMatrix(1,:))); %empty projection matrix
wilksLambda = 1;
for ii=1:length(inputMatrix(1,:));
P_sorted(ii,:) = P(:,index_sorted(ii)); %projection matrix with sorted eigen-
vectors
wilksLambda = wilksLambda*1/(1+E_sorted(ii));
end
for ii=2:length(E_sorted)
wilksLambda(ii) = wilksLambda(ii-1)*(1+E_sorted(ii-1)); %Wilks' Lambda
end
F = (P_sorted*inputMatrix')'; %projection
varianceExplained = E_sorted/sum(E_sorted)*100; %explained variance
%LDA visualization
h1 = figure; hold on;
colors = 'bgrymc';
for ii=1:classes
plot(F((ii-1)*observationsPerClass+1:ii*observationsPerClass,1),F((ii-1)*obs
```

```
ervationsPerClass+1:ii*observationsPerClass,2),'x','Color',colors(ii)) %plot
projections along the two most important dimensions
end
xlabel(['LD1 (',num2str(varianceExplained(1)),'%)']); %label the x-axis
ylabel(['LD2 (',num2str(varianceExplained(2)),'%)']); %label the y-axis
legend(unique(inputLabels)); % plot legend
%--------------------------------------------------------------------------
%LDA testing
unknown_testdata = zscore(randn(1,sensors)+10*rand(1)); %centered and scaled
artifical test data as gaussian distributed quenching efficiencies
unknown_projection = (P_sorted*unknown_testdata')'; %projection of the unknown
test data
predictedLabel = predict(discriminationModel,unknown_testdata); %predicted
class label for the test data
%Prediction visualization
figure(h1);
plot(unknown_projection(:,1),unknown_projection(:,2),'o','Color','black');
%plot projection
legendStr = [(unique(inputLabels));strcat('Predicted as:
',char(predictedLabel))]; %plot legend
legend(legendStr,'Location','NorthEastOutside'); % legend location
%--------------------------------------------------------------------------
%Leave-one-out validation
ldaClassFun = @(xtrain, ytrain, xtest, ytest)confusionmat(ytest,classify(xte
st, xtrain, ytrain),'order',unique(inputLabels)); %prepare leave-one-out vali-
dation
cfMat = crossval(ldaClassFun, inputMatrix, inputLabels, 'partition',
cvpartition(inputLabels, 'leaveout')); %perform leave-one-out validation
cfMat = reshape(sum(cfMat), numel(unique(inputLabels)),
numel(unique(inputLabels))); % summation of confusion matrices for each test
%--------------------------------------------------------------------------
%QQ-Test
observedQuantiles = mahal(discriminationModel, discriminationModel.X,'ClassLab
els', discriminationModel.Y); % Mahalanobis
expectedQuantiles = chi2inv(((1:discriminationModel.NumObservations)-0.5)/dis-
criminationModel.NumObservations,sensors); %expected quantiles
[observedQuantiles,sorted] = sort(observedQuantiles); % sort
h2 = figure;
gscatter(expectedQuantiles, observedQuantiles, discriminationModel.Y(sorted),c
olors,[],[],'off'); %plot
legend(unique(inputLabels)); %plot legend
xlabel('Expected quantile'); %label the x-axis
ylabel('Observed quantile'); %label the y-axis
```

```
line([0 25],[0 25],'color','k'); %plot the bisecting line
%---------------------------------------------------------------------------
%Bartlett test
minusTwoLogV = (observations-classes)*discriminationModel.LogDetSigma; %inter-
mediate result of -2log(V)
for ii=1:classes
minusTwoLogV = minusTwoLogV - (observationsPerClass-
1)*log(det(cov(inputMatrix(((ii-1)*observationsPerClass)+1:ii*observationsPerC
lass,:)))); %calculate -2log(V)
end
degreesOfFreedom = (classes-1)*sensors*(sensors+1)/2; %degrees of freedom
pvalue = 1 - chi2cdf(minusTwoLogV,degreesOfFreedom); %p value
if pvalue>0.05
    disp('Covariance matrices are equal') %accept null hypothesis (not signif-
icant)
else
    disp('Covariance matrices are not equal') %reject null hypothesis (signif-
icant)
end
```

Chapter 5
Correlating Electronic Nose and Field Olfactometer for Industrial Odor Concentration Measurement Using PLS and MLR

Sharvari Deshmukh
National Environmental Engineering Research Institute, India

Nabarun Bhattacharyya
Centre for Development of Advance Computing, India

Arun Jana
Centre for Development of Advance Computing, India

Rajib Bandyopadhyay
Jadavpur University, India

R. A. Pandey
National Environmental Engineering Research Institute, India

ABSTRACT

Industrial odor concentration measurement in continuous mode is a challenging task using olfactometers, as it's expensive and requires human involvement for a prolonged time. This chapter presents the development of an indigenous metal oxide sensor-based electronic nose system for measurement of industrial odor in ou/m3. The results of electronic nose and field olfactometer were correlated using multilinear regression and partial least square regression techniques. The results showed satisfactory prediction by both the models, with RMSE (6.70, and 4.02), RAE (0.29 and 0.16), and NAE (0.89 and 0.96), respectively, for MLR and PLS. The results indicated better performance of PLS compared to MLR. The objective of the present work is to train and employ artificial olfaction system for continuous measurement of obnoxious emissions emitted from industries bypassing involvement of olfactometer.

DOI: 10.4018/978-1-5225-3862-2.ch005

BACKGROUND

Industrial growth is associated with countries' rapid economic growth; however, many such industries have obnoxious emissions associated with it. These odors are due to very low levels of odorant concentration and need to be measured and treated before its emissions into the environment. Therefore, both continuous odor emissions measurement at source point and ambient sources are necessary to maintain the clean surrounding environment and to retain overall sound quality for human dwelling (Chuandong *et.al*, 2016). Odor concentration (ou/m^3) is a cumulative effect of the numerous VOCs present at the industrial site as well as environmental parameters like humidity and temperature. Traditionally odor has been measured using olfactometry. However, prolonged use of olfactometry is not suitable for measurement of industrial odors since it involves human panel. Further, olfactometric measurements are expensive due to the involvement of human panel. Analytical instruments like gas chromatography are expensive, time consuming and depict very little about the odor perceived. These tools basically give the representation of the major odorants and not the overall odor at the industrial site. In the present work pulp and paper industry is chosen as a case study to evaluate the use of electronic nose system for industrial odor concentration measurement. Indian pulp and paper industry is huge with high production of paper quantity and numerous employment generations. However, these industries during pulp manufacturing process emit highly obnoxious gases constituting of dimethyl sulphide [$(CH_3)S$], dimethyl disulphide [$(CH_3)_2S_2$)], methyl mercaptan [$(CH_3)_2SH$)] and hydrogen sulphide [(H_2S)] in different concentration levels (Deshmukh *et.al*, 2014). In this research work a metal oxide sensor based electronic nose system has been developed and trained using field olfactometer for measuring the odor concentration in ou/m^{3}pulp and paper industries. The electronic nose results were correlated with field olfactometer using multilinear regression (MLR) and Partial Least Square regression (PLS). The results on these aspects are presented in this paper.

The motivation and contribution of this chapter is to propose an alternative way to measure industrial odor concentration using an electronic nose system. The chapter presents some onsite industrial odor concentration data collected and measured using an indigenously developed electronic nose system with minimum training sets.

MATERIALS AND METHODS

Developed Electronic Nose Setup and Experimental Procedure

A portable electronic nose system was developed for the measurement of odor concentration (ou/m^3) using commercially available metal oxide sensors. The sensors were procured from FIGARO, USA (Figaro, USA) of TGS series TGS 823, TGS 825, TGS 2610, TGS 2602, TGS 826, TGS 832, TGS 2620, TGS 2611. The TGS sensors are of thick film type, and the sensing element of the TGS sensor is tin dioxide (SnO_2) doped with different impurities. The developed system consisted of (a) stainless steel sample holder (the steel material had no reaction to the sulphurous odorants) (b) programmable suction pump (c) compressed pure air canister for purging of the leftover gases (d) data acquisition system and (c) olfaction software.

The sensors were fixed in a funnel-shaped chamber, the shape of which allowed equal gas exposure to all the sensors. A circular printed circuit board was used for fitting the sensors into the chamber. The chamber had a narrow inlet in the form of Teflon tubing to let the gas collected during sample suction process be exposed to the sensor array. During the purging process, an exhaust fan mounted on the chamber was turned ON to clean the system by removing the left over RSCs. The overall process of the electronic nose can be summarized as follows (1) Suction time = 10 sec, (2) Sample capture = 10 sec, (3) Sampling time = 20 sec, (4) Purging time = 300 sec. All the above operations along with PC – based data acquisition were controlled by specially designed software in LabVIEW® of National Instruments. The developed software helped in reduction of human error while handling a very dynamic thing like gas sample and allowed equal controlled injection of gas samples for repeated studies. Figure 1 shows the schematic of the developed electronic nose system.

Field Olfactometer

Dynamic olfactometry is a common and traditional technique for qualitative & quantitative analysis of odor emissions. It allows measuring the total effect of the odor on human perception using human nose as a detector. The polluted sample is diluted with pure air/ odorless air to reduce the odor concentra-

Figure 1. Electronic nose setup

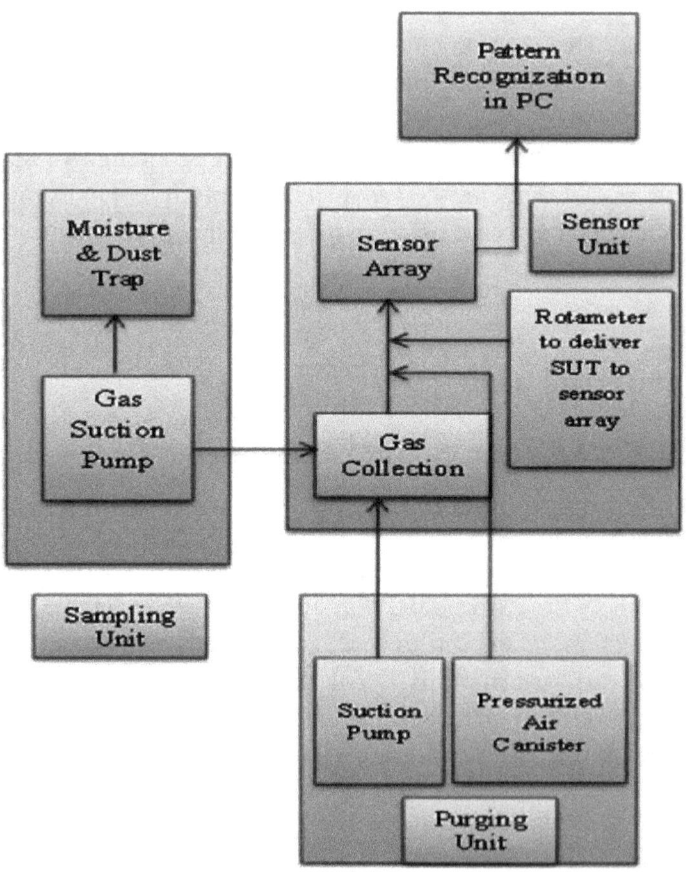

tion to its detection threshold. The odor is diluted with an odorless air in known concentrations, and different samples are prepared. A series of diluted samples are presented to the panelist in descending order. This procedure is followed until 50% of the panelists are unable to distinguish between odorous and odor free sample (McGinley *et.al.,* 2000). This is termed as odor detection threshold and defined as 1 ou/m^{3}of odor concentration. In different countries, it may be termed as 1OU, 1OU$_{Em}$, 1OC or 1 DT^{-1} but conceptually they are all equivalent to 1oum^{-3}. In this study, a portable field olfactometer, NASAL RANGER® Field Olfactometer procured from St Croix Sensory, Inc., Lake Elmo, MN, USA for the measurement of odor concentration at pulp and paper industry site (Figure 2). The field olfactometer is a nasal organoleptic odor measuring device useful for quantifying the odors. It works on the principle of dynamic olfactometry. The field olfactometer dynamically dilutes the ambient air with carbon - filtered air in distinct dilution factors known as 'dilution-to-threshold' factors. It works on Dilution to Threshold (D/T ratios) ratio that is one of the most commonly practiced techniques for odor measurement (Pan *et.al,* 2017). The photograph of the instrument is as shown in Figure 2. The operating principle of the FO compromises of mixing odorous ambient air with odor free filtered air in discrete volume ratios. The discrete volume ratios are called "Dilution-to- Threshold" ratios (D/T ratios). The higher the D/T ratio value assigned to the sample, stronger is the odor (Jae Ho Sohn *et.al.,* 2008) While using the field olfactometer, the human nose is firmly placed inside the nasal mask attached to the Nasal Ranger.

Field-Based Investigation

The industrial samples were analyzed on site to determine the odor concentration using field olfactometer. The samples were collected from the surroundings of the major source points and not exactly at the source points. Four major points were targeted, and the concentrations were measured at these points using the field olfactometer. The details are provided in Table 1. The samples were collected using the sampling unit consisting of moisture and dust trap. From the same sites, gas samples were collected in Tedlar Bags and tested using the developed electronic nose system for correlation studies between field olfactometer and sensor array of the electronic nose.

Figure 2. NASAL RANGER® Field Olfactometer

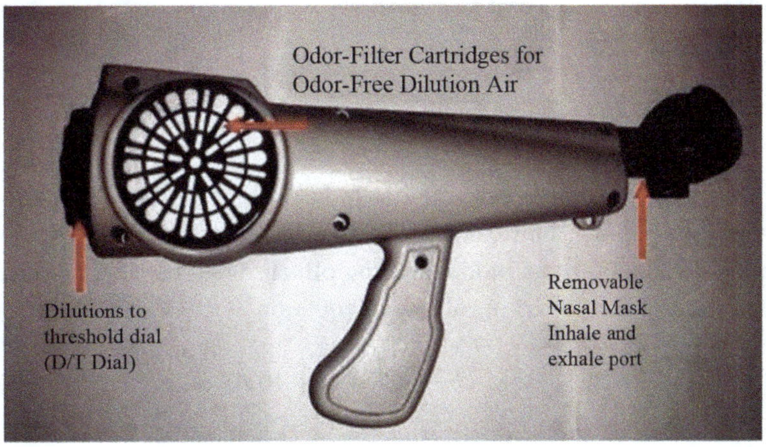

Table 1. Industrial odor concentration (measured using field olfactometer)

Industrial Site Points	Odor Concentration (ou/m³)
Condenser	15-30
Digester Blow	300 – 500
Digester	7 – 15
Evaporator (source surrounding)	400 – 500
Evaporator (Away Source)	30 – 60
Storage Area	100 – 200
Work Environment	4 – 7

Data Analysis

The correlation between the sensor array output and field olfactometer was achieved using Multi Linear Regression (MLR) and Partial Least Squares (PLS). The 40 sets of sensory array data were divided into training (75%) and testing phase (25%). Multiple linear regression (MLR) models the relationship between two or more independent variables and a response variable by fitting a linear equation to observed data. Every value of the independent variable x is associated with a value of the dependent variable y. Whereas, partial least square (PLS) is a technique that reduces the predictors to a smaller set of uncorrelated components and performs least squares regression on these components, instead of on the original data. PLS is particularly useful when your predictors are highly collinear and are redundant responses like that of a sensor array of electronic nose system. To regress the *Y* variables with the explanatory variables X_1...*Xp*, . PLS attempts to find new factors that will play the same role as the *X's*. These new factors often called latent variables or components. Each component is a linear combination of *X1*........*Xp*. To simplify it the relationship between latent variables derived from original individual variable and the output response Y is given by

$$Y = b_0 + b_1 X_1 + \ldots + b_p X_p \quad (1)$$

where b_0 is the regression coefficient for the intercept and *bp* are the regression coefficient calculated for the latent variables.

RESULTS

The results obtained by testing samples collected at industrial site by using electronic nose and field olfactometer were used to develop empirical model using MLR and PLS. The sensor's response in the form of voltage was feed as input to the model and the olfactometer reading acted as the output of the model. The application of MLR resulted in following equation

$$OC = 1.18 - 15.3 * A + 97.8 * B + 37.5 * C - 92.9 * D + 10.4 * E$$

$$-0.6 * F + 46.3 * G - 10.7 * H \quad (2)$$

where *A,B,C,D,E,F,G,H* were different sensors and *OC* the predicted odor concentration. The *P-values* were used as tool to check the significance of each of the coefficients. The smaller the magnitude of *P*, more significant is the corresponding coefficient. The model adequacies were checked by R^2 and predictive error sum of squares (*PRESS*). A model with large R^2 low *PRESS* values is considered to be a good model. The developed model had R^2 value of 0.9510, and *PRESS* value of 2643.81, and R^2_{pred} value of 0.8505. The R^2_{pred} value of the model represents how successfully the model fits the data. The high R^2 for the entire four model shows that the developed models could explain the variations present in the independent variables. Whereas, R^2_{pred} which represents the ability of the derived model to predict the output for unknown samples was also found to be high for all the four models. Fisher's variance ratio *F-value* is calculated as a ratio of mean square regression and mean square residual. It is measure of variance in the data about the mean. Table 2 represents all the model values

Similarly, PLS regression was applied to the data generated from sensor array output and field olfactometer. The results obtained were better compared to MLR. The R^2 value was found to be 0.9576 whereas R^2_{pred} was found to be 0.9242 for 3 components. The *PRESS* value for PLS model was 1692.95. The score plot of PLS is given in Figure 3.

The analysis of variance is table for PLS model is as provided in Table 3.

Table 2. Analysis of variance for MLR

Source	DF	SS	MS	F	P
Regression	8	16804.9	2100.6	74.42	0.00
Residual Error	31	875.0	28.2		
Total	39	17679.9			

Figure 3. The score plot of PLS

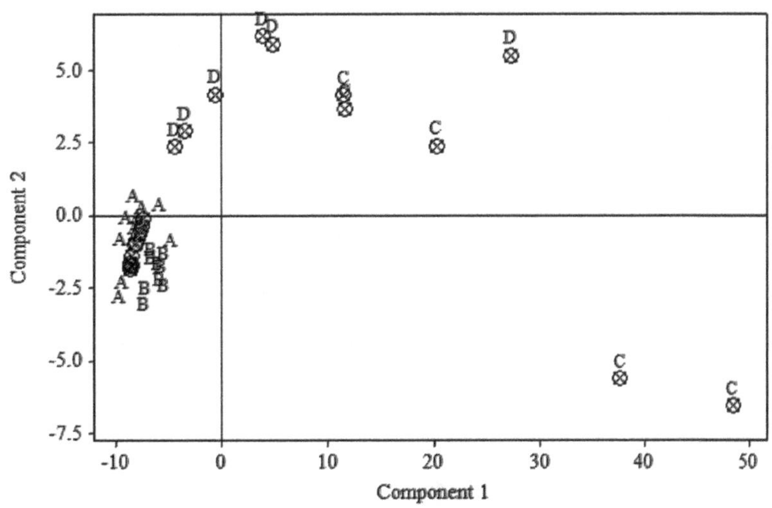

Table 3. Analysis of variance for PLS

Source	DF	SS	MS	F	P
Regression	3	16578.3	5226.10	180.59	0.000
Residual Error	31	875.0	28.2		
Total	39	17679.9			

The performance evaluation of the prediction models implied were verified using root mean square error (RMSE), relative absolute error (RAE) and the Nash-Sutcliffe model efficiency (NSE). The RMSE, RAE and NSE are calculated using the equations given below

$$RMSE = \left[\frac{\sum_{n=1}^{n=N}\left(P-A\right)^{2}}{N}\right]^{-2} \quad (3)$$

$$NSE = 1 - \left[\frac{\sum_{n=1}^{n=N}\left(P-A\right)^{2}}{\sum_{n=1}^{n=N}\left(P-\bar{P}\right)^{2}}\right] \quad (4)$$

$$RAE = \left[\frac{\sum_{n=1}^{n=N}\left(P-A\right)}{\sum_{n=1}^{n=N}\left(P-\bar{P}\right)}\right] \quad (5)$$

where P is the predicted and A is the experimental reading (field olfactometer). Table 4 denotes the values of RMSE, RAE and NSE for MLR and PLS models. RMSE and RAE describe the deviation of model calculations (predicted) from the experimental data. Whereas NSE specifies the quality of how the model data fits to the line of identity with 1 as an optimal value. NSE closer to 1 and RMSE & RAE values closer to 0 indicates good model.

As can be seen from the Table 4 RMSE and RAE values of PLS were comparatively lower than MLR, also NSE for PLS was found to be 0.96 (closer to 1). This indicated that PLS was a better compared to

Table 4. RMSE, RAE and NSE values for MLR and PLS model

	MLR	PLS
RMSE	6.70	4.02
RAE	0.29	0.16
NSE	0.89	0.96

MLR for predicting industrial odor concentration. Figure 4 explains the difference in the results of MLR and PLS model results.

In Figure 4 three lines are used to show the success of the two models. One line is perfect fit; indicating an ideal model (experimental results). The other two lines indicate the prediction capability of MLR and PLS model.

CONCLUSION

In this study, an electronic nose system has been developed to measure the odor concentration (ou/m3) at the industrial site. A field olfactometer has been used to provide training to the developed electronic nose system. The main objective of the study was to provide an alternative to tradition olfactometry procedure and bypass it using machine olfaction system. The results obtained prove that PLS can effectively train the electronic nose system to measure the unknown industrial concentrations. Though the field olfactometer used in this study has limited dilution ratio which could be overcome using a more advance olfactometer. However, the results obtain encourages to explore the electronic nose system further for continuous industrial odor concentration measurement and bypass the human involvement in the process.

Figure 4. Experimental and predicted (MLR and PLS) odor concentration

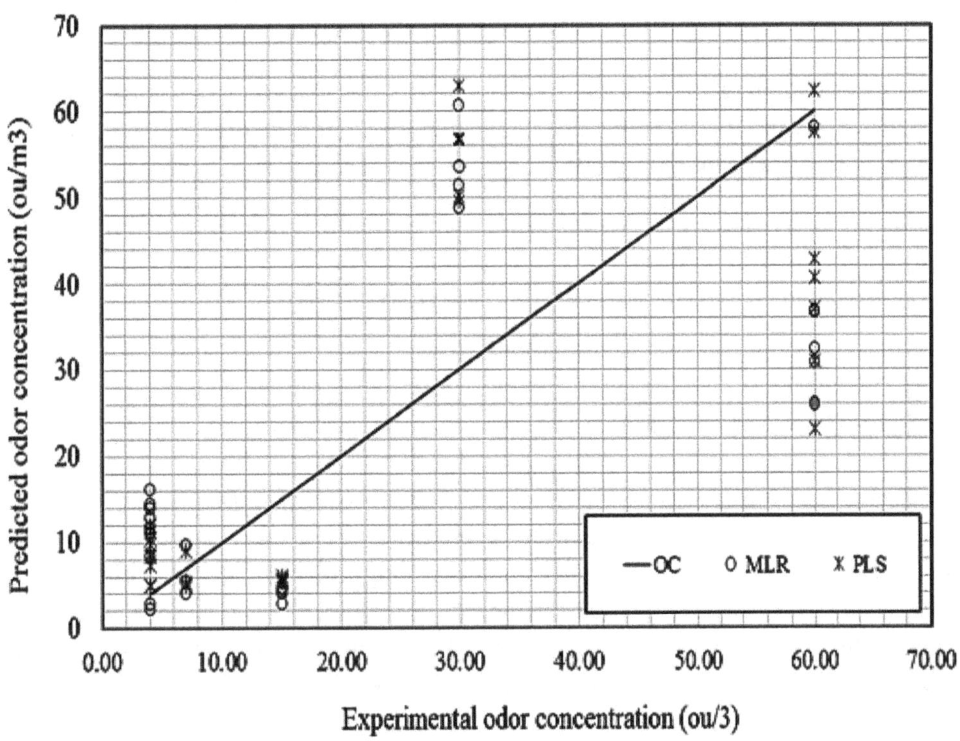

ACKNOWLEDGMENT

The authors are thankful to Director, CSIR-NEERI, Executive Director C-DAC, Kolkata; Prof. E.S.R. Gopal, Emeritus Scientist, IISc, Bangalore their support and guidance throughout the study. The financial support of Department of Electronics and Information Technology, Ministry of Communication and Information Technology, Government of India, and Council of Scientific and Industrial Research, Government of India in carrying out this research work is also acknowledged.

REFERENCES

Deshmukh, S., Jana, A., Bhattacharyya, N., Bandyopadhyay, R., & Pandey, R. A. (2014). Quantitative determination of pulp and paper industry emissions and associated odor intensity in methyl mercaptan equivalent using electronic nose. *Atmospheric Environment*, *82*, 401–409. doi:10.1016/j.atmosenv.2013.10.041

Figaro U. S. A. Inc. (n.d.). Retrieved from http://www.figarosensor.com

McGinley, C. M., & McGinley, M. (2000). A Field odour monitoring and enforcement. *Proceedings of the 22nd Annual Hawaii Water Environment Association Conference.*

Pan, L., Yang, S. X., & DeBruyn, J. (2007). Factor Analysis of Downwind Odours from Livestock Farms. *Biosystems Engineering*, *96*(3), 387–397. doi:10.1016/j.biosystemseng.2006.10.017

Sohn, J. H., Hudson, N., Gallagher, E., Dunlop, M., Zeller, L., & Atzeni, M. (2008). Implementation of an electronic nose for continuous odour monitoring in poultry shed. *Sensors and Actuators. B, Chemical*, *133*(1), 60–69. doi:10.1016/j.snb.2008.01.053

Wu, C., Li, J., Zhao, P., Piringer, M., & Schauberger, G. (2016). Conversion of the chemical concentration of odorous mixtures into odour concentration and odour intensity: A comparison of methods. *Atmospheric Environment*, *127*, 283–292. doi:10.1016/j.atmosenv.2015.12.051

KEY TERMS AND DEFINITIONS

Electronic Nose: A device that works on the lines parallel to human olfaction system using sensors, data acquisition systems, and pattern recognition algorithms.

Industrial Odor: A cumulative obnoxious smell resulting from various activities at industrial site, generally a combination of various chemicals/odorants, smoke, etc.

Multi-Linear Regression: It models the relationship between two or more independent variables and a response variable by fitting a linear equation to observed data.

Olfactometer: An analytical instrument used to measure odor concentration in ou/m3 and odor intensity of the sample under test.

Partial Least Square Regression: It is a technique that reduces the predictors to a smaller set of uncorrelated components and performs least squares regression on these components, instead of on the original data.

Sensors: A device that senses the changes in the environment and sends the signal in suitable form to the intermediate device, usually data acquisition systems.

Volatile Organic Compounds (VOCs): Organic compounds that can be easily vaporized into gases. They generally contain carbon, hydrogen, oxygen, fluorine, chlorine, bromine, sulfur, or nitrogen.

Chapter 6
Electronic Nose Applications

Ping Wang
Zhejiang University, China

Fan Gao
Zhejiang University, China

Xusheng Zhang
Zhejiang University, China

Min Wang
Zhejiang University, China

ABSTRACT

In this chapter, the authors introduce three typical applications of electronic nose, including the diagnosis of lung cancer and gastrointestinal diseases and food freshness detection. Firstly, an electronic nose system intended for lung cancer screening employs semiconductor and electrochemical sensors to compose an array in a chamber to conduct breath analysis with pre-concentration and multiple desorption procedures. Secondly, a design of virtual electronic nose was presented to detect the concentration of hydrogen, methane, and carbon dioxide in breath for the diagnosis of small intestinal bacterial overgrowth. Finally, combined with the metal oxide semiconductor sensors and electrochemical sensors array, the small size electronic nose system is to identify the freshness and classify different kinds of food stored in the refrigerator.

LUNG CANCER DIAGNOSIS

The electronic nose technology, with the advantages of fast response and portable size, has been widely studied in the application of lung cancer diagnosis through breath test. The electric nose has many promising applications. There is still no last word in the research of breath biomarkers relating to lung cancer, and the results from different research team are all a combination of different kinds of volatile organic compounds (VOCs). In the application of electronic nose in detection of lung cancer, many biomarkers need to be detected, including VOCs of both low molecular weight and high molecular weight. Single kind of gas sensors cannot be sensitive to all the biomarkers (Gouma et al., 2011). So we developed an

DOI: 10.4018/978-1-5225-3862-2.ch006

electronic nose system based on mixed gas sensors array and recognition algorithm. The electronic nose must be sensitive enough to many kinds of VOCs to work as an efficient breath diagnostic tool. However, the common electronic nose based on single kind of sensor, such as metal oxide semiconductor (MOS) sensors and conductive polymer (CP), cannot detect all of the biomarkers, and furthermore, it cannot supply all of the useful information and give a reliable detection result. To get better recognition results, an electronic nose based on surface acoustic wave (SAW) combined with MOS sensors was presented in this section.

Design of Electronic Nose System Based on MOS and SAW Sensors

Our electronic nose system is fitted with 8 metal oxide semiconductor (MOS) sensors chosen on the basis of Turner's research papers about breath components like ammonia, acetone, ethanol, and isoprene (Yu et al., 2011). The sensing characteristics of these sensors have been tested in our previous work. The outputs from the sensors array can be affected by the variation of the ambient factors such as temperature, humidity and air velocity. In order to diminish these influences, the sensors array are well packaged in a small chamber made by Teflon. Two micro vacuum pumps are used to drive sampling tubes. The gas flow is monitored by a mass-flow sensor and controlled by solenoid valves. The gas velocity in tubes can also be altered by setting the power supplied to the pumps. The six-way valve is automatically switched to assist the enrichment and desorption unit to fulfil the trace component detection in breath according to a sampling procedure, which is programmed in the micro controller chip, LPC2478. A real-time operating system has been run on this chip to improve the performance of the electronic nose system. The breath sampling structure of electronic nose system based on MOS sensors is presented in Figure 1. A data acquisition software is developed to obtain the filed data and analyze it. The hardware structure of electronic nose system based on MOS sensors is shown in Figure 2.

In order to further improve the diagnosis accuracy of the electronic nose system, we propose a hybrid electronic nose system based on MOS-SAW detection units. The MOS gas sensors are used to detect the VOC molecules with low molecular weight (LMW), and the SAW sensors are adopted for the detection of VOC with high molecular weight (HMW).Thus, the novel combination of these two kinds of gas sensors provides higher sensitivities to more of VOC species in breath than that of using only a single kind

Figure 1. The breath sampling structure of electronic nose system based on MOS sensors

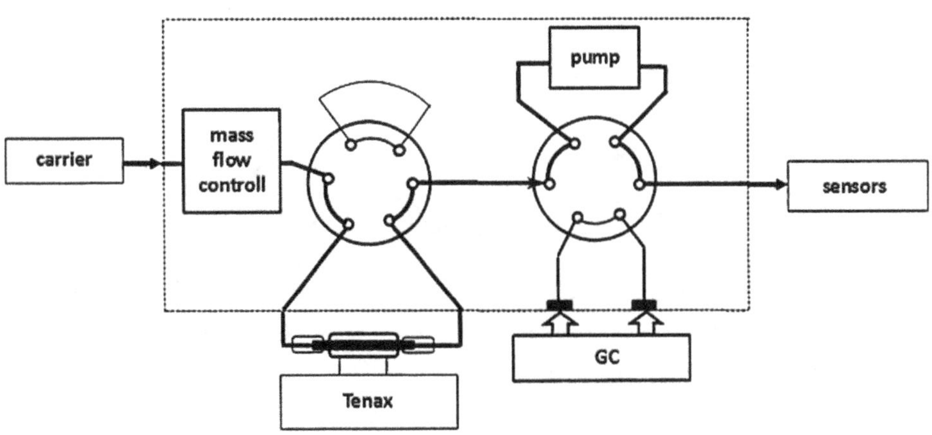

Figure 2. The hardware structure of electronic nose system based on MOS sensors

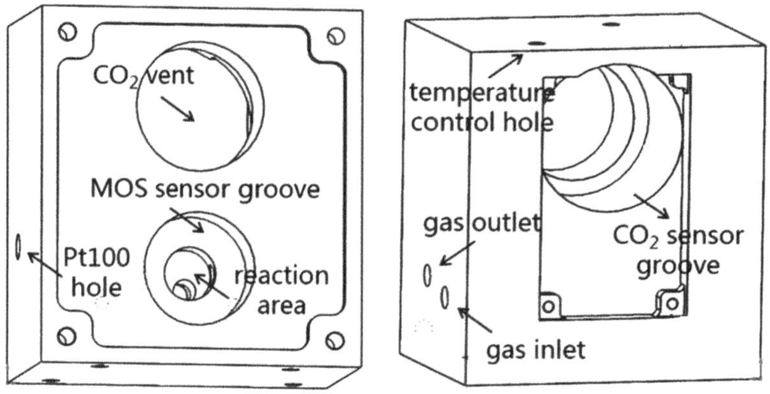

of sensor (Wang et al., 2011). The hybrid electronic nose system is composed of four units, including enrichment unit, MOS detection unit, SAW detection unit and diagnosis unit. The diagram of electronic nose system based on MOS-SAW sensors is shown in Figure 3.

The patients, breath within air bag is pre-concentrated by a Tenax-TA tube, which is equipped in the enrichment unit. After the thermal desorption at 260°C for 3 min with the carrier gas flow of 100 standard cubic centimeter per minute, the enriched VOCs are carried into the MOS detection unit and SAW detection unit, respectively. In the MOS detection unit, nine MOS sensors (Figaro TGS813, TGS822, TGS2600, TGS2602, TGS2620 and Winsen MQ2, MQ3, MQ6, MQ138) are chosen on the basis of

Figure 3. The diagram of electronic nose system based on MOS-SAW sensors

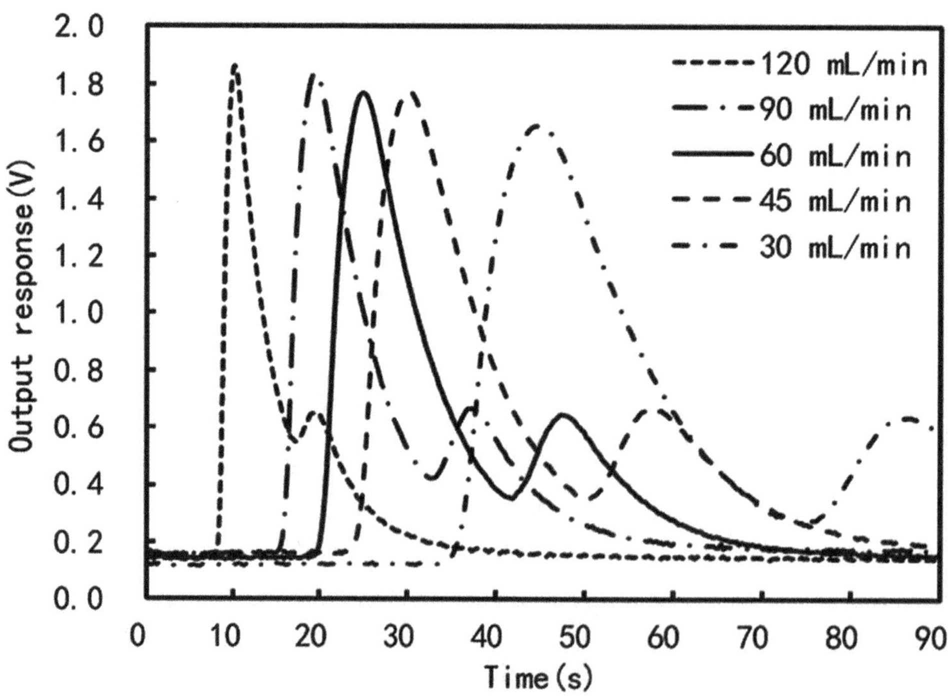

Turner's research papers about the components in breath, like ammonia, acetone, ethanol, methanol and isoprene. The gas chamber structure is shown in Figure 4.

The outputs from this sensors array can be influenced by the variation of the ambient factors such as temperature, humidity and air velocity. In order to diminish these influences, the array is packaged in a small chamber of Teflon. These sensors have high sensitivity to the VOCs with low molecular weights like benzene, styrene, n-hexane, which are considered as the biomarkers of lung cancer. Uncoated SAW resonators and GC separation are adopted in the SAW detection unit. Moreover, the uncoated SAW resonators are much more suitable for analyzing breath than the SAW sensors with delay line or with sensitive film coated. We have developed a SAW resonator using ST-cut quartz, which has a low temperature coefficient. The sensor's central frequency is 250 MHz; the insert loss is 5.96dB. The differential structure is to diminish the environmental influences. When the VOCs come out from the capillary column, they will be condensed on the surface of the temperature controlled SAW resonator, generating a series of frequency changes. The VOCs are captured by the sensor for condensation, so the SAW resonator has a high sensitivity to the high molecular weight VOCs with high boiling point. The signals from MOS-SAW detection units are then recognized by a multi-model diagnosis method. Applying four algorithms, six models were established for diagnosis and tested by leave-one-out cross-validation method (Hunter et al., 2011). The model by artificial neural network (ANN) was selected as the best model to analyze breath samples (Gu & Li, 2014). The MOS gas sensors are applied to detect the VOCs whose molecular weights are lower than 150, e.g., benzene, styrene, n-hexane, while the SAW sensors are adopted for the detection of VOCs with molecular weights higher than 150, e.g., tridecane and tridecanone. As a result, hybrid electronic nose system takes both of the two kinds of sensors' advantages and provides a better diagnostic precision for lung cancer than the system adopting single detecting method we have developed before.

Figure 4. The gas chamber structure of electronic nose system based on MOS-SAW sensors

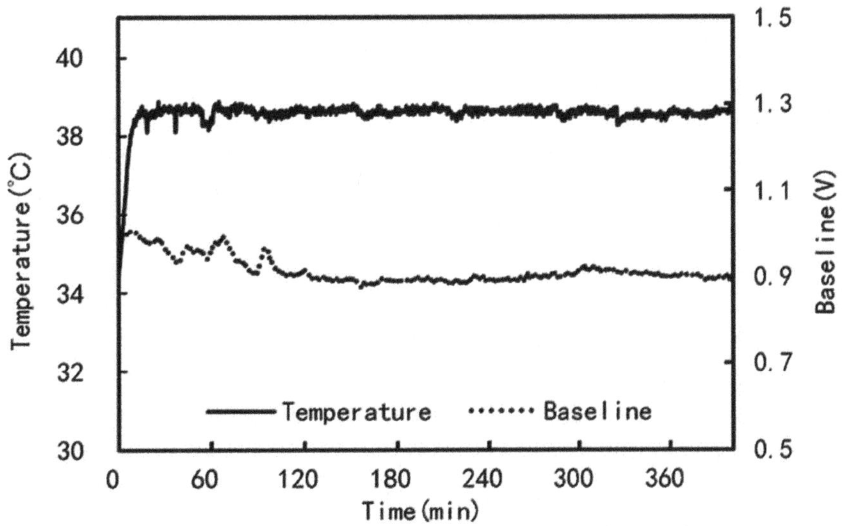

Experiments and Results

We collected 24 samples (sample set A) from normal families to have the breath testing. This samples consist of five chronic smokers, five drinkers, four of them both chronically smoking and drinking, 10 of them neither smoking nor drinking. No sample in sample set A had been diagnosed with lung disease. We also collected 23 samples (sample set B) for the detection of lung disease in hospital, include 9 healthy controls (5 smokers), 5lung Benign Disease (3 smokers) and 9 lung cancer patients (7 smokers).All the volunteers were 41 to 69 years old.

The collections took place in the early morning to ensure that all the volunteers ate nothing for all night long. They were requested to enter an empty room with good ventilating condition. After breathing in that room for 10 minutes, each volunteer was asked to exhale about 70 percent out of lung and the last portion of vital capacity breath was then collected into air bag. Then the air bag is connected to the injection port of our system via silicone tube. This system performs an automatic sampling procedure using the predefined flow rate and total volume. It possesses two desorption methods, full desorption under one constant temperature and gradual desorption under a couple of steady temperatures. The full desorption has been conducted through heating up the EDU to a given temperature, like 240°Cused in subsequent experiments. With the purpose of separation of mixed gas, chromatographic columns have been utilized to improve the discrimination power over conventional electronic nose. For our system, we create a new method to separate the breath during desorption just by each precise match of steady temperature and the volume of carrier gas. This gradual desorption method realizes the separation of sample partly on the hand and reduces the cost of our electronic nose instrument on the other. This new approach is based on the characteristics of Tenax TA's breakthrough volumes for VOCs. Otherwise, the electrochemical sensor NE-NH3 can help quantifying of ammonia utilized for the detection of alimentary diseases.

With the desorption proceeding, nine sensors' signal curves, called smell prints, were displayed on the TFT screen to indicate the sensing results of breath. These signals were transmitted through UART to data acquisition software on PC, and saved in external flash memory simultaneously. Nine curves for each sample are obtained before pattern recognition and it is necessary to perform signal preprocessing. First, the baseline is subtracted from the signal curve and the filter was applied to improve the signal to noise ratio. The next step is feature extraction to get the peak, response time, response peak area, maximum positive slope of the curve and so on. The features of these curves of MOS sensor are shown in Figure 6. In order to acquire a better recognition result, we can take some of them as the features which could be chosen interactively in data acquisition software for certain disease. Normalization is adopted to make compensation for the scale different between the features of all the samples. After the pre-processing, principal component analysis (PCA) performs discrimination from the feature of all samples which is an unsupervised technique as a typical method for reducing the dimensions of data. The result is described on a plot with the axes standing for main principal components (PCs).

The feature of the sensor response are diverse for different samples and different sensors. The smell prints of patients and healthy controls have some differences. It is necessary to state that NE-NH3 has no effective response during desorption, since ammonia which is principally from alimentary tract has been removed during the breath collection. This electrochemical sensor is utilized to satisfy the application for lung cancer. Furthermore, gradual desorption method has been adopted to produce response curves. The typical detection results of the saw detection unit is shown in Figure 6. The first peak contains information about the concentrations of VOCs in breath which have LMW, whereas the second peak represents

Figure 5. The features of the response curves of MOS sensor

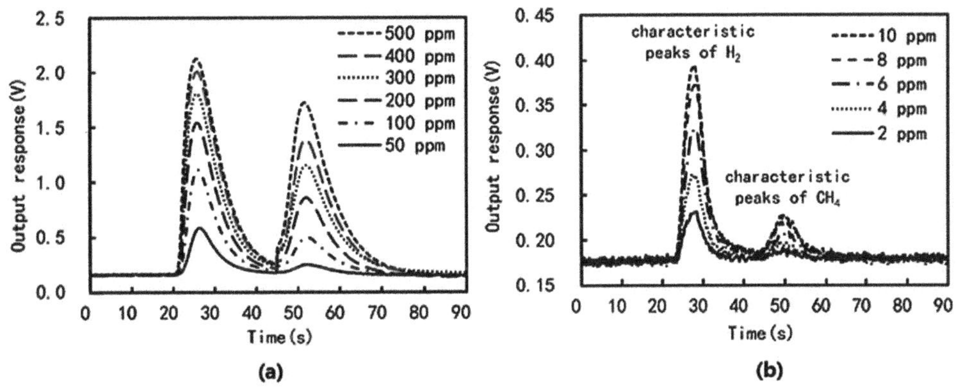

the VOCs with HMW. These two peaks can help providing more information to conduct pattern recognition, especially for the sample set A. The results also reveals the second peak is much lower than the value of the first peak, which is in accordance with the general results for small amounts of biomarkers.

As previously mentioned, data pre-processing includes three steps. First step is to subtract the baseline. Then weighted mean and medium filtering, as the second step, are applied to be the foundation of feature extraction. Finally, the feature of sensor response are extracted before recognition. The SNR for signals are improved which makes the feature extraction more accurate. We select the first two main PCs whose contribution rates are 53.08% and 31.96% respectively. The PCA result of the sample set A is shown in Figure 7. From the result, it can be known that the total accuracy is about 83.3%.88.9% of lung cancer patients are correctly identified and 77.8% of healthy controls are well clustered. The misclassification rate is 16.67 percent and there is no crossover region between the two categories. The result above from the full desorption method will not satisfy the need of a future healthcare system. Thus, the PCA result of the sample set B with gradual desorption method whose false positive rate is 11.1%,while all the patients have been correctly identified. The accumulation contribution rate for first two PCs is 90.7%, which has been obviously improved from that full desorption method. Through the comparison, we know that a better discrimination rate could be obtained employing the gradual desorption approach without increasing any cost for such electronic nose instrument which will be crucial for the home application. However, it takes more time for the breath testing than time used in the occasion of full desorption under the same conditions. Hence, the sample set A has been analyzed utilizing the gradual desorption method and the recognition result.

Although the subjects in this sample set are all healthy people, they have been classified into the correct categories. Current evidence support that most lung diseases, especially lung cancer, are related to chronically smoking. It is observed that the category of smoking has the farthest distance from that of non-smoking. Between the 'Neither' and 'Drinking' categories, there is crossover region mainly for the biomarkers' choice which are consider to originated from alimentary tract. Maybe a more precise gradual desorption procedure will be able to distinguish these two categories completely. Two response signals of one sample obtained from the MOS and SAW detecting units are filtered by appropriate filters. In the following step, features extracted from these two sets of signals are normalized to make compensation for the scale difference between the features of all the samples. We employ PCA, ANN, linear discriminant analysis (LDA), and partial least squares (PLS) to establish diagnostic models. The method of leave-

Figure 6. The detection result of the SAW detection result

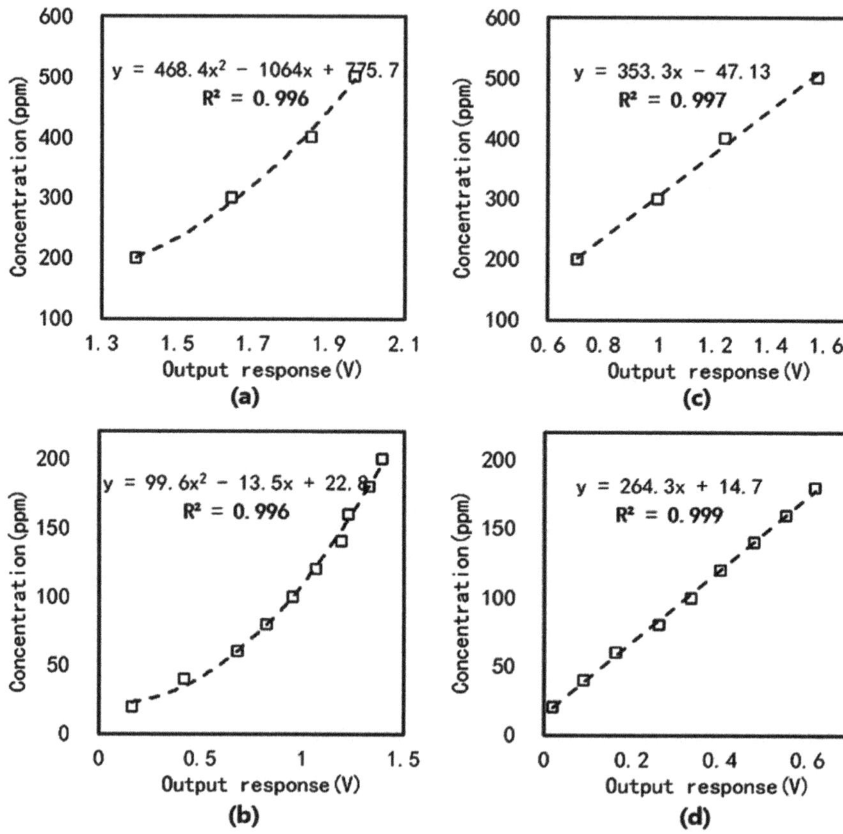

one-out cross-validation is adopted to test these models. The model with the highest sensitivity and selectivity will be selected to validate the ability of this electronic nose system to diagnose lung cancer.

Benzene, styrene and n-hexane are selected to calibrate the MOS detection unit as the low molecular weight VOCs, and tridecane and tridecanone are selected as the high molecular weight VOCs for the SAW detection unit. All of the five VOCs given above are considered as part of the lung cancer biomarkers. The MOS detection unit is calibrated by benzene, styrene and hexane with five concentrations, i.e., 0.01 ppm, 0.1 ppm, 0.2 ppm, 0.3 ppm and 0.4 ppm. Similarly, the SAW detection unit is calibrated by tridecane and tridecanone with five concentrations, i.e., 1 ppb, 3 ppb, 5 ppb, 7 ppb and 10 ppb.

The samples range from male to female, smoker to nonsmoker and alcohol drinker to nondrinker with ages from 30 to 80. All the patients and controls are asked to eat nothing and ban smoking for 12h (test in the morning), and avoid high-fat dinner the day before the measurement. The subjects use clean water to gargle half an hour before the test. Then, 110 features were extracted from these two responses for each sample, and six diagnostic models are established: LDA, ANN, PLS, PCA-LDA, PCA-ANN and PCA-PLS. Because of the nonlinear relationship between the diagnostic result and the signal features, both the sensitivity and selectivity (lower than 70%) of the models based on PCA, LDA and PLS cannot satisfy the requirements very well. Finally, ANN is selected to build the diagnostic model in application. Three kinds of ANN diagnostic models are established separately: Model based on features of MOS detection unit, the one based on features of SAW detection unit and another one based on features of both MOS and

Figure 7. The PCA result of set A

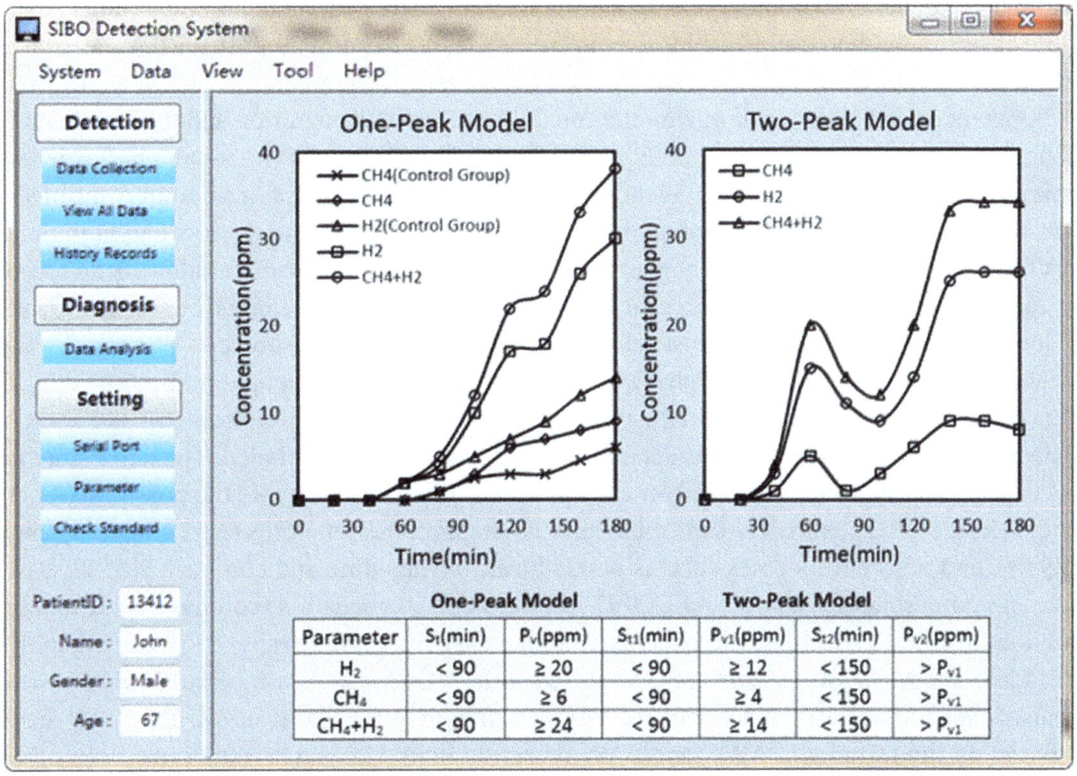

SAW detection units. The accuracies of each model obtained through the leave-one-out cross-validation method are compared. The sensitivity and selectivity to model I are both not good enough for diagnosis. Although the sensitivity to model II is 91.49% which is much higher than that of the former, the electivity is only 57.14% even much lower than 69.05%. The model III, which is established on the features of both MOS and SAW detection units, has the highest sensitivity of 93.62%. Besides, the selectivity substantially is improved to 83.37% satisfying the requirements of breath testing. A separate analysis of smokers and nonsmokers had been done by the GCMS before the hybrid electronic nose system was established. The result shows that some specific VOCs are higher due to the tobacco smoking, such as 6, 10- dimethyl-2-hendecanone, but smoking does not adversely affect the accuracy of the breath test. So this separate analysis was not performed by the hybrid electronic nose system.

Our study shows that exhaled breath of patients with lung cancer/diseases has some features that can be identified using our electronic nose based on mixed sensors array and multiple methods. During the study, we have also made comparison between lung benign diseases, like pneumonia and pneumoconiosis, with healthy controls. The false positive rate is about 40% which should be improved in further research to fulfil the requirements of the electronic nose system. Currently, as the primary reason for the high false positive rate, it should be considered that there are few efficient approaches to wipe off the breathing gas from the anatomical dead space. During the experiment, the VOC biomarkers of low concentrations have been hidden by the background gas which comes from the ambient air and even from the local cavity. As a solution, we have developed new gas sensors to enlarge the array and a novel breath collector for in situ application. Besides, more precise gradual desorption procedure will be required to receive better discrimination. Another point to note here is that the precise gradual desorption procedure is based on the choice of biomarkers from diseases. The main cost for our electronic nose instrument are the fabrication of mixed sensors array with a small chamber, the fabrication of EDU, and the development of embedded system with the capability of smart recognition and the interfaces for applications. Because of the long term stability and suitable sensitivity, the instrument has excellent cost performance for the home lung cancer detection without additional reagent or air source during the test.

It can be seen that both the detection units have a low detecting limit to the low and high molecular weight VOCs, respectively, and the combination of these two kinds of sensors can adapt to the complexity of lung cancer diagnosis through breath analysis. The 89 breath samples came from 42 healthy persons and 47 lung cancer patients. There are strict criteria when selecting the subjects: The healthy controls were selected from the synchronous physical examination of healthy people and the healthy family members of the lung cancer patients. In the physical examination, both the spirometry and chest X-rays were performed. If any abnormality was detected, the subject would be excluded. The family members of the lung cancer patients were also asked for chest X-rays examination to avoid the possibility of chronic bronchitis. Besides, the healthy controls must have no respiratory symptoms (e.g., cough, dyspnea and hemoptysis), and no acute illnesses such as peptic ulcer, oxyhepatitis and coronary heart disease. The patients with comorbidities such as AECOPD, peptic ulcer, oxyhepatitis, coronary heart disease and diabetes were excluded. Besides, the patients who had surgery, radiotherapy, and chemotherapy were also excluded. The MOS gas sensors are used to detect the VOCs with low molecular weight, while the SAW sensors are adopted for the detection of VOCs with high molecular weights. The ANN diagnostic model based on the features of MOS and SAW detection units has reached the sensitivity of 93.62% and the selectivity of 83.37%. Otherwise, the results also indicate that the hybrid electronic nose system based on MOS-SAW sensors has significant advantage over electronic nose consisting of single kind of gas sensors in breath testing for lung cancer. For the sake of application into large-scale screening, the further work will be on reduction of testing time and the size of the instrument. Therefore, the technology of fast gas chromatography could be introduced into hybrid electronic nose system, and different kinds of sensors would be integrated to form an entire structure and conduct more efficient analysis.

GASTROINTESTINAL DISEASES DIAGNOSIS

Alveolar air from the lungs of a person can be analyzed for, but not limited to, the noninvasive diagnosis of a wide variety of conditions including the stomach infections related to a high incidence of ulcers, enzymatic deficiencies and small intestinal bacteria overgrowth (SIBO) (Yang et al., 1998). Hydrogen (H2) and methane (CH4) are produced in the digestive system primarily only by the bacterial fermentation of carbohydrates (Gasbarrini et al., 2007). The generation of H2 and/or CH4 will result in the reabsorption of these gases into the blood stream from the site of their digestion, and they will appear in the expired air (Riordan et al., 1995). The work presented here has focused on the establishment of virtual electronic nose with diagnosis model for the detection of H2 and CH4 in breath sample, containing a sufficient volume of air representative of true alveolar air, which takes advantage of noninvasive procedure, accurate data and high repeatability.

Design of Virtual Electronic Nose

The virtual electronic nose separates the components by the basic principle of gas chromatography, and its schematic diagram as shown in Figure 8. The gas chamber is cleaned by using room air as the carrier gas, which is pumped through the system by an internal circulating pump. The sample is pumped into the sample loop after filtering water, while the three three-way valves are in full lines. H_2 and CH_4 are separated from all other reducing gases by the chromatographic column, and are carried past a carbon dioxide (CO_2) sensor and a metal oxide semiconductor (MOS) sensor, while the three three-way valves are in dashed lines. The flow valve and flowmeter control and measure the flow rate accurately all the time. Within 50 seconds after introducing the sample, the instrument displays sample H_2, CH_4, and CO_2 concentrations on the alphanumeric backlit data display. Then, the instrument activates the correction feature and displays the correction factor (Alveolar CO_2/Sample CO_2) and the corrected values for H_2 and CH_4. Figure 9 shows the material object of the virtual electronic nose.

The gas detection chamber, which is made of metal material for good thermal inertia, consists of MOS and CO_2 sensor groove mainly as shown in Figure 10. The effect to the sensor output caused by temperature drift and breath condensation is inhibited by the temperature sensor (Pt100) and the temperature control module for the gas detection chamber. There is a cylindrical tank inside the gas detection chamber as reaction area just covering the sensitive material of the MOS sensor. The gas flows vertically to the MOS sensor surface from top of the reaction area through the gas inlet, and exhaust through the gas outlet which is connected with the side wall to ensure that the MOS sensor could output the response quickly and recover the baseline shortly.

Experiments and Results

Figure 11 shows the temperature control of the gas chamber and baseline drift of MOS sensor. The temperature of gas chamber is 38 ± 0.5 degrees Celsius after warming up for two hours, and the base-

Figure 8. The schematic diagram of virtual electronic nose system

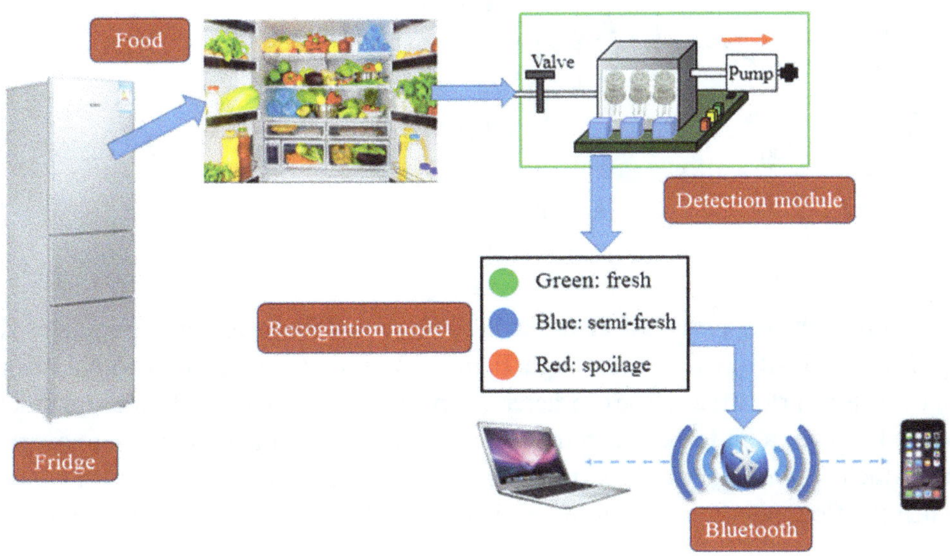

Figure 9. The material object of virtual electronic nose system

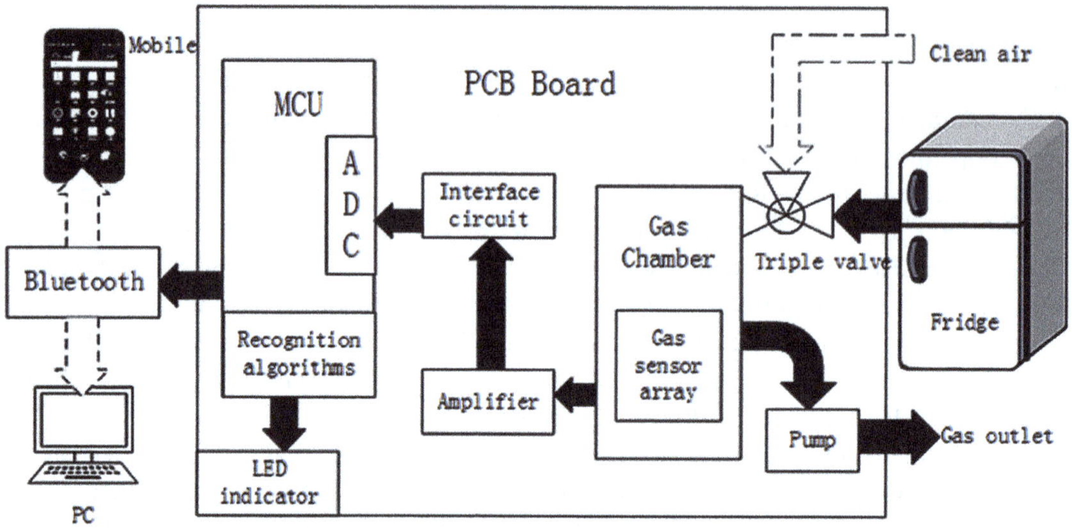

Figure 10. The gas detection chamber of virtual electronic nose system

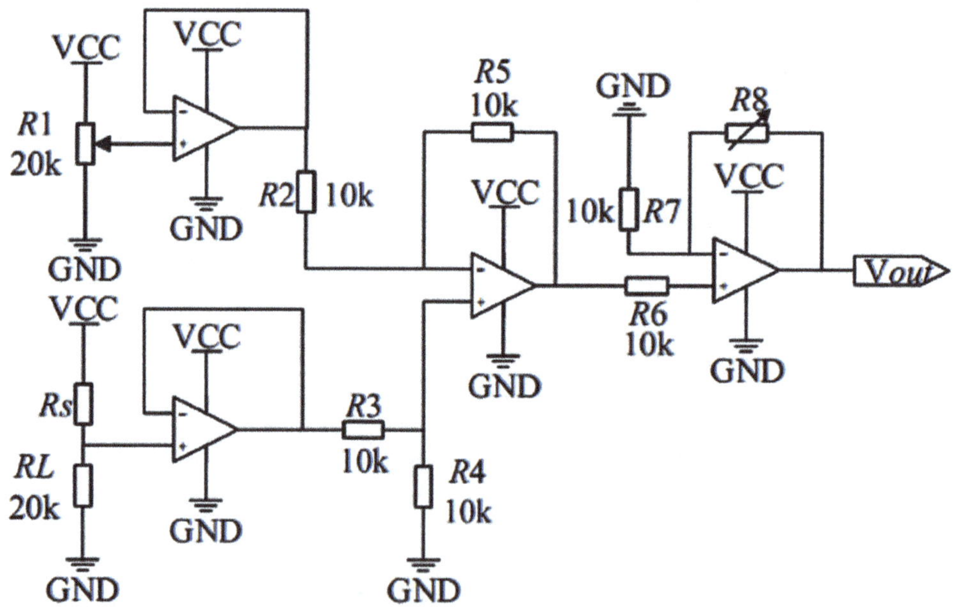

line of sensor output reaches a stable state basically. The virtual electronic nose regulates flow rate of the carrier gas automatically in a suitable range. Figure 12 shows the impact on output response of the MOS sensor by different flow rates. The characteristic peaks of H2 and CH4 show more quickly and the output response becomes a bit larger when the flow rate is faster, while there is more overlap between the two peaks which is hard to distinguish from each other. The characteristic peaks of H2 and CH4 are separated when the flow rate is small, but more time is needed for returning to the baseline, resulting

Figure 11. The temperature control of the gas chamber and baseline drift of MOS sensor

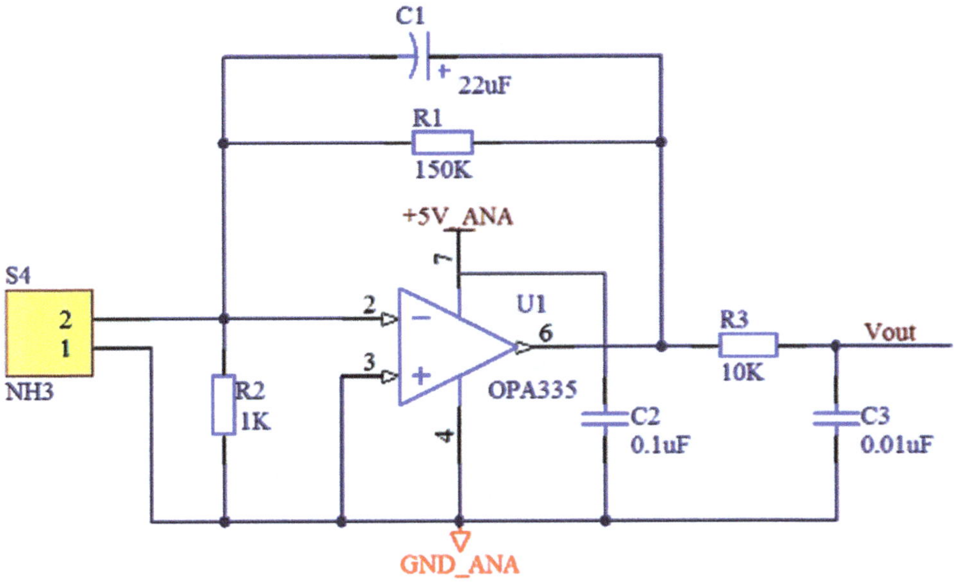

in longer detection cycle for one single sample, which is not conducive to clinical analysis with large samples. Considering the accuracy and detection cycle, the instrument is allowed to enter the calibration and detection state when the flow rate of carrier gas is in the range of 60 ± 3 mL/min.

The output response of the MOS sensor with high and low concentration gradient both have a typical two-peak feature, as shown in Figure 13 (a) and Figure 13 (b), respectively. The detection range of H2 and CH4 of the virtual electronic nose was set up as 0-500 ppm with a resolution of 1 ppm, according to the highest concentration of patients and the required sensitivity. The fitting formulas are obtained with good relativity (R2 > 0.99) after analyzing the correlation between the values of characteristic peaks and the concentration gradient, as shown in Figure 14 (a) (high concentration gradient of H2), Figure 14 (b) (low concentration gradient of H2), Figure 14 (c) (high concentration gradient of CH4) and Figure 14 (d) (low concentration gradient of CH4), respectively. The concentration of H2 and CH4 can be calculated after the output response of the MOS sensor is brought into the formula.

Figure 15 shows the software interface with two diagnostic models for SIBO. Bacteria are ordinary not present in significant numbers in the small intestine of healthy people. Therefore, when a challenge dose (e.g. lactose) is ingested, the level of H2 and/or CH4 will rise significantly within one to two hours only if the sugar is not digested, and reaches the colon (control group). The one-peak model has mono-phasic pattern in breath H2 or/and CH4 with a large increase (Pv) of at least 20/6/24 ppm with start time before 90 minutes (St). The two-peak model has biphasic pattern in breath H2 or/and CH4 levels two peaks with an early increase (Pv1) of at least 12/4/14 ppm within the first 90 minutes (St1), followed by a much larger increase (Pv2).

The breath test of H2 and CH4 as a diagnostic method for SIBO has greater reliability and acceptability than the blood test, according to most reports in the literature. The virtual electronic nose measures H2 and CH4 preferably from a single sample of alveolar air, increasing the accuracy of the breath test. The

Figure 12. The impact on output response of the MOS sensor by different flow rates

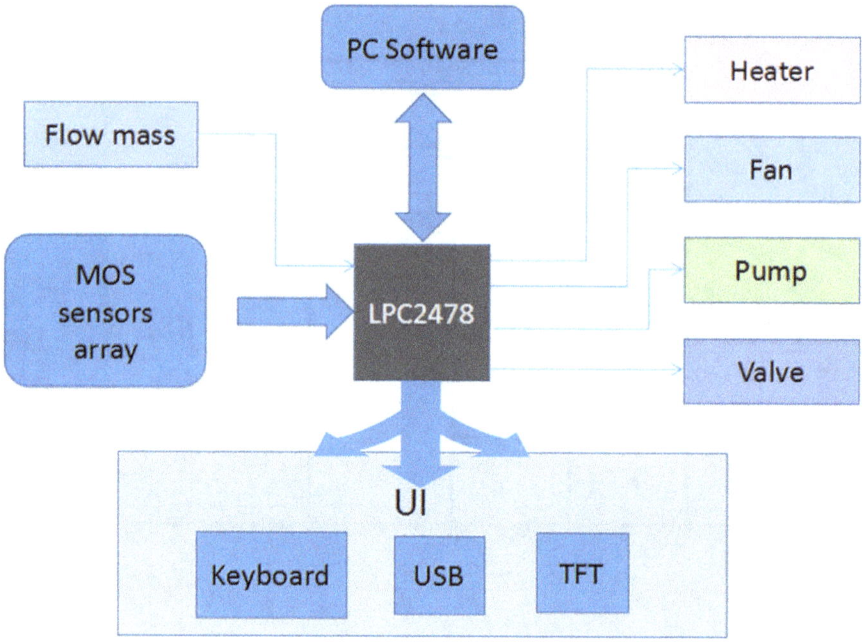

Figure 13. Gas concentration gradient of virtual electronic nose system: (a)-high concentration, (b)-low concentration

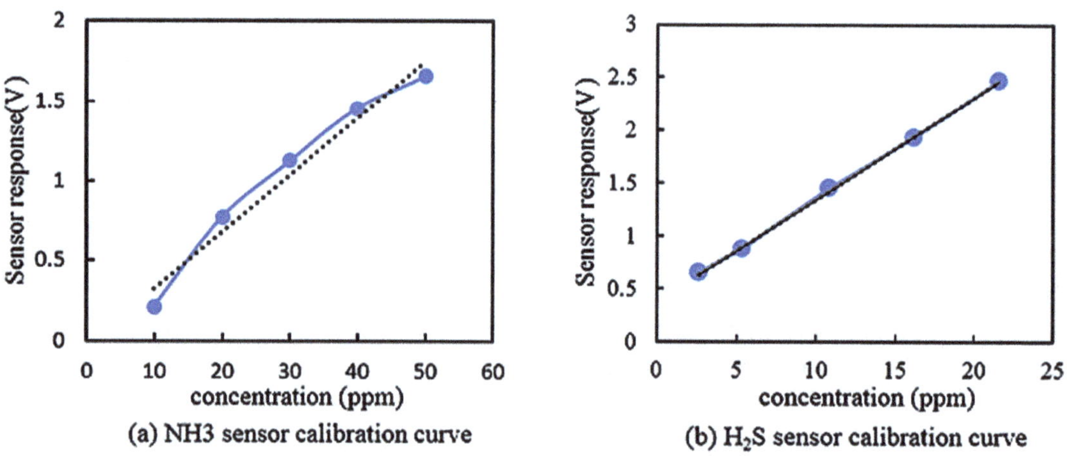

sample was pushed into chromatographic column for separation and every component was detected by the MOS sensor. The concentration of H2 and CH4 were achieved after calibrated calculation, and can be corrected by CO2. The detection range of H2 and CH4 is 1-500 ppm and the resolution is 1 ppm. The diagnosis model for SIBO was established and validated by local patients. This method has the advantages of high specificity, good repeatability, and noninvasive, so that can be used for the detection of increasing volume of clinical and domestic patients.

Figure 14. The correlation analysis of the concentration gradient: (a)-H2, high, (b)-H2, low, (c)-CH4, high, (d)-CH4, low

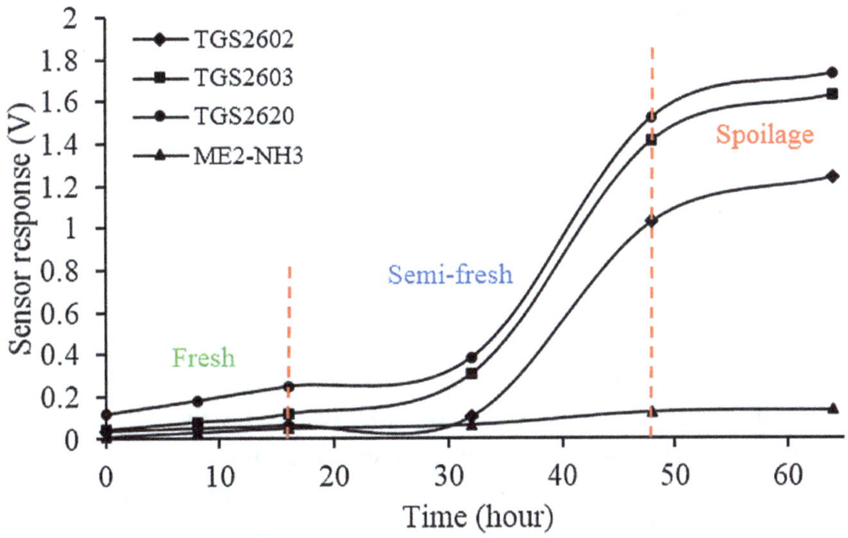

Figure 15. The Software interface with diagnostic models for SIBO

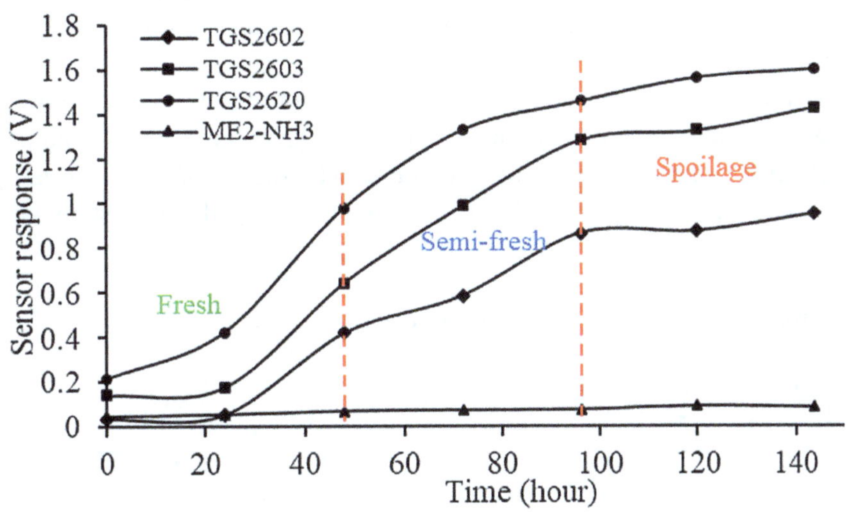

FOOD FRESHNESS DETECTION IN THE REFRIGERATOR

With the development of society and the improvement of people's living standards, the requirement for the quality of life is increasingly high. The food stored in the refrigerator mainly includes meat, vegetables and fruits. Due to the effects of microbes and storage environments, these food will be varying degrees of corruption. Eating such spoiled and contaminated food can cause severe health problems. To avoid such risk and guarantee the public health, there are lots of research that concentrate on food freshness (Haugen, 2001). The traditional food freshness detection methods were human sensory evaluation, chemical detection, microbiological testing and so on. These methods had many shortcomings such as complicated steps, long consuming time, experimental environment requirements (Berna, 2010). With the development of gas sensor and patter recognition technology, electronic nose has become a rapid, non-invasive and precise detection methods on food freshness compared with traditional detection methods (Shen, et al., 2001). The electronic nose system is to be built with special detection chamber and MOS sensors array, using LDA and PCA method to identify the freshness and classify different kinds of food (Hong & Wang, 2012).

Design of Virtual Electronic Nose

The current research of food freshness is mainly about meat (Eom et al., 2014) and fish (Hasan et al., 2012). While for vegetables and fruits, there is little research on the freshness adopting the technology of electronic nose most of their experimental objects are foods at the room temperature. In this article, electronic nose is mainly used to detect the freshness of food stored in the refrigerator. Figure 16 shows the overall design of system. The gas produced by food corruption is pumped out of the refrigerator by a vacuum mini-pump. A detection sensors array based on the chemical sensitivity of MOS and electrochemical sensor is designed. Meanwhile, a special gas chamber is designed for the low temperature circumstance of fridge. The detected results can be transmitted to computer and mobile by Bluetooth communication technology. The small electronic nose can be implemented to detect the food freshness of refrigerator with low cost, fast and simple features. Pork, leek and banana are selected to verify the performance of system. The detected information will be calculated and displayed as one of three grades: Fresh, Semi-fresh and Spoilage.

For demands of refrigerator storing meat, fresh fruits and vegetables, the detection system is mainly composed of three parts including gas sampling chamber, MCU control module and data transmission module. The gas chamber is used to gather the gas produced by food spoilage in the refrigerator and the gas sensors array is embedded in chamber. For controlling the gas path and system hardware circuit, a

Figure 16. The overall design of the food freshness detection system

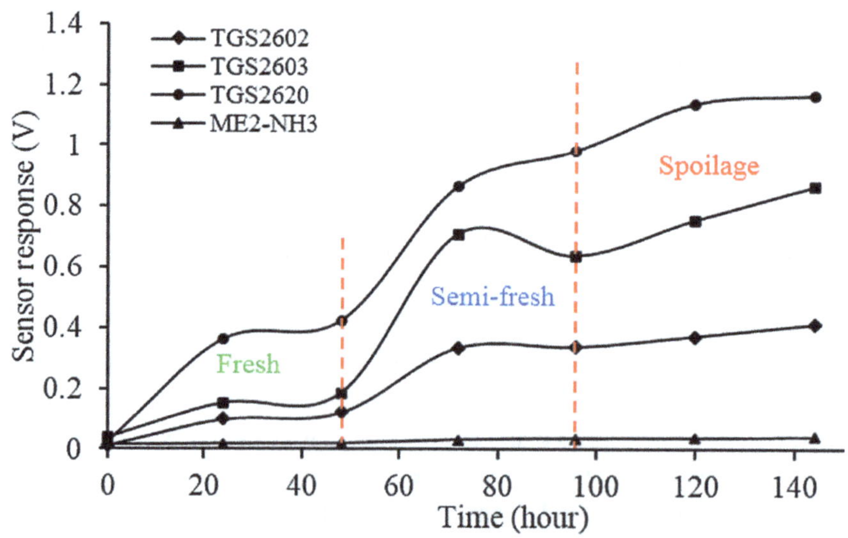

low power microcomputer chip is used. And the data transmission module use Bluetooth technology to communicate with PC and mobile.

Figure 17 shows the block diagram of fridge food freshness detection system. When beginning the measurement, the fresh food is placed in the refrigerator. Opening up the triple valve and the gas in the refrigerator will be pumped into the gas chamber by a mini-pump. The sensors array embedded in chamber is sensitive to gas produced by refrigerator food and transfer the odor into voltage. The voltage go through amplifier and interface circuit and finally enter into the ADC channels of MCU. Then the microcomputer acquire the voltage change generated by gas sensors array and threshold is set to establish a freshness discriminant model. Using human sensory evaluation as the standard to judge the freshness of refrigerator food. Warning signal will be sent for the users with the LED director when it is detected that the food is not fresh according to the simple recognition algorithms model. Meanwhile, the data of gas sensors array is transmitted to PC and mobile by Bluetooth for constructing more complex and intelligent discriminant model in the future. After end of the measurement, closing the triple valve and the clean air will be pumped into the gas chamber to clean the whole gas path.

In order to ensure the release of cryogenic gas from refrigerator stored food will not affect the sensors' response and normal life, we designed the gas detection chamber which can control the gas temperature at room environment before it pumped into sensors array. The material of gas chamber is resin and is made by the technology of 3D print. The sensors array is embedded in the gas chamber. During food stored in the refrigerator, some components will be decomposed and release odors due to the reaction of bacteria and enzymes. Protein is decomposed into corrosive amines, further producing ammonia, hydro-

Figure 17. The block diagram of fridge food freshness detection system

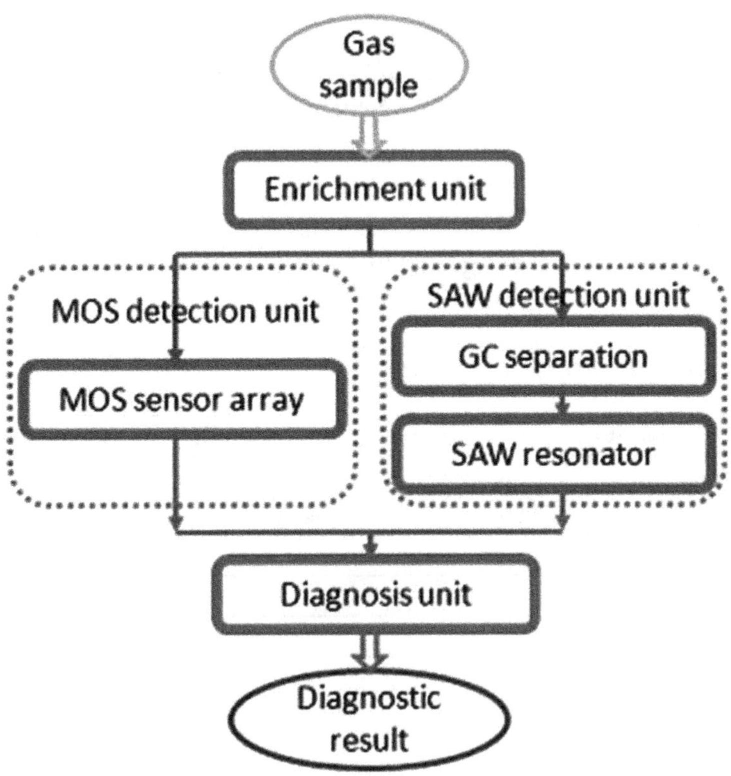

gen sulfide, ethanethiol and so on. Fat is decomposed into fatty acids, and further producing aldehydes and uric acid odor. Carbohydrates are broken down into alcohols, ketones, aldehydes and carboxylic acid gases. The corruption process of vegetables mainly produces organic volatile gases, while fruits produces ethanol when spoiled. In the process of spoilage, these gases' concentrations will increase exponentially. In this paper, three kinds of food including meat, vegetables and fruits are selected as the object of study. So corresponding with this, three kinds of corrosive characteristic gases including ammonia, hydrogen sulfide, and ethanol are selected as the object of detect (Liao et al., 2010). Using these gas sensors allows us to detect these gases and judge the food's freshness stored in the refrigerator.

For the selected corrosive characteristic gases, considering the cost and size of the sensor detection array, the following four kinds of sensors are used to set up the sensor detection array. The specific model, the corresponding sensitive gas and detection range are shown in Table 1.

Sensor S1~S3 are the MOS sensors manufactured by Figaro, sensor S4 is an electrochemical sensor manufactured by Winsen Electronics Technology. The sensitivity shows that three MOS sensor have a good response to ammonia, hydrogen sulfide, ethanol. And because of its broad spectrum characteristics, MOS sensor also have special response to some other gases, such as toluene, methane. While these gas will not appear in the detecting environment when using in the refrigerator, and therefore the sensors array still has good selectivity. In order to improve the accuracy of the detection, we add an electrochemical sensor that only has response to the ammonia to the sensors array. For MOS sensor, the relationship between sensor resistance (Rs) and gas concentration (C) can be expressed by the formula (1). The coefficient α and A is constant related to the type of gas.

$$Rs = A \cdot C^{-\alpha} \tag{1}$$

For electrochemical sensor, the output current is proportional to the ammonia concentration. For MOS sensor, in order to eliminate the influence of baseline drift and improve the detection sensitivity, the sensor detection circuit selects the differential detection circuit including the Wheatstone bridge and the OPA2333 amplifier. The core detection circuit design is shown in Figure 18.

As can be seen from Figure 18, the output voltage of V_{out} is subtracted from the baseline caused by the partial pressure by adjusting R1, as a result expanding the detected dynamic range. We can get the relationship between the final output voltage and the sensor input voltage by the formula (2).

Table 1. The characteristics of MOS sensors array

Serial Number	Sensor Model	Characteristics of Sensitivity	Detection Range (ppm)
S1	TGS2602	Ammonia, Toluene	1-30
S2	TGS2603	Hydrogen sulfide, Trimethylamine	1-10
S3	TGS2620	Ethanol, Methane	50-5000
S4	ME2-NH3	Ammonia	0-50

Figure 18. Detection circuit of MOS sensor

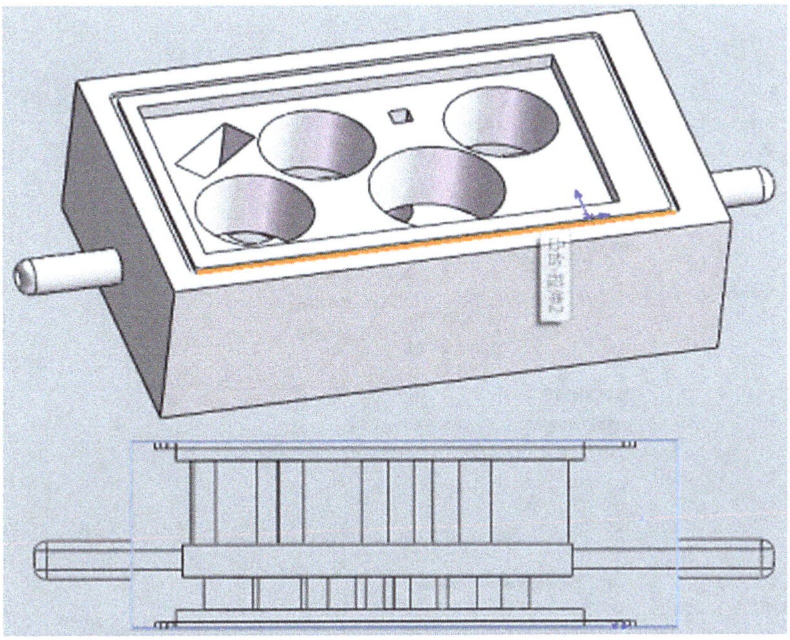

$$\begin{cases} V_{out} = \left(V_s - V_1\right)\left(1 + \dfrac{R_8}{R_7}\right) \\ V_s = \dfrac{V_{cc}R_L}{\left(R_s + R_L\right)} \end{cases} \tag{2}$$

V1 is the input voltage of resistance R1 after setting output voltage to zero. Vs is the input voltage of MOS sensor. Rs is the sensor resistance exposing to different kinds of gases.

For electrochemical sensor, the output of sensor is the microampere level current. In order to ensure the precise measurement of the sensor current value, we use a current feedback amplifier circuit to amplify the resolution. While the amplifier also amplify the noise at the same time. Therefore, we included a noise filter to remove the noise and then amplified the signal. There was a small amount of noise that could not be removed with the noise filter, so we used an RC filter as a secondary filter. Figure 4 shows the detection circuit of electrochemical sensor.

As shown in the Figure 19, we can get the relationship between the output voltage and the sensor input current by the formula (3).

$$V_{out} = I_w * R_1 \tag{3}$$

I_w is the input current of electrochemical sensor. R1 and C1 form a noise filter, and R3, C3 is an RC filter as s secondary filter. From this formula, we can conclude that the output voltage is proportional

Figure 19. Detection circuit of electrochemical sensor

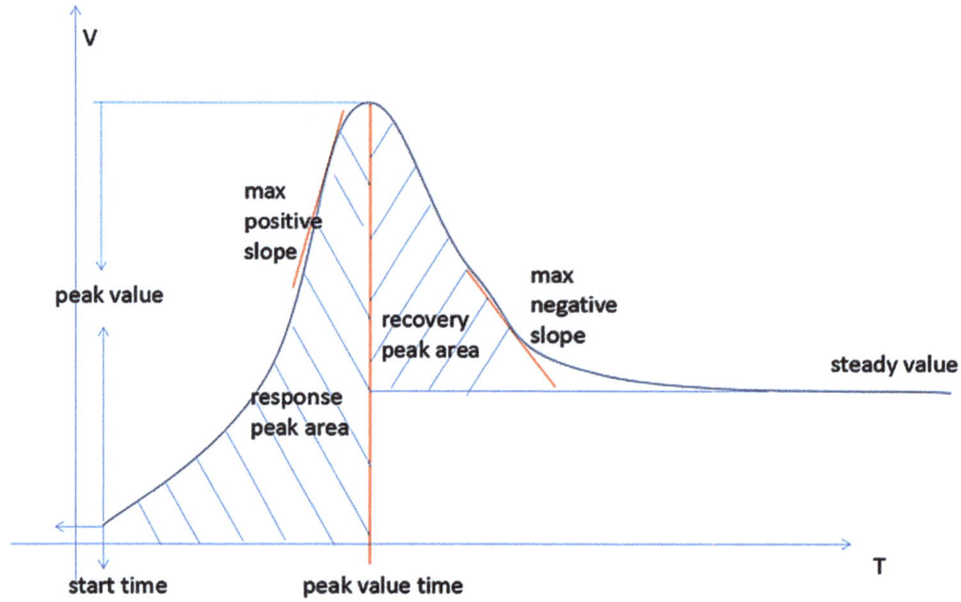

to the ammonia concentration. And the output voltage can be changed by adjusting the feedback resistor R1.

Experiment and Results

The food stored in the refrigerator is mainly meat, vegetables and fruits. After several groups of experiments, we choose pork as a representative of meat, leek as a representative of vegetables and banana as a representative of fruits taking into accout the speed of corruption and odor distribution. Placing the food in the refrigerator and detecting the output response of the sensors array. Based on the test results, the indentification model of food freshness will be established.

There are ammonia and hydrogen sulfide standard gas in our laboratory. The MOS sensor of sensors array is calibrated on the basis of this standard gas. Due to the standard gas concentration is very high, we use the method of syringe extraction to dilute the standard gas. For ammonia, the standard gas concentration is 1999 ppm and is used to calibrate the TGS2602 sensor. Using syringe to extract standard ammonia and clean air, and configuring 500 mL ammonia concentration of 10 ppm, 20 ppm, 30 ppm, 40 ppm and 50 ppm. For hydrogen sulfide, the standard gas concentration is 54 ppm and is used to calibrate the TGS2603 sensor. Using syringe to extract standard hydrogen sulfide and clean air, and configuring the 500 mL hydrogen sulfide concentration of 2.7 ppm, 5.4 ppm, 10.8 ppm, 16.2 ppm, 21.6 ppm and 27 ppm. Performing the calibration experiments, the sensor calibration curve is obtained of Figure 20.

It can be safely concluded from the Figure 5 that the TGS 2602 sensor has a good linearity for ammonia at a concentration of 10 ppm to 50 ppm and TGS2603 sensor has a good linearity for hydrogen sulfide at a concentration of 5.4 ppm to 27 ppm. The detection range of TGS2602 and TGS2603 sensor is much higher than the range showed in the datasheet. And we don't calibrate the ethanol due to the laboratory conditions.

Figure 20. The calibration curve of MOS sensor

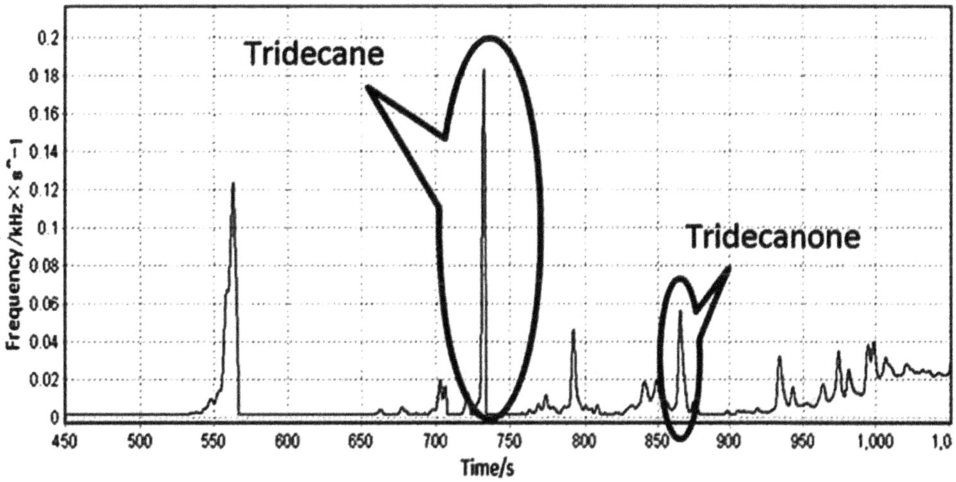

The experimental materials were fresh pork, leek and watermelon brought from the market and we weighed 100g each for the refrigerator storage. The experimental steps are as follows.

1. Opening the refrigerator door at night to get rid of the smell of the refrigerator.
2. During the day, supplying the power for refrigerator. Waiting for the temperature inside the refrigerator stable and the detecting system warming up, the sensors array's response of the empty refrigerator are measured and used as the detecting baseline.
3. Placing the food needed to be measured in the refrigerator and measure the response of sensors array after 1.5 hours.
4. After the measurement is over, removing the food from refrigerator and ventilating half an hour. Place other food into the refrigerator and continue the measurement.

We repeat the experiments on pork, leek and banana in the process of food from fresh to corruption. Finally, we can get discrimination results of freshness and classification according to the testing data.

Sensory evaluation about the freshness of the pork, leek and banana stored in the refrigerator made by several evaluators is shown in Table 2. Based on the food features in Table 2, evaluators scoring the food freshness under different conditions. Taking the human sensory evaluation as the standard, and comparing with the results of our system judgment to calculate the system accuracy. Human sensory evaluation has the advantages of quickness, simplicity, no instrument and no fixed experimental place. However, the evaluate results have the problem of non-quantification, lack of precision, subjectivity and so on (Benedetti et al., 2008). So it is necessary for experienced and well-trained personnel to be qualified the job of human sensory evaluation.

Figure 21-23 illustrates the records of each sensors responses for recognition pattern which can divide into three models: fresh, semi-fresh and spoilage. We judged the food to be semi-fresh or spoilage if the sensor peak value was more than the threshold that we had set in advance. For electrochemical sensor, it only responses to ammonia. There is none or a very small amount of ammonia in the process of corruption of vegetables and fruits. Therefore, electrochemical sensor is used to discriminate the freshness

Table 2. Sensory evaluation and results

Food Kinds	Stored Time(h)	Sensory Features	Evaluation Result
Pork	24	red, good flexibility, no ammonia odor	Fresh
	36	dark red, poor flexibility, slight ammonia odor	Semi-fresh
	48	gray red, no flexibility, pungent ammonia odor	Spoilage
Leek	48	green, brittle, no corruption odor	Fresh
	96	dark green, soft, slight corruption odor	Semi-fresh
	120	brown, atrophy, pungent corruption odor	Spoilage
Banana	48	green-yellow or yellow, slight hard	Fresh
	96	getting dark, slightly sour odor, getting soft	Semi-fresh
	120	black and rotten, juice flowing out, strong alcohol odor	Spoilage

of meat and can be used to judge whether there is meat when the food is mixed together. We weighed 100g pork, leek and banana and put it into the refrigerator for experiments.

Figure 21 shows the pork's peak value response curves of sensors array. As is shown in the figure, the MOS sensor is peak response value is quite small less than 300mV when the pork is fresh. When the MOS sensor peak value is more than 300mV, we judge the pork is under semi-fresh. While if the peak value of MOS sensor is more than 1000mV, there is no doubt that the pork must be spoilage in a certain degree. For electrochemical sensor, the sensor peak value is smaller compared to MOS sensor due to the different amplifying circuit. We judge the pork to be semi-fresh if the electrochemical sensor peak value is more than 50mV, and to be spoilage if the sensor peak value is more than 100mV. Combined with the MOS sensor and electrochemical sensor, we can confirm the freshness of meat stored in the refrigerator with a quite high accuracy.

Figure 22 shows the leek's value response curves of sensors array. The decomposition of vegetables is mainly hydrogen sulfide. The electrochemical sensor is not sensitive to the hydrogen sulfide. We can see from the figure that the peak value of electrochemical sensor almost keep constant and the peak value

Figure 21. The pork's peak value response curves of sensors array

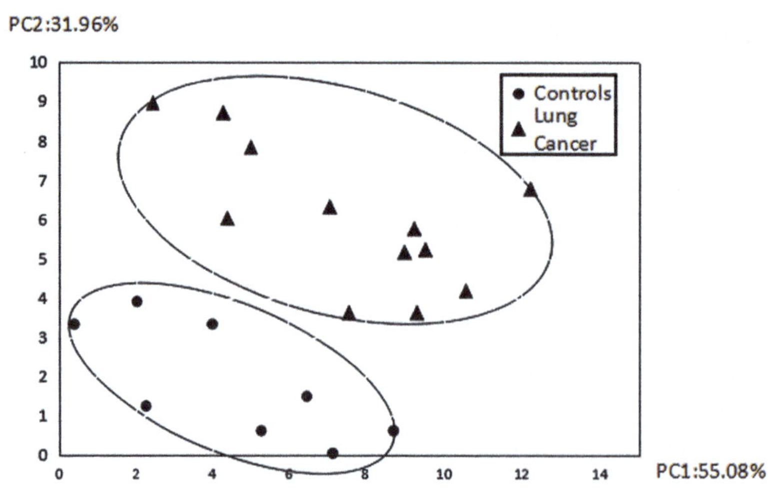

Figure 22. The leek's peak value response curves of sensors array

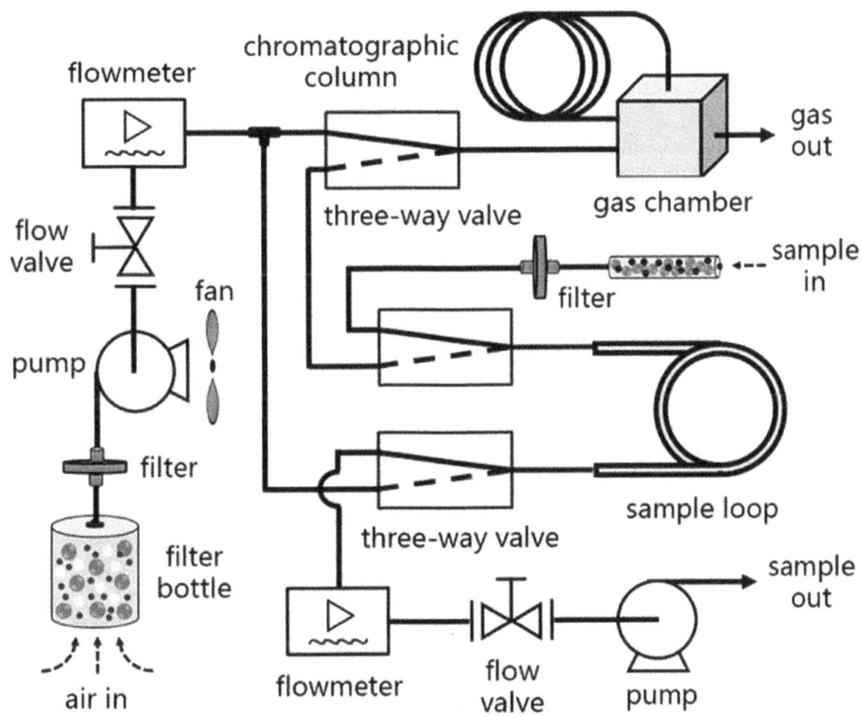

maybe the noise produced by the amplifying circuit. So we judge the freshness of leek just according to the responses of MOS sensor. And with the characteristic of three MOS sensor, we can discriminate the leek's degree of corruption.

Figure 23 shows the banana's peak value response curves of sensors array. The peak value response of banana is similar to the leek about the electrochemical sensor. The decomposition of vegetables is mainly ethanol in the process of corruption. So we do not consider the electrochemical sensor when discriminate the freshness of banana. While the peak value of three MOS sensor is varying greatly at same time. As a result, we choose the different threshold for the three MOS sensor when judging the freshness of banana (Torri et al., 2010).

By analyzing the detecting results of three kinds of food, we can preliminary establish a refrigerator food freshness detection model by setting the threshold (Vainionpaa et al., 2004). And combining the MOS sensor and electrochemical sensor, detection accuracy can be improved greatly. However, most of the food stored in the refrigerator is no single but mixed. In the next experiments, we will detect the sensor response of mixed food stored in the refrigerator.

This article proposed a food freshness detection system that can be used in the refrigerator. The current test results have shown that the food freshness could be detected fast and simply based on electronic nose system and the electronic nose system could be widely used in detecting freshness of food stored in refrigerator. However, as the actual refrigerator internal environment is more complicated, our research is only a preliminary exploration. There are still a number of changes to be met include drift, humidity influence, sensor aging and so on. In the near future, with a better sensors array and improved recognition algorithms, identification of spoiled food may be more precise and intelligent.

Figure 23. The banana's peak value response curves of sensors array

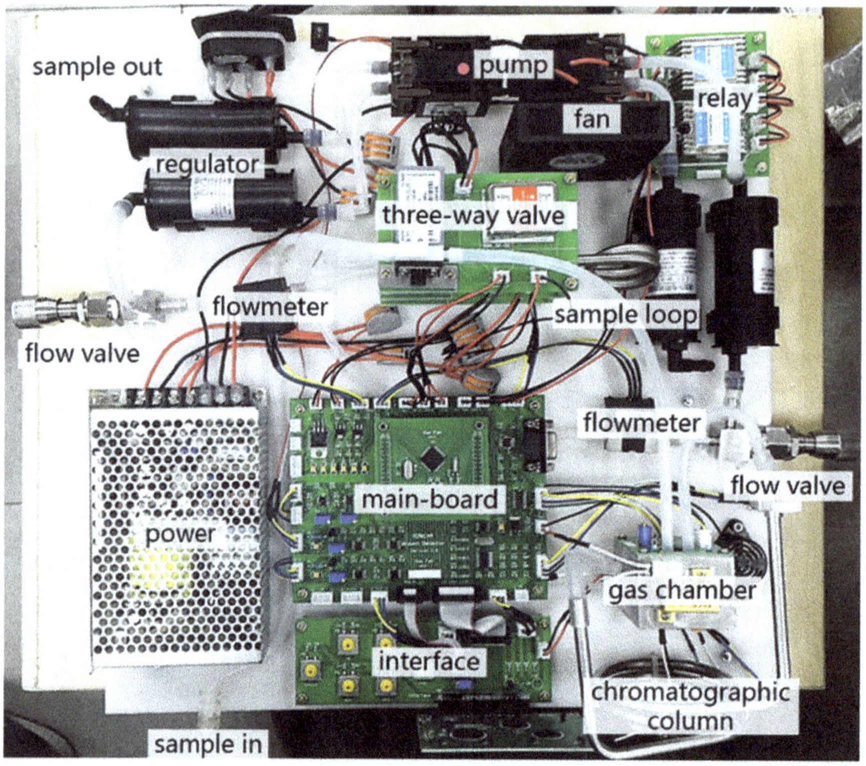

FUTURE RESEARCH DIRECTIONS

A non-linear discriminant model (NLDM) is going to be proposed, which will contribute to optimization of the electronic nose design. The detection units of detection system consist of MOS sensors and SAW sensor. In total, 138 feature values will be extracted from responses of detection units. In the present research, an NLDM algorithm is developed to analyze the multivariate data. For clinical diagnosis of different patient, physicians often take different factors into account. Using that for reference, when analyzing a certain breath sample, NLDM determines which feature values could enter the model by assigning different diagnostic importance to each value. Comparing with ANN used before, NLDM will show better discriminant capacity and low complexity.

We will measure the combination of carcinoembryonic antigen (CEA), squamous cell carcinoma antigen (SCC) and neural specific enolase (NSE) in exhaled breath condensate (EBC). In addition, it has been proved they can be potential markers in early lung cancer. A modified EBC collecting device will be used, which makes it possible to collect proteins with large molecular weight. Chemiluminescence immunoassay will be used to detect concentrations of CEA, SCC and NSE in EBC. Fasting venous blood samples are collected from patients and healthy individuals for comparing studies. A significantly higher concentration level of each marker will be found in lung cancer patients than healthy controls. Markers in EBC have a higher positive rate and were more specific to histologic types than markers in serum. Tumor markers in EBC may have a better diagnostic performance for LC than those in serum.

Gold nanoparticle enhancement strategy will be employed, which significantly amplifies the sensitivity of Love-wave immunosensor for detection of tumor markers in exhaled breath condensate. Coating antibodies are immobilized onto the sensitive area of Love-wave sensors through protein A. Markers to be detected are captured by coating antibodies. The specificity of immunosensor results from the specific recognition between antibody and antigen. In addition, the detecting antibodies are labeled by gold nanoparticles. A "sandwich" immunoassay is performed on a Love-wave sensor. Subsequently, gold staining solution is introduced, which results in catalyzed deposition of gold (III) on to the gold nanoparticles. The deposition amplifies the mass loading on immunosensor and makes it possible to detect tumor markers in EBC. Those limits of detection of the measurement system will be lower than the corresponding cut-off values, and the described Love-wave immunosensor has potential to detect lung cancer in clinic.

A detecting instrument will be designed for detecting of exhaled breath condensate, and a software running on iPad is being developed, which implement real-time detection of tumor markers in EBC. An instrument will be designed to detect responses of the above Love-wave immunosensors. Automatic inject system, control circuit and mobile software are planned to implement mechanized detection which could reduce the experimental error brought about by manual operation. Simultaneously, remote monitoring of detection procedure and data-sharing among multi-platform have been achieved. The detection results of sensors measured by the present instrument is consistent with those measured by commercial network analyzer. The higher levels of CEA, NSE and SCC could be detected in EBC of patients than in EBC of controls, which suggests the promising application of this instrument for early detection of lung cancer.

CONCLUSION

In the present phase, our electronic nose system concentrates on the detection for the respiratory healthy. Furthermore, we are planning apply also the health test for alimentary tract. It can be concluded that our electronic nose system has great potential for lung cancer and SIBO detection through breath testing with a wireless sensors network and communication techniques. The electronic nose is an instrument, which comprises a sampling system, an array of chemical gas sensors with differing selectivity, and a computer with an appropriate pattern-classification algorithm, capable of qualitative and/or quantitative analysis of simple or complex gases, vapors, or odors. Recently, lung cancer has the continuously greatest incidence and mortality among all cancers. Because lung cancer is almost asymptomatic in the early stages, it is always detected in advanced stage when the curative intervention is no longer possible. It is obviously that screening is the only systematic means of finding lung tumors at early stage. However, the present detection tools for lung cancer is not available to a general high-risk population. Breath test is a non-invasive and low-cost detecting method which has potential to be a screening tool. Additionally, exhaled breath air originates in the respiratory system, which makes it more specific to lung cancer than any other cancers. The present research is devoted to exploring lung cancer markers in exhaled gas and condensate, developing detecting instruments and detecting technology which would meet the requirements of physical examination. It is expected that the early diagnosis of lung cancer will be realized after developing a detection technology which can be applied conveniently in regular medical checkup.

REFERENCES

Benedetti, S., Buratti, S., Spinardi, A., Mannino, S., & Mignani, I. (2008). Electronic nose as a non-destructive tool to characterise peach cultivars and to monitor their ripening stage during shelf-life. *Postharvest Biology and Technology*, *47*(2), 181–188. doi:10.1016/j.postharvbio.2007.06.012

Berna, A. (2010). Metal Oxide Sensors for Electronic Noses and Their Application to Food Analysis. *Sensors (Basel)*, *10*(4), 3882–3910. doi:10.3390/s100403882 PMID:22319332

Eom, K. H., Hyun, K. H., & Sen, L. (2014). The Meat Freshness Monitoring System Using the Smart RFID Tag. *International Journal of Distributed Sensor Networks*, *3*, 1–9.

Fleischer, M. (2011). Solid State and Spectroscopic Chemical Sensors and Applications. *Sensor Letters*, *9*(2), 706–709. doi:10.1166/sl.2011.1597

Gasbarrini, A., Lauritano, E. C., Gabrielli, M., Scarpellini, E., Lupascu, A., Ojetti, V., & Gasbarrini, G. (2007). Small Intestinal Bacterial Overgrowth: Diagnosis and Treatment. *Digestive Diseases (Basel, Switzerland)*, *25*(3), 237–240. doi:10.1159/000103892 PMID:17827947

Gouma, P., Prasad, A., & Stanacevic, S. A. (2011). Selective nanosensor device for exhaled breath analysis. *Journal of Breath Research*, *5*(3), 037110. doi:10.1088/1752-7155/5/3/037110 PMID:21896971

Hasan, N. U., Naveed, E., & Waleed, E. (2012). Meat and fish freshness inspection system based on odor sensing. *Sensors (Basel)*, *12*(11), 15542–15557. doi:10.3390/s121115542 PMID:23202222

Haugen, J. E. (2001). Electronic Noses in Food Analysis. Headspace Analysis of Foods and Flavors. *Headspace Analysis of Foods and Flavors*, *488*, 43–57. doi:10.1007/978-1-4615-1247-9_4

Hong, X., & Wang, J. (2011). Discrimination and Prediction of Pork Freshness by Electronic nose. *Computer and Computing Technologies in Agriculture V*, *370*, 1–14.

Hunter, G. W., Xu, J. C., Biaggilabiosa, A. M., Laskowski, D., Dutta, P. K., Mondal, S. P., ... Dweik, R. A. (2011). Smart sensor systems for human health breath monitoring applications. *Journal of Breath Research*, *5*(3), 037111. doi:10.1088/1752-7155/5/3/037111 PMID:21896970

Korotkova, O. V., Salinas, Y. E., Vasilieva, E. A., Kozlov, A. V., Yashina, N. V., Loginov, I. A., & Dalin, M. V. (2013). Hydrogen respiratory test: Pilot examinations for evaluation of the small intestinal colonization by normal microflora. *Bulletin of Experimental Biology and Medicine*, *154*(6), 818–820. doi:10.1007/s10517-013-2064-x PMID:23658932

Lin, E. C., & Massey, B. T. (2016). Scintigraphy Demonstrates High Rate of False-positive Results From Glucose Breath Tests for Small Bowel Bacterial Overgrowth. *Clinical Gastroenterology and Hepatology*, *14*(2), 203–208. doi:10.1016/j.cgh.2015.07.032 PMID:26241509

Lupascu, A., Gabrielli, M., Lauritano, E. C., Scarpellini, E., Santoliquido, A., Cammarota, G., ... Gasbarrini, A. (2005). Hydrogen glucose breath test to detect small intestinal bacterial overgrowth: A prevalence casecontrol study in irritable bowel syndrome. *Alimentary Pharmacology & Therapeutics*, *22*(11-12), 1157–1160. doi:10.1111/j.1365-2036.2005.02690.x PMID:16305730

Mendoza, E., Crismatt, C., & Matos, R. (2007). Diagnosis of small intestinal bacterial overgrowth in children: The use of lactulose in the breath hydrogen test as a screening test. *Biomedica*, *27*(3), 325–332. doi:10.7705/biomedica.v27i3.195 PMID:18320098

Omatu, S., & Yano, M. (2014). Electronic nose System by Using Neural Networks. *Neurocomputing*, *172*(C), 311–318.

Pimentel, M., Chow, E. J., & Lin, H. C. (2003). Normalization of lactulose breath testing correlates with symptom improvement in irritable bowel syndrome. a double-blind, randomized, placebo-controlled study. *The American Journal of Gastroenterology*, *98*(2), 412–419. PMID:12591062

Reis, J. C., Morais, M. B., & Oliva, C. A. G. (2007). Breath hydrogen test in the diagnosis of environmental enteropathy in children living in an urban slum. *Digestive Diseases and Sciences*, *52*(5), 1253–1258. doi:10.1007/s10620-006-9288-9 PMID:17372830

Riordan, S. M., Mciver, C. J., Duncombe, V. M., & Bolin, T. D. (1995). Bacteriologic analysis of mucosal biopsy specimens for detecting small-intestinal bacterial overgrowth. *Scandinavian Journal of Gastroenterology*, *30*(7), 681–685. doi:10.3109/00365529509096313 PMID:7481532

Saad, R. J., & Chey, W. D. (2014). Breath testing for small intestinal bacterial overgrowth: Maximizing test accuracy. *Clinical Gastroenterology and Hepatology*, *12*(12), 1964–1972. doi:10.1016/j.cgh.2013.09.055 PMID:24095975

Shen, N., Moizuddin, S., Wilson, L., Duvick, S., White, P., & Pollak, L. (2001). Relationship of electronic nose analyses and sensory evaluation of vegetable oils during storage. *Journal of the American Oil Chemists' Society*, *78*(9), 937–940. doi:10.1007/s11746-001-0367-z

Singh, H., Raj, V. B., Kumar, J., Durani, F., Mishra, M., Nimal, A. T., & Sharma, M. U. (2016). SAW mono sensor for identification of harmful vapors using PCA and ANN. *Process Safety and Environmental Protection*, *102*, 577–588. doi:10.1016/j.psep.2016.05.014

Sung, H. J., Paik, C. N., & Chunget, W. C. (2015). Small Intestinal Bacterial Overgrowth Diagnosed by Glucose Hydrogen Breath Test in Postcholecystectomy Patients. *Journal of Neurogastroenterology and Motility*, *21*(4), 545–551. doi:10.5056/jnm15020 PMID:26351251

Torri, L., Sinelli, N., & Limbo, S. (2010). Shelf life evaluation of fresh-cut pineapple by using an electronic nose. *Postharvest Biology and Technology*, *56*(3), 239–245. doi:10.1016/j.postharvbio.2010.01.012

Vainionpaa, J., Smolander, M., Alakomi, H. L., Ritvanen, T., Rajamäki, T., Rokka, M., & Ahvenainen, R. (2004). Comparison of different analytical methods in the monitoring of the quality of modified atmosphere packaged broiler chicken cuts using principal component analysis. *Journal of Food Engineering*, *65*(2), 273–280. doi:10.1016/j.jfoodeng.2004.01.025

Wang, D., Yu, K., & Wang, Y. (2011). An Electronic Nose Based on Hybrid MOS-SAW Sensors for Detection of Different Biomarkers of Lung Cancer. American Institute of Physics.

Wilson, A. D., & Baietto, M. (2009). Applications and Advances in Electronic-Nose Technologies. *Sensors (Basel)*, *9*(7), 5099–5148. doi:10.3390/s90705099 PMID:22346690

Yang, C. Y., Chang, C. S., & Chen, G. H. (1998). Small-intestinal bacterial overgrowth in patients with liver cirrhosis, diagnosed with glucose H2 or CH4 breath tests. *Scandinavian Journal of Gastroenterology, 33*(8), 867–871. doi:10.1080/00365529850171549 PMID:9754736

ADDITIONAL READING

Costa, M. B., Azeredo, I. L. Jr, Marciano, R. D., Caldeira, L. M., & Bafutto, M. (2012). Evaluation of small intestine bacterial overgrowth in patients with functional dyspepsia through H2 breath test. *Arquivos de Gastroenterologia, 49*(4), 279–283. doi:10.1590/S0004-28032012000400009 PMID:23329223

Gu, Y., & Li, Q. (2014). Application of the new pattern recognition system in the new electronic nose to detecting Chinese spirits. *Chinese Physics B, 4*, 330–334.

Liao, Y. T., Fan, Y. X., & Cheng, F. (2010). On-line prediction of fresh pork quality using visible/near-infrared reflectance spectroscopy. *Meat Science, 86*(4), 901–907. doi:10.1016/j.meatsci.2010.07.011 PMID:20728281

Walters, B., & Vanner, S. J. (2005). Detection of bacterial overgrowth in IBS using the lactulose H2 breath test: Comparison with 14C-D-xylose and healthy controls. *The American Journal of Gastroenterology, 100*(7), 1566–1570. doi:10.1111/j.1572-0241.2005.40795.x PMID:15984983

Yu, K., Wang, Y., Yu, J., & Wang, P. (2011). A Portable Electronic Nose Intended for Home Healthcare Based on a Mixed Sensors array and Multiple Desorption Methods. *Sensor Letters, 9*(2), 876–883. doi:10.1166/sl.2011.1635

KEY TERMS AND DEFINITIONS

Artificial Neural Network: The neural network is an operational model that consists of a large number of nodes (or neurons). The network itself is usually an approximation of some algorithm or function of the nature, or it may be an expression of a logical strategy.

Electronic Nose: A system consisting of gas sensors array and appropriate pattern recognition algorithm and is used to identify different kinds of smell that the identification results is consistent with human sensory evaluation. The electronic nose can be implemented to disease diagnosis, food safety detection, and so on.

Food Freshness: When food is stored in the refrigerator, food will be a certain degree of corruption due to the reaction of bacteria and enzymes. The food freshness can be displayed as one of three grades: fresh, semi-fresh, and spoilage.

Food Freshness Detection System: A small electronic nose system is to be built with special detection chamber and gas sensors array, using pattern recognition algorithm to identify the freshness of food and classify different kinds of food stored in the refrigerator.

Metal Oxide Semiconductor Sensor: The conductivity of the metal oxide semiconductor sensor changes when the semiconductor adsorbs the measured gas. The response can be usually affected by the environment gas.

Small Intestinal Bacterial Overgrowth: Small intestinal bacterial overgrowth is a disorder of excessive bacterial growth in the small intestine. Unlike the colon (or large bowel), which is rich with bacteria, the small bowel usually has fewer than 10,000 organisms per millilitre.

Surface Acoustic Wave: Surface acoustic wave is a kind of elastic wave propagating on the surface of a solid. The surface characteristics can be used for gas detection.

Virtual Electronic Nose: The electronic nose, which is based on the gas chromatography technology, causes its responses to have two components: the appearance time and the response intensity. The difference of physical and chemistry characteristics, different components need difference time to pass through the capillary, so the appearance time of the peak can be used for determining material.

Chapter 7
Electronic Noses in Food Analysis

Runu Banerjee Roy
Jadavpur University, India

Rajib Bandyopadhyay
Jadavpur University, India

Bipan Tudu
Jadavpur University, India

Nabarun Bhattacharyya
Centre for Development of Advanced Computing, India

ABSTRACT

Food quality evaluation is a tedious job as it can be sensed either by its flavor, taste, or appearance. Sensory evaluation for food quality determination is very complex and depends on biological sensor systems and is subject to high variability depending on taster's mood, physical state, etc. Moreover, as it is purely subjective, the observations may vary for different tasters. For food industries the instrumental means for quality monitoring has significant appreciations which are repeatable, accurate, and reliable. Electronic nose is an array of sensors that senses based on aroma of samples. For the last two decades, electronic nose has been applied in several domains of applications in food analysis, for example, food quality monitoring based on seasonal effect, ageing, geographical origin, fermentation, etc. This chapter aims to focus on different domains of application of electronic nose in food analysis.

INTRODUCTION

Olfaction or odour sensing is considered to be a conversion of chemical information into an electrical signal. This chemical information carried through is essentially a code in live organic system and the perception with discrimination is by stimulation of the olfactory nerve. In artificial systems, type, intensity, occurrence of odours individually or collectively are to be considered for coding which is possible through a model made *a priori* to be exploited for the purpose. Certain types of sensations known as

DOI: 10.4018/978-1-5225-3862-2.ch007

chemesthetic ones such as burning, piquancy, prickling, stinging, tingling etc. stimulate the trigeminal nerve and 'pungency' is the result of such sensations.

Mechanism of human olfaction is a special chemical process affected by odorants or odour molecules which interact with special receptors in the olfactory epithelium located at the top of the nasal cavity (Rosenblith, 1961). These are molecules with widely variant structures each of which is responsible for a characteristic odour designated as the signature of the chemical (Amoore, 1952). Minor variations in the structure can lead to large change in the odour quality and signature. Human olfactory system can detect around 10000 complex odours because of the existence of the range of receptor types in the nose (Buck and Axel, 1991). Size and shape of molecules of the chemicals are correlated to its odour quality (Amoore and Venstrom, 1967). There are other postulates of odour quality as well as quantity (Wright, 1966). Odour threshold specified in parts per billion (ppb) is determined by the number of molecules excited. In terms of ppb, the threshold of hydrogen sulphide i.e. rotten egg is 1.1 ppb (Chastrette, 1997). Threshold is also determined / correlated by the molecular weight, volume, hydrogen bonding ability, cross-sectional area, partition co-efficient and many other such parameters (Breen, 1992). Science of olfaction in humans involving physiology, anatomy, bio-physics, molecular biology (Serby and Chobar, 1992) has demonstrated that, like all other sensing systems, excitation produces an electrical signal which is transmitted for exciting the brain.

The Objective of this chapter is to find the application of electronic nose for food quality determination. The quality of food is dependent on several issues like climate and soil where the crops developed, the processing techniques those are employed to convert from raw to edible, storage conditions and ageing of food samples, fermentation conditions if the food is taken after fermentations etc. Electronic nose is an instrument which is eligible to detect food quality considering above conditions. This chapter is an attempt to find out the applications of electronic nose in all these areas.

The next section describes basic architecture of electronic nose, sensor technologies of electronic nose and use of electronic nose in different applications. After that the chapter focuses to the main focus of this chapter – application of electronic nose in food quality evaluation. The different sections under this header explain application of electronic nose for food quality evaluation based on different features. A generalized discussions and recommendation is there followed by a future scope of research. The chapter concluded with a conclusion.

Electronic Nose

Since time immemorial, five physiological sensory organs of the humans have been in use for detection and identification of parameters relevant and significant to societal and individual perspective. Sensing and perception through these organs are done through biological, physiological, chemical processes involving finally the brain as the Central Processing Unit (Pearce et al., 2003). With rapid growth in electronic hardware technology and spurt of software-based computation and processing, standardized identification through artificial sensors like electronic eyes, electronic ears etc. have been quite common since the eighties of the last century. The sensor technology of artificial olfaction had its beginnings with the invention of the first gas multisensor array in 1982 (Persaud and Dodd, 1982). Advances in aroma-sensor technology, electronics, biochemistry and artificial intelligence made it possible to develop devices capable of measuring and characterizing volatile aromas released from a multitude of sources for numerous applications. These devices, known as electronic noses, were engineered to mimic the mammalian olfactory system within an instrument designed to obtain repeatable measurements, allowing identifications and classifications of aroma mixtures while eliminating operator fatigue Electronic nose has generated a lot of interest over the past decade because of its versatile applicability in terms of identification and classification of agro-based and processed food and consumer items. Its contribution towards the development of solution methodology in the areas of environmental studies, medical diagnostics and biological processes (Jinks and Laing, 2001) is being recognized now. It is an instrument that uses array(s) of gas sensors, electronic hardware and processors for detection and discrimination of odours from the ensemble (Gardner and Bartlet, 1999). Electronic nose, as a complete instrument, would therefore, consist of a few specific modules as shown in Figure 1 (Hatfield et al, 1994).

Sensor Technologies for Electronic Nose

The human sense of smell is intensely subjective as it is dependent on one's state of mind and physique. Synthetic sensors/sensor arrays in electronic nose can detect and identify odours objectively keeping the sensing principle same as in humans such as chemical interaction in the ionic, atomic or molecular level. These sensors are basically chemosensors and the principles in various types of them are (1) conductometric or resistive, (2) capacitive, (3) potentiometric, (4) gravimetric, (5) calorimetric, (6) optical and (7) amperometric where the output obtained on interaction with the odour molecules are in the form of changes in (1) resistance / conductance, (2) capacitance, (3) emf / voltage, (4) mass / piezoelectricity, (5) temperature, (6) refractive index / wavelength / intensity of radiation, (7) electrical current respectively. Devices used are metal oxide semiconductors (MOS), MOSFET's, Quartz Crystal Microbalance (QCM), (Organic) Conducting Polymers (CP), Surface Acoustic Wave Devices (SAWD), Surface Plasmon Resonance Devices (SPRD), Flourescent Dyes and so on.

A good sensor should fulfill a number of criteria. First, the sensor should have highest sensitivity to the target group of chemical compound(s) intended for detection and with a threshold of detection similar to that of the human nose. To be most useful with diverse detection capabilities, electronic nose sensors should have relatively low selectivity in order to be sensitive to a wide number of different chemical compounds. Sensors should be capable of operating at relatively low temperatures when necessary, have short calibration and training requirements, fast recovery time between runs and maintenance procedures

Figure 1. Block diagram representation of electronic nose system

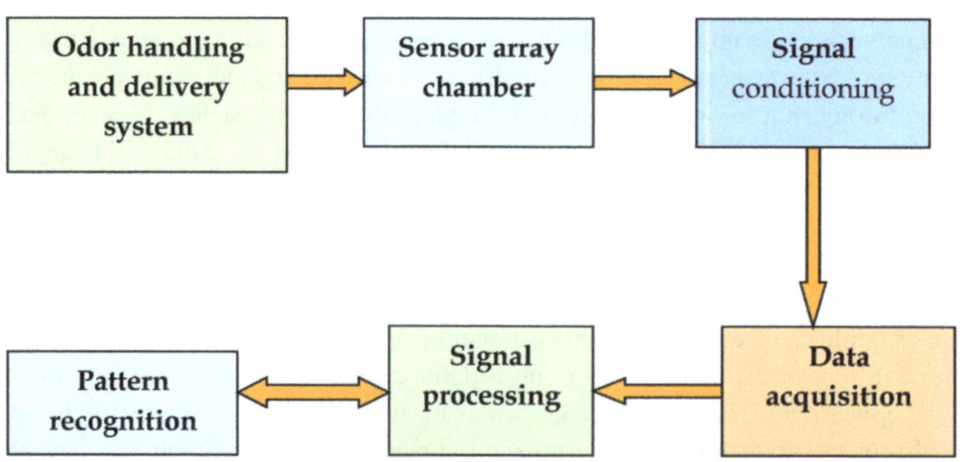

to maintain low operating costs. They must also have short recording and analysis times, particularly when used as on-line systems, and high sensor array stability. The commonly used sensors for electronic nose and their principles of detection are summarized in Figure 2.

Applications of Electronic Nose

Electronic nose has generated a lot of interest over the past decade because of its capability in terms of identification and classification of odours of consumer and other products. The food and beverage industries are currently the largest beneficiary of this technology, even though its use is no less in fragrance and wine industries. A vivid coverage on electronic nose applications for food and beverages is available in the book titled "Measurement and Control in Food Processing" by Manabendra Bhuyan (CRCPress, 2006). An air-quality monitor electronic nose has already traveled to space on board NASA's space shuttle flight STS 95 and was designed to monitor the ambience of the spacecraft.

The different applications of electronic nose are in environmental monitoring, medical diagnosis, food and beverages quality estimation, narcotic detection, explosive detection, quality monitoring of cosmetics and fragrance. Among all these fields, the major application of electronic nose is in food analysis. Recent results showed the utility of an electronic nose to smell chemicals related to lung, melanoma, prostatic, breast and pancreatic cancers. In general, electronic nose has wide applications in agriculture, biomedical, automobile and aerospace, environmental, food, military, pharmaceutical, explosive and in various other fields (Bhattacharyya *et al.*, 2008a; Brezmes *et al.*, 2001; Di Natale *et al.*, 2001; Boilot *et al.*, 2000; Young *et al.*, 2003; Capua *et al.*, 2009).

Figure 2. Electronic nose sensors

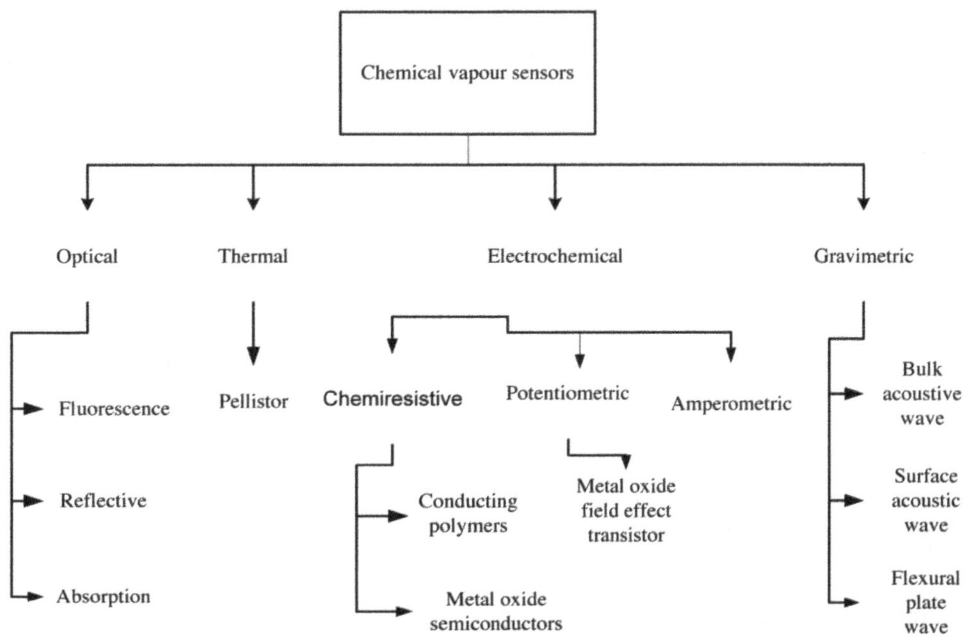

Applications of Electronic Nose in Food Quality Evaluation

Considerable work has already been carried out on meat, grains, coffee, mushrooms, cheese, sugar, fish, beer and other beverages, as well as on the odour quality evaluation of food packaging material. Food quality evaluation using electronic nose may be classified based on their original quality. More specifically, say for tea, quality may vary due to seasonal effect, based on their geographical origin. Electronic nose has successful applications in determining food quality evaluation based on this (Luo et al, 2017). Another major application of electronic nose in food quality is to detect additives/adulterates used in food for ripening of fruits, fresh appearance of food sample etc. Food quality also varies depending on aging. Literature survey finds many applications of determining the age of food samples or how aging affects in quality determination process. Final quality of different food products like tea, wine etc. depends on fermentation process. Quality determination at different fermentation level is a crucial task and electronic nose can be applied for that. The aim of this chapter is to explain these different types of applications alongwith analysis techniques.

BACKGROUND

With the advent of electronic nose it becomes easy to analyze the quality of food sample for different applications based on the odor. Already few commercially available electronic noses are there in the market. Among all the applications, food industries are mostly benefitted with this technology. Again in food analysis, there are several issues related to food quality and most of the issues can be resolved with electronic nose. The introduction of electronic noses in the area of food is envisaged for quality control, process monitoring, freshness evaluation, shelf-life investigation and authenticity assessment. Electronic nose has several significant features, like, the relatively fast assessment of headspace, a quantitative representation of volatile component present and low cost sensors which can be easily integrated in current production processes. The application of electronic nose is expanding rapidly and a considerable number of significant achievements are already published relevant for food industry. The major important task of quality determination process using electronic nose is to correlate the sensor response with an appropriate benchmark and to calibrate the instrument for unknown sample detection. The common methodology used to prepare the correlation model, either quality assigned by expert panellist is considered or high end analytical instrumental analysis of sample using gas chromatogram or GC-MS provides the benchmark data. Several researchers have attempted to determine food quality based on geographical origin, same origin but from different batch of production, degree of fermentation, ageing of sample, spoilage detection and adulterate detection in different food and beverages using electronic nose. The following article will focus on each type of application with detail methodology.

ELECTRONIC NOSE IN FOOD ANALYSIS

Food Quality Based on Geographical Origin

For agro-food sectors, food quality certification is one of the major important tasks. It is proved that for several agro products quality varies for different geographical location due to climate, soil etc. and hence the market price has been set accordingly. Consumers are increasingly concerned about the origin of foods, so the geographical origin of foods has been a major topic of debate and extensive research. Electronic nose has been successfully applied for determining geographical origin of olive oil (casale et al, 2010, Melucci et al, 2016), rice(Han et al, 2016), saffron (Tahri et al, 2015),tea (Luo et al, 2017, Kovács et al, 2010), etc. In all the cases it is found that the food quality varies depending on geographical origin and the difference in quality can be detected by electronic nose. Table 1 listed few applications of electronic nose in determining the quality of some other food products based on geographical origin. In the Table1 the electronic nose model, sensor material, the standard reference data with which the electronic nose response correlated are also listed.

Table 1. Applications of electronic nose for determining the quality of food based on geographical origin

Application	Sensing Element	Enose Model	Data Processing Algorithm	Correlated With	Reference
Determination of the geographical origin of extra virgin olive oils	The Heracles II was equipped with two columns working in parallel mode, A single comprehensive chromatogram was created by joining the chromatograms obtained with the two columns	FGC E-nose Heracles II AlphaMos, Toulouse, France	PCA, LDA, PLS-DA	SPME/GC–MS	Melucci et al,2016
identifying the geographical origin of licorice roots	MOS sensors	ISE Nose 2000 (ISE, Pisa, Italy	DFA	HS-SPME-GC–MS	Russo et al, 2014
geographical origin of commercial red ginseng concentrates	MS-EN	SMart Nose300; SMart Nose, Marin-Epagnier, Switzerland	DFA	Liquid chromatogram-mass spectrometry LC-MS/MS	Moon et al, 2014
Geographical origin identification of propolis	MOS sensors	Alpha M.O.S. Heracles, France	DFA,PCA	gas chromatography-mass spectrometry and olfactometry (GC-MS-O)	Cheng et al, 2013
to predict the geographical origin of Tempranillo wines produced in Australia and Spain	Samples were analyzed on the Chemical Sensor (HP 4440, Hewlett Packard) equipped with headspace sampler (HP 7694, Model G 1290A)	Mass spectrometry electronic nose (MS-EN)	PCA, PLS-DA, SLDA	Mass spectrometry(MS)	Cynkar et al,2010
To determine the geographical origin of Liguria extra virgin olive oil	MS-EN	MS-EN	Kennard and Stone algorithm is used to fuse the information from multivariate sensors	UV–visible and NIR spectroscopy	Casale et al, 2010
Prediction of Geographical origin of Sauvignon Blanc wines	array of 12 semiconducting metal oxide sensors	FOX 3000 Enose (Alpha M.O.S, Toulouse, France), MS-Enose	LDA,SDDA	GC–MS	Berna et al, 2009

Food Quality Based on Fermentation Level

Food quality for different food stuffs and agro products may vary depending on fermentation level. Different food stuffs like sauce, pickle, bread, beer, wine, yogurt etc are prepared by fermentation process. Again some agro products like tea, coffee, cocoa are processed from raw grains by the process of fermentation. Fermentation in food processing is the process of converting carbohydrates to alcohol or organic acids using microorganisms—yeasts or bacteria—under anaerobic conditions. During fermentation sugar content chemically converted into ethanol producing alcoholic compound such as wine, beer, cider. Similar process takes place during leavening in bread, preparation of yogurt etc. In case of tea, coffee etc the grassy smell of leaves is transformed into floral smell in fermentation stage only which is very crucial. Here fermentation is basically oxidation process where typical temperature, humidity are to be maintained. Now with the slight variation of fermentation, quality of food sample may vary significantly. Electronic nose is able to successfully discriminate the fermentation stages and can declare optimum fermentation condition for a particular application considering food quality.

Electronic nose has some specific advantages over conventional analytical techniques such as fast measurements, possibility of easy automation, relatively simple and inexpensive instrumentation. As well as the measurement technique is non-destructive, the continuous monitoring at all the fermentation stages is possible with electronic nose. The electronic nose has a good potential for the assessment of the state and operation conditions of a fermentation process. In the review article (Peris and Escuder-Gilabert, 2013), author described several area of applications where electronic nose is used to monitor fermentation process. Table 2 describes the application of electronic nose for quality detection of different food stuffs during fermentation.

Food Quality Based on Shelf-Life Investigation (Freshness Evaluation)

Another burning issue related to food quality is shelf-life determination of a food product. Food industries have to declare shelf-life of a processed food on its package as expiry date of the product for the sake of consumption. Examples of foods that have to be consumed preferably within dates include canned, dry and frozen foods under certain storage conditions. That means the quality of food sample deteriorates beyond the stipulated time mentioned and sometimes there may be a chance for food poisoning. Food manufacturers are responsible for determining the shelf life and to label accordingly alongwith mentioning storage condition. The purpose of shelf life is to help consumers make safe and informed about the use of foods. Food shelf life should be considered valid only if the product is bought intact and not damaged. Several research has already been carried out in determining shelf-life of various food samples. With aging of certain food stuff, microbial spoilage cause the generation of some volatile organic components (VOCs) due to these microorganisms (Boothe and Arnold, 2002).

For the storage of meat, the electronic nose can be applied for assessing quality, detection of off-flavours and spoilage identification. In this direction one of the study reported the changes in the head-space of vacuum packaged beef, vaccinated with Salmonella typhimurium, was evaluated using a metal oxide (MOS) based enose (Balasubramanian et al. 2008). In another report the volatile components of pork and meat products such as sausages were also studied (Nurjuliana et al. 2011). For the detection of Salmonella contamination in packaged beef steaks an olfactory system has been considered using neural network classifiers (Khot et al. 2012). Meat spoilage detection was also explored for the case of beef fillets stored aerobically at different isothermal conditions (0, 4, 8, 12, and 16 °C) (Papadopoulou et al.

Table 2. Application of electronic nose during fermentation of different food samples

Application	Sensing Element	Enose Model	Data Processing Algorithm	Correlated With	Reference
Flavor quality of Chinese-type soy sauces	MOS sensors	PEN3 e-nose	pattern recognition software (Win Muster)	GC-MS combined with HS-SPME	Gao et al, 2017
Flavor quality of natural pickled wax gourd	MOS sensors	PEN3 e-nose	PCA	GC-MS	Wu et al, 2017
Fermentation on characteristics of doughs obtained with durum wheat	MOS sensors	PEN2 e-nose	analysis of variance (ANOVA) and Fisher's least significant difference (LSD)	GC	Balestra et al, 2015
The quality of 26 fermented cocoa beans	MS-based e-nose	MS-based e-nose	PCA, HCA, SIMCA	MS, HS-SPME-GC-MS	Tran et al., 2015
Detect the optimum fermentation time of black tea	QCM sensors	Array of eight QCM sensors	PCA	UV-vis Spectrophotometer	Sharma et al, 2015
Classification of three different fermentation degrees of tea	porphyrins, metalloporphyrins materials	odor imaging sensor array	LDA, LOOCV	GC-MS	Chen et al., 2013
Prediction of optimum fermentation time	MOS sensor array	Developed enose set up with Figaro sensors	TDNN, m-TDNN, SOM	Used prediction models	Bhattacharya et al., 2008b

2013). In another study, for measuring lavou quality changes of refined chicken fat during controlled oxidation, an enose, consisting of 18 MOS gas sensors has been used and as a prediction model PLSR was utilized (Song et al. 2013). An et al, 2017 showed that low dose of e-beam irradiation and vacuum packaging is beneficial for safety and shelf life extension without affecting the sensory characteristics of smoked duck meat by monitoring flavor profiles during the different storage intervals employing electronic nose.

The age of fruits (ripeness or maturity level) determines the shelf life and future rate of quality loss due to changes in lavour, firmness and color. For assessing fruit quality and detection of ripening, several studies have been done with the application of e-nose (Brezmes et al., 2000); Costa et al, 2003). Some experiments have been performed for apples (Di Natale,2001; Saevels,2004; Li et al, 2005), peaches (Benedetti et al, 2008; Zhang, 2008; Su, 2013) and tomatoes (Berna, 2004 ; Gomez, 2008). Gomez et al., 2006, investigated the volatile production change of ripeness stages (unripe, half- ripe, full-ripe, and over-ripe) of tomato and mandarin, using an e-nose with ten different MOS sensors. Brezmes et al., 2001 used an e-nose system, consisting of 21 MOS sensors array, in order to assess the ripeness state of apples during their shelf-life.

Another major application area for shelf-life detection is milk and milk products. It has already been proved that standard microbiological methods, including the Moseley keeping-quality test, have very little value as predictors of shelf life (Bishop and White, 1986). The major problem of the Moseley keeping-quality test is the extended time period required to perform testing. Another problem is the relatively poor correlation between microbial counts and actual shelf life (Urbach and Milne, 1988; Vallejo-Cordoba and Nakai, 1994). In Table 3 few applications of electronic nose has been listed for shelf-life detection of several food products like fish, meat, milk etc.

Table 3. Application of electronic nose for shelf-life detection of different food samples

Application	Sensing Element	Enose Model	Data Processing Algorithm	Correlated With	Reference
Effect of vacuum packaging on the shelf-life of silver carp	MOS Sensors	ISENSO INTELLIGENT., China	PCA	Sensory data	Kachele et al., 2017
Monitoring the spoilage of beef fillets	QCM Sensors	Libra enose	MLP,PLS,ANFIS, SVM	FTIR, Raman spectroscopy, Hyper spectral imaging	Kodogiannis, 2017
For prediction of banana quality indices at different shelf-life stages	Array of MOS senors	Hanwei Electronics Co., Ltd., Henan, China	PLS, SVR, MLR	Chemical analysis of quality indices	Sanaeifar et al, 2016
Rapid assessment of shelf-life of fried potato wedges	Array of MOS senors	TGS Figaro sensors	PCA, Fuzzy logic analysis	HS-SPME/GC-MS	Chatterjee et al., 2014
Change in aromatic composition of white truffles during presevation	QMB sensor array	QMB sensor array	PCA	GC–FID, GC-MS	Pennazza et al., 2013
Predictive modelling for freshness decay and shelf life of European sea bass	Mos sensors	PEN 2 model,WMA	PCA	Chemical analysis of volatiles	Limbo et al., 2009
To predict the shelf life of pasteurized and homogenized reduced-fat milk and whole-fat chocolate milk	MS-EN	mass spectrometry-based electronic-nose	PCA, PLS	gas chromatograph/mass spectrometer	Marsili, 2000
To determine the shelf life of milk	MOS sensor chambers	Fox 4000 (ALPHA MOS, Toulouse FR)	PCA, Radar representation	microbial count by using the 3M™-Petrifilm™ test	Labreche et al., 2005

Food Quality Based on Purity (Adulterate/Additive Detection)

Food additives are added to food in order to preserve colour, taste and flavor. All type of processed food have some kind of food additives. For example, vinegar, citric acid, tartaric acid, malic acid, fumaric acid, and lactic acid are used as preservatives in several food products for acidic taste. Ascorbic acid is used as several vegetable and fruit juices, fortified wines, baby foods as colour retention agent. There are a huge number of additives used for several purposes in processed food. However, there must be a specified limit of adding food additives to get technological benefit and in turn the market price of the product.

Food adulteration is the process in which the quality of food is lowered either by the addition of inferior quality material or by extraction of valuable ingredient. A food item is said to be adulterated when the added substance is injurious for human health. Imitation of food products is considerable risk for producers, distributors, retailers and consumers from the point of business, price, health. Flavour of several food products is one of the key attributes of their quality and authenticity. In the case of some foods, the aroma of a product is specific enough to discriminate an original product from its adulterated counterpart. Electronic nose is a rapid and powerful technique, which is able to determine the aroma of a product without additional special sample preparation. Food adulterants are used to get more economical value with low quality products and may be used for all kind of foods. Pesticides and food pathogens are two major issues for food borne illness. Electronic nose is capable of detecting the traces of pesticides as well as detection of these pathogens originating from different sources are possible using electronic nose(Canhoto and Magan, 2005; Arrora et al., 2006). Table 4 describes the application of electronic nose for detection of different food additives/adulterants in various food stuffs.

DISCUSSIONS AND RECOMMENDATIONS

The sense of smell has a fundamental role in human development and biosocial interactions. Food qualities are mainly decided based on flavour of food. From ancient times spices are most valuable things for cooking to enhance flavour of food. The good quality tea, coffee, wine, fruit juices, cooked food, ripened fruit all are characterised by aroma of them. Thus aroma characteristics have contributed immensely to the many commercial and agro products. Human smell sense is severely limited as a practical application due to high subjectivity. An electronic nose has a number of significant features- the relatively fast assessment of headspace, a quantitative representation of a gas, repeatable measurement and cheap sensors which can be easily integrated in current production processes. These cause the use of electronic nose for food applications as well as in other area of applications. In the past few years, there have been notable achievements in the use of electronic nose relevant to the food industries.

Despite of the above features, still there is limited use of electronic nose in industry because there is no universal electronic nose that can identify any gas samples. This is due to the selectivity and sensitivity limitation of electronic nose for specific analyte. A proper selection of an appropriate electronic nose system for a particular application must involve an evaluation of systems on a basis of typical case. The key considerations for electronic nose development for a particular application is the (i) suitable material selection, (ii) the selectivity and sensitivity range of individual sensor for particular target analyte gases, (iii) repeatability check with same sample, (iv) appropriate data analysis method use and response

Table 4. Application of electronic nose for food additives/adulterants detection in different food samples

Application	Sensing Element	Enose Model	Data Processing Algorithm	Correlated With	Reference
Additive detection in fruit juice	MOS sensors	PEN2 E-nose, AIRSENSE Analytics GmbH, Schwerin, Germany	LDA, PLSR and SVM	for quantitative monitoring, SVM, RF, ELM and PLSR were applied to establish regression models between E-nose signals and the amount of food additives in fruit juices.	Qiu and Wang, 2017
Differentiation of fresh and frozen pork	Metal oxide sensors	Alpha M.O.S., Toulouse, France	PCA, ANN	ultra-fast gas chromatography (UFGC)	Górska-Horczyczak et al., 2017
To detect mislabeling and adulteration of PDO Oscypek cheese.	Metal oxide sensors	FOX 3000 E-Nose (Alpha M.O.S, Toulouse, France	PCA, LDA, SIMCA, SVM	For SPME-MS analysis, a Hewlett-Packard HP5890II coupled to a Hewlett-Packard 5971 quadrupole mass spectrometer was used	Majcher et al, 2015
Saffron adulteration detection	Metal oxide sensors	Assembled enose setup with sensor array of six metal oxide semiconductors, MQ-type (HANWEI Electronics Co., Ltd., Henan, China	PCA, LDA, PLSR, ANN including MLP, SOM, LVQ, and ART, fuzzy logic,	Pure samples are considered and known adulterates have been mixed	Heidarbeigi et al, 2015
Determining both organoleptic adulterations of processed tomato	Metal oxide sensors	EOS835 (Sacmi Imola scarl, Italy)	PCA, Box-plot, KNN	GC-MS	Concina et al, 2009

interpretation. Other operational considerations for electronic nose are (i) portability, (ii) low cost, (iii) easy to handle. The instrument must be trained with a known set of data which is quite easy compared to complex analytical instruments. Several abbreviations used within the chapter those are listed with full forms in the Table 5.

FUTURE RESEARCH DIRECTIONS

The major scope of research in the direction of electronic nose is in sensor development for typical applications. Commercially available sensors for electronic nose are most generalized and not appropriate for all applications. Thus searching appropriate material and employing suitable sensor development technique a devoted electronic nose can be developed for a particular application. In order to choose sensor material the knowledge about the target gas, its chemical structure and nature is very much important. Major importance must be given to the design and development of drift free sensors that can be used reliably over long period without deviating from the performance. Consequently, the internal drift influencing factors like crystalline structure variation, grain size variations, grain boundary effects, uniformity in the

Table 5. List of abbreviations used in this chapter

Abbreviation	Full Term
ANFIS	Adaptive Neural Fuzzy Inference System
ART	Adaptive resonance theory
ANN	Artificial Neural Network
CA	Cluster
DFA	Discriminant function analysis
ELM	Extreme learning machine
FTIR	Fourier transform infrared spectroscopy
GC–FID	Gas Chromatogram with Flame Ionization Detector
GC–MS	Gas chromatogram- mass spectrometry
HCA	Hierarchical cluster analysis
k-NN	k-nearest neighbors algorithm
LDA	Linear discriminant analysis
LVQ	Learning vector quantization
LOOCV	leave-one-out cross-validation
m-TDNN	Multiple-time delay neural network
MLR	Multiple linear regression
MLP	Multi layer perceptron
MOS	Metal oxaide semiconductor
NIR	Non infrared
PCA	Principal component analysis
PLS-DA	Partial least squares- discriminant analysis
PLSR	Partial least square regression
PNN	probabilistic neural network
QMB	Quartz micro balance
RF	Random forest
SPME	Solid phase-microextraction
SOM	Self-organizing map
SVM	Support vector machine
SVR	Support vector regression
TDNN	Time delay neural network
UV	Ultra violet
SLDA	Stepwise discriminant analysis
SIMCA	Soft independent modeling of class analogy
SDDA	Stepwise diagonal discriminant analysis

dopant concentrations, perfect contact materials, thickness of the sensing elements should be addressed. Moreover, sensor response must not be influenced by humidity, ambient temperature and atmosphere. Along with these the performance of other parts of the instrument like data acquisition system must be handled carefully in order to get better accuracy in measurement. Data analysis and presentation is one of the major task in this regard. So exploration of classification techniques and prediction tools must improve the measurement process in different applications.

CONCLUSION

A fundamental problem for any food industry is the relationship between perceived quality and the presence of natural flavour compounds in the product. On one hand, methodologies for estimation of such perceived quality being human-dependent, problems like subjectivity, non-repeatability and lack of reproducibility create huge problem towards standardization of quality. On the other hand, chemical analysis methods anchored on chromatographic techniques are cumbersome for online deployment due to high cost, sample preparation complexities and non-availability of expert manpower. Electronic nose is the solution for all these problems as most of the food qualities are categorized on the basis of flavor. In the present chapter, application of electronic nose in various field of study for food quality analysis has been described. From the study of different applications with different types of sensors employing various data analysis methods it can be said that electronic nose is a handy tool to monitor quality of food.

Though aroma is a major important characteristic for food quality but taste is another important feature. When we perceive food, we actually recognize the flavor of food which is amalgamation of smell as well as taste. In human physiology smell and taste are influenced by each other. If due to cold, nose gets blocked that is the smell sense becomes weak, and then we cannot feel the taste of food properly. Smell and taste sense have cross correlation effects. Towards the direction of overall quality of food, combined analysis of electronic nose and electronic tongue has shown quite improved result than single sensor system. In the area of food analysis combination of electronic sensor responses has opened a new research direction.

REFERENCES

Amoore, J. E. (1952). The stereochemical specificities of human olfactory receptors, perfumes. *Essent Oil Rec.,* (43), 321-323, 330.

Amoore, J. E., & Venstrom, D. (1967). Correlations between stereochemical assessment and organoleptic analysis of odorous compounds. Olfaction and Taste, (2), 3 – 17.

An, K.-A., Arshad, M. S., Jo, Y., Chung, N., & Kwon, J.-H. (2017). E-Beam Irradiation for Improving the Microbiological Quality of Smoked Duck Meat with Minimum Effects on Physicochemical Properties During Storage. *Journal of Food Science, 82*(4), 865–872. doi:10.1111/1750-3841.13671 PMID:28267865

Arora, K., Chand, S., & Malhotra, B. D. (2006). Recent developments in bio-molecular electronics techniques for food pathogens. *Analytica Chimica Acta, 568*(1-2), 259–274. doi:10.1016/j.aca.2006.03.078 PMID:17761267

Balasubramanian, S., Panigrahi, S., Logue, C. M., Doetkott, C., Marchello, M., & Sherwood, J. S. (2008). Independent component analysis-processed electronic nose data for predicting salmonella typhimurium populations in contaminated beef. *Food Control, 19*(3), 236–246. doi:10.1016/j.foodcont.2007.03.007

Balestra, F., Laghi, L., Taneyo Saa, D., Gianotti, A., Rocculi, P., & Pinnavaia, G. (2015). Physico-chemical and metabolomic characterization of KAMUT® Khorasan and durum wheat fermented dough. *Food Chemistry, 187*, 451–459. doi:10.1016/j.foodchem.2015.04.041 PMID:25977050

Benedetti, S., Buratti, S., Spinardi, A., Mannino, S., & Mignani, I. (2008). Electronic nose as a non-destructive tool to characterise peach cultivars and to monitor their ripening stage during shelf-life. *Postharvest Biology and Technology, 47*(2), 181–188. doi:10.1016/j.postharvbio.2007.06.012

Berna, A.Z., Lammertyn, J., Saevels, S., Natale, C.D., Nicolaï, B.M. (2004). Electronic nose systems to study shelf life and cultivar effect on tomato aroma profile. *Sens. Actuat. B: Chem.*, (97), 324–333.

Berna, A. Z., Trowell, S., Clifford, D., Cynkar, W., Cozzolino, D., Berna, A. Z., ... Cozzolino, D. (2009). Geographical origin of Sauvignon Blanc wines predicted by mass spectrometry and metal oxide based electronic nose. *Analytica Chimica Acta, 648*(2), 146–152. doi:10.1016/j.aca.2009.06.056 PMID:19646576

Bhattacharya, N., Tudu, B., Jana, A., Ghosh, D., Bandhopadhyaya, R., & Bhuyan, M. (2008b). Preemptive identification of optimum fermentation time for black tea using electronic nose. *Sensors and Actuators. B, Chemical, 131*(1), 110–116. doi:10.1016/j.snb.2007.12.032

Bhattacharya, N., Tudu, B., Jana, A., Ghosh, D., Bandhopadhyaya, R., & Saha, A. (2008a). Illumination heating and physical raking for increasing sensitivity of electronic nose measurements with black tea. *Sensors and Actuators. B, Chemical, 131*(1), 37–42. doi:10.1016/j.snb.2007.12.031

Bhuyan, M. (2006). *Measurement and Control in Food Processing*. Taylor & Francis. doi:10.1201/9781420006797

Bishop, J. R., & White, C. H. (1986). Assessment of dairy product quality and potential shelf lifesa review. *Journal of Food Protection, 49*(9), 739–753. doi:10.4315/0362-028X-49.9.739

Boilot, P., Hines, E. L., John, S., Mitchell, J., Lopez, F., Gardner, J. W., ... Gonogora, M. A. (2000). Detection of bacteria causing eye infections using a neural network based electronic nose system. In J. W. Gardner & K. C. Persaud (Eds.), *Electronic nose and olfaction 2000* (pp. 189–196). Bristol, UK: IOP Publishing.

Boothe, D. D. H., & Arnold, J. W. (2002). Electronic nose analysis of volatile compounds from poultry meat samples, fresh and after refrigerated storage. *Journal of the Science of Food and Agriculture, 82*(3), 315–322. doi:10.1002/jsfa.1036

Breen, H. (1992). Odor recognition and second messenger signaling in olfactory receptor neurons. *Seminars in Cell Biology*, (5): 25–32. PMID:8186393

Brezmes, J., Llobet, E., Vilanova, X., Orts, J., Saiz, G., & Correig, X. (2001). Correlation between electronic nose signals and fruit quality indicators on shelf-life measurements with pinklady apples. *Sensors and Actuators. B, Chemical, 80*(1), 41–50. doi:10.1016/S0925-4005(01)00867-X

Brezmes, J., Llobet, E., Vilanova, X., Saiz, G., Correig, X., (2000). Fruit ripeness monitoring using an electronic nose. *Sens. Actuat. B: Chem.,* (69), 223–229.

Buck, L., & Axel, R. (1991). A novel multigene family may encode odorant receptors, a molecular basis for odor recognition. *Cell, 65*(1), 175–187. doi:10.1016/0092-8674(91)90418-X PMID:1840504

Canhoto, O., & Magan, N. (2005). Electronic nose technology for the detection of microbial and chemical contamination of potable water. *Sensors and Actuators, B: Chemical, 106*(1), 3-6.

Capua, E., Cao, R., Sukenik, C. N., & Naaman, R. (2009). Detection of triacetone triperoxide (TATP) with an array of sensors based on non-specific interactions. *Sensors and Actuators. B, Chemical, 140*(1), 122–127. doi:10.1016/j.snb.2009.04.045

Casale, M., Casolino, C., Oliveri, P., & Forina, M. (2010). The potential of coupling information using three analytical techniques for identifying the geographical origin of Liguria extra virgin olive oil. *Food Chemistry, 118*(1), 163–170. doi:10.1016/j.foodchem.2009.04.091

Chastrette, M. (1997). Trends in structure – odor relationships. *SAR and QSAR in Environmental Research, 6*(3-4), 215–254. doi:10.1080/10629369708033253 PMID:9487700

Chatterjee, D., Bhattacharjee, P., & Bhattacharyya, N. (2014). Development of methodology for assessment of shelf-life of fried potato wedges using electronic noses: Sensor screening by fuzzy logic analysis. *Journal of Food Engineering, 133*, 23–29. doi:10.1016/j.jfoodeng.2014.02.009

Chen, Q., Liu, A., Zhao, J., & Ouyang, Q. (2013). Classification of tea category using a portable electronic nose based on an odor imaging sensor array. *Journal of Pharmaceutical and Biomedical Analysis, 84*, 77–83. doi:10.1016/j.jpba.2013.05.046 PMID:23810847

Cheng, H., Qin, Z. H., Guo, X. F., Hu, X. S., & Wu, J. H. (2013). Geographical origin identification of propolis using GC-MS and electronic nose combined with principal component analysis. *Food Research International, 51*(2), 813–822. doi:10.1016/j.foodres.2013.01.053

Concina, I., Falasconi, M., Gobbi, E., Musci, M., Mattarozzi, M., Pardo, M., ... Sberveglieri, G. (2009). Early detection of microbial contamination in processed tomatoes by electronic nose. *Food Control, 20*(10), 873–880. doi:10.1016/j.foodcont.2008.11.006

Costa, G., Noferini, M., Fiori, G., Orlandi, A., & Miserocchi, O. (2003). Non-destructive technique to assess internal fruit quality. *International Conference on Quality in Chains. An Integrated View on Fruit and Vegetable Quality*, (604), 571–576. doi:10.17660/ActaHortic.2003.604.68

Cynkar, W., Dambergs, R., Smith, P., & Cozzolino, D. (2010). Classification of Tempranillo wines according to geographic origin: Combination of mass spectrometry based electronic nose and chemometrics. *Analytica Chimica Acta, 660*(1-2), 227–231. doi:10.1016/j.aca.2009.09.030 PMID:20103167

Di Natale, C., Macagnano, A., Martinelli, E., Paolesse, R., Proietti, E., D'Amico, A., (2001). The evaluation of quality of post-harvest oranges and apples by means of an electronic nose. *Sens. Actuat. B: Chem.,* (78), 26–31.

Di Natale, C., Olafsdottir, G., Einarsson, S., Martinelli, E., Paolesse, R., & D'Amico, A. (2001). Comparison and integration of different electronic noses for freshness evaluation of cod-fish fillets. *Sensors and Actuators. B, Chemical, 77*(1-2), 572–578. doi:10.1016/S0925-4005(01)00692-X

Gao, L., Liu, T., An, X., Zhang, J., Ma, X., & Cui, J. (2017). Analysis of volatile flavor compounds influencing Chinese-type soy sauces using GC–MS combined with HS-SPME and discrimination with electronic nose. *Journal of Food Science and Technology, 54*(1), 130–143. doi:10.1007/s13197-016-2444-0 PMID:28242911

Gardner, J. W., & Bartlet, P. N. (1999). *Electronic Noses, Principles and Applications.* Oxford, UK: Oxford University Press.

Gómez, A.H., Hu, G., Wang, J., & Pereira, A.G. (2006). Evaluation of tomato maturity by electronic nose. *Comput. Electron. Agricult.,* (54), 44–52.

Gómez, A.H., Wang, J., Hu, G., & Pereira, A.G. (2006). Electronic nose technique potential monitoring mandarin maturity.*Sens. Actuat. B: Chem.,* (113), 347–353.

Gomez, A. H., Wang, J., Hu, G., & Pereira, A. G. (2008). Monitoring storage shelf life of tomato using electronic nose technique. *Journal of Food Engineering, 85*(4), 625–631. doi:10.1016/j.jfoodeng.2007.06.039

Górska-Horczyczak, E., Horczyczak, M., Guzek, D., Wojtasik-Kalinowska, I., & Wierzbicka, A. (2017). Chromatographic fingerprints supported by artificial neural network for differentiation of fresh and frozen pork. *Food Control, 73*, 237–244. doi:10.1016/j.foodcont.2016.08.010

Han, H. J., Lee, S. H., Moon, J. Y., Park, S., Dong, H., & Noh, B. S. (2016). Discrimination of the cultivar, growing region, and geographical origin of rice (Oryza sativa) using a mass spectrometer-based electronic nose. *Food Science and Biotechnology, 25*(3), 695–700. doi:10.1007/s10068-016-0121-8

Hatfield, J. V., Neaves, P., Hicks, P. J., Persaud, K., & Travers, P. (1994). Towards an integrated Electronic Nose using conducting polymer sensors. *Sensors and Actuators. B, Chemical, 18*(1-3), 221–228. doi:10.1016/0925-4005(94)87086-1

Heidarbeigi, K., Mohtasebi, S. S., Foroughirad, A., Ghasemi-Varnamkhasti, M., Rafiee, S., & Rezaei, K. (2015). Detection of adulteration in saffron samples using electronic nose. *International Journal of Food Properties, 18*(7), 1391–1401. doi:10.1080/10942912.2014.915850

Jinks, A., & Laing, D. G. (2001). The analysis of odor mixtures: Evidence for a configurational process. *Physiology & Behavior, 72*(1-2), 51–63. doi:10.1016/S0031-9384(00)00407-8 PMID:11239981

Kachele, R., Zhang, M., Gao, Z., & Adhikari, B. (2017). Effect of vacuum packaging on the shelf-life of silver carp (Hypophthalmichthys molitrix) fillets stored at 4 °C. *Lebensmittel-Wissenschaft + Technologie, 80*, 163–168. doi:10.1016/j.lwt.2017.02.012

Khot, L. R., Panigrahi, S., Doetkott, C., Chang, Y., Glower, J., Amamcharla, J., ... Sherwood, J. (2012). Evaluation of technique to overcome small dataset problems during neural-network based contamination classification of packaged beef using integrated olfactory sensor system. *Lebensmittel-Wissenschaft + Technologie*, *45*(2), 233–240. doi:10.1016/j.lwt.2011.06.011

Kodogiannis, V. S. (2017). Application of an Electronic Nose Coupled with Fuzzy-Wavelet Network for the Detection of Meat Spoilage. *Food and Bioprocess Technology*, *10*(4), 730–749. doi:10.1007/s11947-016-1851-6

Kovács, Z., Dalmadi, I., Lukács, L., Sipos, L., Sza'ntai-Ko˝hegyi, K., Kókai, Z., & Fekete, A. (2010). Geographical origin identification of pure Sri Lanka tea infusions with electronic nose, electronic tongue and sensory profile analysis. *Journal of Chemometrics*, *24*(3-4), 121–130. doi:10.1002/cem.1280

Labreche, S., Bazzo, S., Cade, S., & Chanie, E. (2005). Shelf life determination by electronic nose: Application to milk. *Sensors and Actuators, B: Chemical 106*(1), 199-206.

Li, C., Heinemann, P., Irudayaraj, J., & Peterson, D. (2005). Detection of apple defects using an electronic nose and znose. *ASAE Annual International Meeting.*

Limbo, S., Sinelli, N., Torri, L., & Riva, M. (2009). Freshness decay and shelf life predictive modelling of European sea bass (Dicentrarchus labrax) applying chemical methods and electronic nose. *Lebensmittel-Wissenschaft + Technologie*, *42*(5), 977–984. doi:10.1016/j.lwt.2008.12.011

Luo, D., Chen, J., Gao, L., Liu, Y., & Wu, J. (2017). Geographical origin identification and quality ontrol of Chinese chrysanthemum flower teas using gas chromatography–mass spectrometry and olfactometry and electronic nose combined with principal component analysis. *International Journal of Food Science & Technology*, *52*(3), 714–723. doi:10.1111/ijfs.13326

Majcher, M. A., Kaczmarek, A., Klensporf-Pawlik, D., Pikul, J., & Jeleń, H. H. (2015). SPME-MS-Based Electronic Nose as a Tool for Determination of Authenticity of PDO Cheese, Oscypek. *Food Analytical Methods*, *8*(9), 2211–2217. doi:10.1007/s12161-015-0114-x

Marsili, R. T. (2000). Shelf-life prediction of processed milk by solid-phase microextraction, mass pectrometry, and multivariate analysis. *Journal of Agricultural and Food Chemistry*, *48*(8), 3470–3475. doi:10.1021/jf000177c PMID:10956135

Melucci, D., Bendini, A., Tesini, F., Barbieri, S., Zappi, A., Vichi, S., ... Gallina Toschi, T. (2016). Rapid direct analysis to discriminate geographic origin of extra virgin olive oils by flash gas chromatography electronic nose and chemometrics. *Food Chemistry*, *204*, 263–273. doi:10.1016/j.foodchem.2016.02.131 PMID:26988501

Moon, J. Y., Han, H. J., Park, S. W., Dong, H., Han, K.-Y., Kim, H. J., ... Noh, B. S. (2014). Discrimination of the origin of commercial red ginseng concentrates using LC-MS/MS and electronic nose analysis based on a mass spectrometer. *Food Science and Biotechnology*, *23*(5), 1433–1440. doi:10.1007/s10068-014-0196-z

Nurjuliana, M., Che Man, Y. B., Mat Hashim, D., & Mohamed, A. K. S. (2011). Rapid identification of pork for halal authentication using the electronic nose and gas chromatography mass spectrometer with headspace analyzer. *Meat Science*, *88*(4), 638–644. doi:10.1016/j.meatsci.2011.02.022 PMID:21420795

Papadopoulou, O., Panagou, E. Z., Mohareb, F., & Nychas, G.-J. (2013). Sensory and microbiological quality assessment of beef fillets, using a portable electronic nose in tandem with support vector machine analysis. *Food Research International*, *50*(1), 241–249. doi:10.1016/j.foodres.2012.10.020

Pearce, T. C., Schiffman, S. S., Nagle, H. T., & Gardner, J. W. (Eds.). (2003). Handbook of Machine Olfaction, Electronic Nose Technology. Wiley – VCH, Weinheim (FDR).

Pennazza, G., Fanali, C., Santonico, M., Dugo, L., Cucchiarini, L., Dachà, M., ... Mondello, L. (2013). Electronic nose and GC-MS analysis of volatile compounds in Tuber magnatum Pico: Evaluation of different storage conditions. *Food Chemistry*, *136*(2), 668–674. doi:10.1016/j.foodchem.2012.08.086 PMID:23122112

Peris, M., & Escuder-Gilabert, L. (2013). On-line monitoring of food fermentation processes using electronic noses and electronic tongues: A review. *Analytica Chimica Acta*, *804*, 29–36. doi:10.1016/j.aca.2013.09.048 PMID:24267060

Persaud, K. C., & Dodd, G. (1982). Analysis of discrimination mechanisms in the mammalian olfactory system using a model nose. *Nature*, *299*(5881), 352–355. doi:10.1038/299352a0 PMID:7110356

Qiu, S., & Wang, J. (2017). The prediction of food additives in the fruit juice based on electronic nose with chemometrics. *Food Chemistry*, *230*, 208–214. doi:10.1016/j.foodchem.2017.03.011 PMID:28407902

Rosenblith, W. A. (Ed.). (1961). *Sensory Communication*. New York: John Wiley.

Russo, M., Serra, D., Suraci, F., Di Sanzo, R., Fuda, S., & Postorino, S. (2014). The potential of e-nose aroma profiling for identifying the geographical origin of licorice (Glycyrrhiza glabra L.) roots. *Food Chemistry*, *165*, 467–474. doi:10.1016/j.foodchem.2014.05.142 PMID:25038700

Saevels, S., Lammertyn, J., Berna, A. Z., Veraverbeke, E. A., Di Natale, C., & Nicolaï, B. M. (2004). An electronic nose and a mass spectrometry-based electronic nose for assessing apple quality during shelf life. *Postharvest Biology and Technology*, *31*(1), 9–19. doi:10.1016/S0925-5214(03)00129-7

Sanaeifar, A., Mohtasebi, S.S., Ghasemi-Varnamkhasti, M., Ahmadi, H., (2016). Application of MOS based electronic nose for the prediction of banana quality properties. *Measurement: Journal of the International Measurement Confederation,* (82), 105-114.

Serby, M. J., & Chobar, K. L. (1992). *Science of Olfaction*. New York: Springer Verlag. doi:10.1007/978-1-4612-2836-3

Sharma, P., Ghosh, A., Tudu, B., Sabhapondit, S., Baruah, B. D., Tamuly, P., ... Bandyopadhyay, R. (2015). Monitoring the fermentation process of black tea using QCM sensor based electronic nose. *Sensors and Actuators. B, Chemical*, *219*, 146–157. doi:10.1016/j.snb.2015.05.013

Song, S., Yuan, L., Zhang, X., Hayat, K., Chen, H., Liu, F., ... Niu, Y. (2013). Rapid measuring and modelling flavour quality changes of oxidised chicken fat by electronic nose profiles through the partial least squares regression analysis. *Food Chemistry*, *141*(4), 4278–4288. doi:10.1016/j.foodchem.2013.07.009 PMID:23993616

Su, M., Zhang, B., Ye, Z., Chen, K., Guo, J., Gu, X., & Shen, J. (2013). Pulp volatiles measured by an electronic nose are related to harvest season, TSS concentration and TSS/TA ratio among 39 peaches and nectarines. *Scientia Horticulturae*, *150*, 146–153. doi:10.1016/j.scienta.2012.10.020

Tahri, K., Tiebe, C., Bougrini, M., Saidi, T., El Alami El Hassani, N., El Bari, N., ... Bouchikhi, B. (2015). Characterization and discrimination of saffron by multisensory systems, SPME-GC-MS and UV-Vis spectrophotometry. *Analytical Methods*, *7*(24), 10328–10338. doi:10.1039/C5AY01986A

Tran, P. D., Van de Walle, D., De Clercq, N., Winne, A. D., Kadow, D., Lieberei, R., ... Van Durme, J. (2015). Assessing cocoa aroma quality by multiple analytical approaches. *Food Research International*, *77*, 657–669. doi:10.1016/j.foodres.2015.09.019

Urbach, G., & Milne, T. (1988). The concentration of volatiles in pasteurized milk as a function of storage time and storage temperaturesa possible indicator of keeping quality. *Australian Journal of Dairy Technology*, *43*, 53–58.

Vallejo-Cordoba, B., & Nakai, S. (1994). Keeping-quality assessment of pasteurized milk by multivariate analysis of dynamic headspace gas chromatographic data. 1. Shelf life prediction by principal component regression. *Journal of Agricultural and Food Chemistry*, *42*(4), 989–993. doi:10.1021/jf00040a029

Wright, R. H. (1966). Odor and molecular vibration. *Nature*, *209*(5023), 571–572. doi:10.1038/209571a0 PMID:5921179

Wu, Z., Zhuang, B., Weng, P., & Zhang, X. (2016). Fermentation Quality Characteristics and Flavor Formation Changes during the Process of Pickled Wax Gourd in Eastern Zhejiang. *International Journal of Food Properties*, *19*(2), 409–419. doi:10.1080/10942912.2015.1027775

Young, R. C., Buttner, W. J., Linnell, B. R., & Ramesham, R. (2003). Electronic nose for space program applications. *Sensors and Actuators. B, Chemical*, *93*(1-3), 7–16. doi:10.1016/S0925-4005(03)00338-1 PMID:14584511

Zhang, H., Wang, J., & Ye, S. (2008). Predictions of acidity, soluble solids and firmness of pear using electronic nose technique. *Journal of Food Engineering*, *86*(3), 370–378. doi:10.1016/j.jfoodeng.2007.08.026

Chapter 8
Application of Electronic Nose Systems on Animal-Source Food:
An Overview

Ambra Rita Di Rosa
University of Messina, Italy

Francesco Leone
University of Messina, Italy

ABSTRACT

Electronic nose, which is designed to perceive artificially the odour-active molecules in a sample head-space, has seen an increased use in the food industry as a rapid and reliable tool for quality assessment, classification, and authentication of several food items. The use of chemometrics and pattern recognition methods, together with gas sensors, emerged to be a very powerful analytical approach. In this chapter, an overview of the recent achievements in the field of electronic nose applications on animal-source food is given. Moreover, the authors deal with the recent research trends to overcome the actual sensor shortcomings, including sensor fusion techniques and their applications to evaluate animal-source foods and novel electronic nose systems.

BACKGROUND

In order to grant an adequate level of quality and safety, all food products need to be subjected to several analysis steps. In addition, to compete effectively in the marketplace, food companies must produce foods that meet consumers' expectation of safe, nutritious and high-quality food products (Di Rosa et al., 2017). Moreover, attention must be paid to the government regulations and to the policies and standards of international organizations (Nielsen, 2010).

DOI: 10.4018/978-1-5225-3862-2.ch008

Generally, food quality is specified in terms of traceable origin, known chemical composition, commensurate physical properties, satisfactory sensory evaluation, safety and health safeguards with respect to microbiological and toxic contamination, and is influenced by the processing and storage (Borras et al., 2015). Once food has been authenticated, the main technique for quality assessment, from the consumer point of view, is the sensory analysis, which is based on the evaluation of the attributes perceptible by the five sense organs, and allows establishing the organoleptic profile of the diverse products (Piana et al., 2004). This method, however, suffers from several disadvantages, being considered time-consuming and expensive, subjective (depending upon the professional acumen of the personnel involved) (Kiani et al., 2016), inconsistent and unpredictable, due to various human factors (Banerjee et al., 2016), and requires a panel of skilled assessors. Consequently, despite the extreme importance of aroma as an indicator of quality and product conformity, monitoring of food products is mainly performed via physicochemical measurements, due to the lack of reliable odour assessing instruments (Ampuero and Bosset, 2003). Conventional standard methods for gas analysis usually rely on the utilization of precision laboratory instruments like headspace gas chromatography (HGC) and two-dimensional gas chromatography using different types of detectors, for example, mass spectrometry coupled to olfactometry (Wardencki et al., 2009). These methods, providing an accurate approach for the analysis of type and concentration of the single component in a mixture of substances, may be considered a useful sensorial tool for food odour analysis. However, they are laborious and time-consuming, require considerable analytical skills, involve a lot of tedious and complex pre-treatment of samples, and use many hazardous organic reagents (Huang et al., 2015).

In the last two decades, much research has been performed to substitute the perception of human senses with artificial systems, providing signals related to the sensory attributes (Di Rosa et al., 2017). In this regard, advances in sensor technology, microelectronics and artificial intelligence led to the development of instruments such as the electronic nose (E-nose), capable of measuring and characterizing the odour of various products (Wilson and Baietto, 2009; Ghasemi-Varnamkhasti et al., 2011; Cubero et al., 2011). E-noses have been used in various research fields; however, the most attention has been paid to the food industry. Particularly, this technology has been employed for quality control of raw and manufactured products: process, freshness and maturity monitoring, shelf-life investigations, authenticity assessments of premium products and microbial pathogen detection (Schaller et al., 1998). In conclusion, the aim of this chapter is to give an overview of the most important, recent achievements in the field of E-nose applications, with regards to the assessment of animal source food products.

ANIMAL SOURCE FOOD

The use of animal source foods in human diet has a long history, going back at least 5 million years ago (Larsen, 2003). Nutritionally, they are energy-dense and an excellent source of protein (Neumann et al., 2002). The proteins in these foods are considered the highest quality available, as they contain a full complement of essential amino acids, in the right proportion, and are readily digested (Millward, 1999). Further, animal foods are an efficient source of micronutrients, offered in abundant and bioavailable form, such as calcium, iron, zinc, vitamin A, vitamin B-12 and riboflavin (Neumann et al., 2002). In addition, animal foods are in general more distinctive in flavour and texture, and often more palatable than foods of vegetable origin (Howe, 1950). For these reasons, the development of direct, rapid, precise, and accurate determination of several target compounds, with minimal or no sample preparation and reagent consumption, is an actual food industry need. Moreover, the traditional analysis are performed off-line by wet chemical assays, often involving enzymatic reactions or separation techniques (Peris and Escuder-Gilabert, 2013). In this sense, the E-nose technology is a promising approach. Table 1 summarizes the most important, recent contributions in this field.

CASE STUDIES FOR ANIMAL SOURCE FOODS E-NOSE ANALYSIS

Meat

Meat is one of the most consumed food product in the world. Food companies are constantly in seeking of novel evaluation techniques for rapid, real-time assessment of livestock, carcasses and meat products, in order to ensure high-quality meat (Ojha et al., 2016).

Although fresh meat is almost unscented, and the only odour detected by consumers is described as "bloody" (Khan et al., 2015), a complex series of reactions, leading to a number of compounds responsible for meat aroma, occur during cooking process. Maillard reaction, between reducing sugars and amino acids, and thermal degradation of lipid are the main routes to flavour compounds. Several hundred volatiles, derived from lipid degradation, have been found in cooked meat; however, aldehydes are believed to be the most useful in determining the specific aromas of different meat species (Mottram, 1998). Nurjuliana et al. (2011), for example, used an E-nose based on a single, uncoated, high quartz surface acoustic wave (SAW) sensor, to discriminate between pork and other types of meat. Principal component analysis (PCA) was applied for data interpretation. According to the authors, the volatile

Table 1. E-nose applications on animal source food

Category	Object of Investigation	E-Nose Model	Number of Sensors and Technology	Multivariable Data Analysis	References
Meat	Rapid identification of pork for halal authentication	zNose™	1 SAW	PCA	Nurjuliana et al., 2011
	Analysis of pork adulteration in minced mutton	PEN2	10 MOS	PCA, step-LDA, CDA, PLS, MLR, BPNN	Tian et al., 2013
	Monitoring aerobically packaged beef fillets spoilage	LibraNose	8 QMB	PCA, DFA, SVMs, SVR	Papadopoulou et al., 2013
	Characterization and differentiation of Italian dry-cured hams	PEN2	10 MOS	PCA	Laureati et al., 2014
	Rapid prediction of ochratoxin A-producing strains of *Penicillium*	ISE Nose 2000	12 MOS	DFA	Lippolis et al., 2016
Honey	Botanical origin identification and quality determination	FOX 4000	18 MOS	PCA, DFA, LS-SVMs	Huang et al., 2015
	Geographical origin identification of propolis	Heracles	GC-based	PCA	Cheng et al., 2013
Egg	Monitoring storage time and quality attributes	Laboratory-made	8 MOS	PCA, LDA, BPNN, GANN	Yongwei et al., 2009
	Prediction of TVB-N content	PEN3	10 MOS	PCA, SVR, BPNN	Liu and Tu, 2012
	Discriminating eggs from different poultry species	FOX 3000	18 MOS	PCA	Wang et al., 2014

continued on following page

Table 1. Continued

Category	Object of Investigation	E-Nose Model	Number of Sensors and Technology	Multivariable Data Analysis	References
Dairy products	Monitoring of rancidity of milk	Laboratory-made	5 MOS	PCA	Capone et al., 2001
	Milk shelf life determination	FOX 4000	18 MOS	PCA	Labreche et al., 2005
	Analysis of volatile flavour composition in milk	PEN2	10 MOS	PCA	Ai et al., 2015
	Effects of UV-C and HHP treatments on flavour of cow milk	iNose	14 MOS	PCA, LDA	Hu et al., 2016
	VOCs profile during milk fermentation	PEN3	10 MOS	PCA, LDA	Pan et al., 2014
	Classification of Pecorino cheeses	EOS 507	6 MOS	PCA, ANN	Cevoli et al., 2011
	Consumer acceptability of ovine cheese	EOS 835	6 MOS	PCA	Branciari et al., 2012
	Validation of aroma authenticity	S3	6 MOS	PCA	Sberveglieri, 2016
	Discriminate between cheese varieties	Laboratory-made	5 QMB	PCA, CA	Pais et al., 2012
	Butter type differentiation	PEN2	10 MOS	PCA, PLS-DA	Lorenzen et al., 2013
Fish	Study of grass carp quality predictive model	Laboratory-made	8 MOS	PCA, max-SNR	Guohua et al., 2012
	Freshness assessment	Laboratory-made	8 MOS	k-NN, SVMs	Guney and Atasoy, 2013
	Penaeus orientolis prawn freshness rapid determination	Laboratory-made	8 MOS	PCA, SR, DCSSR	Wei et al., 2015
	Rapid freshness analysis of mantis shrimps	Laboratory-made	8 MOS	PCA, SR	Jiang et al., 2016
	Rapid evaluation of fish freshness	Laboratory-made	9 colorimetric sensors	PCA, ANN	Huang et al., 2011
	Detection of off-flavours in catfish	AromaScan 32S	32 CP	PCA	Wilson et al., 2013
	Study of fish species	Laboratory-Made	8 MOS	Binary tree structured algorithm	Guney and Atasoy, 2015
	Species discrimination	FOX 4000	18 MOS	PCA, DFA	Zhang et al., 2012
	Water quality classification and monitoring	PEN3	10 MOS	LDA	Adnan et al., 2014
	Classification of fish sauces	Laboratory-made	8 MOS	PCA, LVQNN	Nimsuk, 2014
	Effect of fermentation by *Aspergillus oryzae* on properties of fish sauce	PEN3	10 MOS	Loading analysis	Sun et al., 2016
	Influence of heating temperature on volatile compounds	PEN 3	10 MOS	PCA, LDA	Sun et al., 2013

profile of pork contained a higher proportion of heptanal, which was considered the major compound influencing the discrimination. The ability of the E-nose in the identification of different meat species is relevant to consumers for several reasons. The fallout from fraudulent substitution will possibly lead to economic losses, jeopardize the health of consumers, who may have specific food allergies, and emotional disturbance due to religious beliefs. Other authors, aiming to detect meat adulteration, used a portable E-nose to predict pork content in minced mutton (Tian et al., 2013). The E-nose showed an excellent prediction ability, with R^2 higher than 0.9092 and RMSE lower than 10.94%.

Due to improper handling and storage, raw meat perish extremely fast. Enzymes and bacteria will decompose the main ingredients, producing many volatile organic compounds (VOCs), which are responsible for off-flavours and product rejection. At the same time, meat colour becomes dark; texture turns to obscure and elasticity declines during bacterial spoiling process (Huang et al., 2014). Moreover, consumption of spoiled meat could cause serious health hazards (Gram et al., 2002). E-nose is one of the approaches being looked into for "sniffing" out the VOCs emitted from food products due to the action of microorganisms (Balasubramanian et al., 2016). Papadopoulou et al. (2013) evaluated the performance of an E-nose in monitoring aerobically packaged beef fillet spoilage at different storage temperatures (0, 4, 8, 12, and 16 °C). Qualitative interpretation of the E-nose data was based on samples discrimination in three quality classes, namely fresh, semi-fresh and spoiled, using support vector machines (SVMs) classification and regression models.

Along with cooking for immediate consumption, meat can be processed for long-term preservation. Some of the most well-known meat products are those manufactured using pork meat, such as dry-cured hams. Parma, San Daniele and Toscano are the main consortia for the production of Italian dry-cured hams, labelled with Protected Designation of Origin (PDO). Laureati et al. (2014) evaluated the aromatic profile of the three products using a portable E-nose, based on 10 metal-oxide semiconductor (MOS) sensors. The experimental results showed that the volatile fraction of San Daniele ham was perceived by W1C, W3C and W5C sensors, which are sensitive to aliphatic compounds, aromatic compounds and slightly polar compounds. In agreement, alcohols compose the 40% of the total volatile fraction of San Daniele ham. Moreover, the odour fingerprint of Parma ham, characterized by esters, present in a much higher extent than in San Daniele ham (Gaspardo et al., 2008), and by sulphur compounds (Hinrichsen and Pedersen, 1995), was perceived by W1W and W2W, sensitive to sulphur-organic compounds.

E-nose technology have also been used to detect ochratoxigenic species during the seasoning processes for dry-cured meats (Lippolis et al., 2016). Although the fungal colonization contributes to improvement of the quality, some uncontrolled fungal development may also occur, causing either undesirable alteration and/or synthesis of mycotoxins, such as Ochratoxin A (OTA). OTA is a strong nephrotoxic agent and it has been shown to be teratogenic, mutagenic, hepatotoxic and immunosuppressive to animal species (Sorrenti et al., 2013). Therefore, it is crucial to prevent and monitor possible contamination of meat, to avoid undesirable negative economic impact and a potential health hazard to consumers. Lippolis et al. (2016) successfully applied an E-nose, coupled with discriminant function analysis (DFA), for rapid prediction of OTA-producing and non-producing *Penicillium* strains, with an overall recognition percentage of 73%.

Finally, several attempts to apply gas sensor technology for the detection of boar taint have been reported (Vestergaard et al., 2006; Haugen, 2006; Ampuero and Bee, 2006). Boar taint is an unpleasant odour perceived in pork during cooking and eating, described as "animal", "urine", "faecal" and/or "sweet" (Gunn et al., 2004). The results from these studies show significant correlation between sensor signals and levels of skatole and androstenone, which are considered the main responsible of boar taint (Andresen, 2006). However, there is still a need for research and development in this field in order to end up with a successful application.

Honey

Honey is defined as the natural sweet substance produced by *Apis mellifera* bees from the nectar of plants, from secretions of living parts of plants or from excretions of plant-sucking insects on plants. Bees collect it, transform it by combining with specific substances of their own, deposit it, dehydrate it, store it and leave it in honeycombs to ripen and mature (Spiteri et al., 2015). Honey is a very nutritious food product, which contains water, saccharides, amino acids, minerals, vitamins, enzymes, phenols, organic acids, pigments, volatile oils and over 100 varieties of aromatic substances (Gan et al., 2016). Its composition depends on several factors, such as floral source used to collect the nectar, geographical origin, climatic conditions and seasonality. This high variability highlights the importance of honey authenticity, a fundamental requirement for all food products that can be easily adulterated (Perna et al., 2012). Moreover, monofloral honeys have a higher commercial value than multifloral; therefore, it is important to be able to identify its botanical origin (Oroian et al., 2015).

For honey floral classification, traditionally a melissopalynology analysis is used. This method consists of counting down the number of pollens grains of a honey sample and calculating the respective percentages of nectariferous pollens. These are then used to identify the botanical origin and the overall pollen spectrum, which may allow determining the geographical origin of honeys. This technique is quite laborious, time-consuming and requires a high-skilled and trained technician (Dias et al., 2015). Another major limitation of this method is fraud addition of pollen for authenticity (Tiwari et al., 2013). Analytical and quantitative methods such as high-performance liquid chromatography and high-performance anion-exchange chromatography are also routinely performed (Wang et al., 2010; Cozzolino et al., 2011).

The sensory properties are also very important for honey characterization, in order to distinguish and assess the declared botanical origin (de Sousa et al., 2016). The colours of the honeys defined by visual observations are noticeably different, and they range from almost colourless to dark brown. Further, odour, taste and aroma are characteristic for each individual honey type (Bertoncelj et al., 2011). This allows the botanical origin of a honey to be identified, thus establishing the quality criteria, as well as revealing and quantifying certain defects (Piana et al., 2004). However, there is a need for detection methods that are non-invasive, faster, more manageable and suitable for wide-scale sample screening. In the past decade, E-noses have seen increased use, and coupled with chemometrics have been shown to be useful tools for honey classification (Ulloa et al., 2013). Recently, Huang et al. (2015) employed a system equipped with 18 MOS sensors for identifying the botanical origin of different honey samples. The authors established a discriminant model, based on least square support vector machines (LS-SVMs), which showed 100% overall accuracy.

Moreover, gas sensor technology was applied on other honeybee products. Propolis, for example, is a glue-like substance, composed of plant resins, bee waxes and pollens, used by bees to protect their hives. Since various biological activities of propolis were found, it is used as a healthy food (Chang et al., 2002). Cheng et al. (2013), taking into account the volatiles composition, were able to distinguish between propolis coming from different geographical areas.

Egg

Eggs are considered tasty and wholesome nutritious food, being a rich and balanced source of essential amino and fatty acids, as well of minerals and vitamins (Surai and Sparks, 2001). In addition, food companies are interested in high-quality egg products in a liquefied form, extensively used for pasta and bakery products preparation (Suman et al., 2007), due to unique functional proprieties such as gelling and foaming (Karoui et al., 2006). Freshness makes a major contribution to the quality of eggs and for this reason; the non-destructive detection of storage periods has been considered an important aspect (Karoui et al., 2006).

In recent years, many papers concerning with egg quality inspection appeared in literature. Computer vision (CV) have been used to detect dirty shells (excretions and blood), broken shells and odd sharps (Di Rosa et al., 2017). At the same time, some researchers attempted to identify volatile components that contribute to the egg flavour and aroma, working with different extraction and analytical techniques (steam distillation, solvent extraction, purge and trap, etc.) (Yongwei et al., 2009). Several aldehydes, aromatic compounds and sulphur compounds were identified in greatest concentrations (Warren et al., 1995). In particular, methylsulphide compounds are closely related to deterioration and perception of unacceptable odours in whole eggs (Brown et al., 1986).

An alternative strategy to sense the global profile of organic volatiles emitted by eggs may potentially be achieved using artificial olfactory systems. Yongwei et al. (2009), for example, used an E-nose to monitor egg storage time under cool (4°C) and room temperature. According to the authors, the volatile fraction changed quickly during room-temperature storage, while, under cool storage condition, the difference in volatiles between fresh and 1-week old eggs was not obvious, causing overlapped clusters by both PCA and LDA (linear discriminant analysis). An important freshness index of egg is total volatile basic nitrogen (TVB-N) content, which includes measurement of trimethylamine, dimethylamine and other compounds associated with the spoilage process (Ocaño-Higuera et al., 2011). Liu and Tu (2012) successfully employed a portable E-nose, coupled with support vector regression (SVR), for predicting TVB-N content in eggs. The experimental results showed the average prediction accuracy, RMSE and MRE (Mean Relative Error) to be 94.62%, 0.09% and 0.05%, respectively. Hence, the E-nose can be considered an effective tool for monitoring egg freshness.

Moreover, a variety of eggs from different species of birds is commercially available in different parts of the world, from the petite quail eggs to the very large ostrich eggs (Sinanoglou et al., 2011). From these different varieties, various odour characteristics and chemical proprieties, such as fatty acids composition, can be obtained. However, it is difficult for consumers to identify counterfeit eggs using common sense and experiences, suggesting that the identification of volatile compounds might be an effective way to distinguish different eggs (Wang et al., 2014). In agreement, Wang et al. (2014) achieved a clear separation between different species of avian eggs using an E-nose device equipped with 18 MOS sensors and coupled with PCA.

Dairy Products

Fresh, good-quality milk is characterized by three features: a characteristic, subtle, and delicate flavour; a pleasant mouth-feel, since it is an emulsion of fat globules dispersed in a colloidal aqueous solution; and a slightly sweet and salty taste, due to the presence of lactose and salts (d'Acampora Zellner et al., 2011). The analysis of milk aroma is an especially complex problem, as its heterogeneous nature makes difficult to isolate flavour-active chemicals, based on general proprieties like polarity or volatility (Capone et al., 2001). In this sense, the E-nose, relying on a non-separative working mechanism, can be seen as a powerful tool for milk quality evaluation. Capone et al. (2001), for example, developed an array of five SnO_2 gas sensors, in order to obtain the selective discrimination of UHT (ultra-high temperature) and pasteurised milk, at different rancidity levels. A dynamic PCA analysis allowed achieving a correlation among spoilage and ageing days. Further, Labreche et al. (2005) showed that the E-nose measurements could be used to detect bacteria growth in milk and to determine its shelf life; therefore, gas sensor devices can be used for milk freshness control. E-nose technology have also been used to analyse the volatile flavour composition of raw whole and skim milk, according to their different fat contents (Ai et al., 2015).

In addition, as almost all milk types, being highly perishable goods, are submitted to heat treatments to eliminate potential pathogens and spoilage organisms, but producing thermally derived off-flavour compounds, its volatiles evaluation is routinely performed. Novel non-thermal technologies, such as high hydrostatic pressure (HHP) and UV-C radiation, have shown the potential to replace, at least partially, the traditional preservation processes. Hu et al. (2016) used an E-nose to distinguish the flavour attributes and compare the flavour changes of fresh cow milk after single and multiple-cycle HHP and UV-C treatments.

Milk; however, is also a raw material used in the manufacture of several food products (Gomes, 2016). Traditionally, fermentation was primarily applied to improve and prolong food preservation. Moreover, the chemical transformation of lactose, citrate, milk fat, and proteins in dairy food raw materials, contribute to flavour development (d'Acampora Zellner et al., 2011). Pan et al. (2014) employed an E-nose for monitoring the VOCs emitted during milk fermentation by *Lactobacillus pentosus.* Anyway, one of the most widely studied fermented products is cheese, which has more than 500 varieties (Singh et al., 2003). Cheese manufacture consists on the addition of salt, an enzyme coagulant and a culture of a starter organism to milk (d'Acampora Zellner et al., 2011). Moreover, ripening or maturation during storage is responsible to large biochemical, microbial, chemical, and physical changes, leading to a great number of flavour compounds (Gomes, 2016). So far, E-noses have been massively used for investigating the volatile composition of various cheeses. Cevoli et al. (2011) used an E-nose based on an array of six MOS sensors, jointly with artificial neural networks (ANNs), to classify Pecorino cheeses according to their ripening time and manufacturing techniques. In addition, a comparison with gas chromatography-mass spectrometry (GC-MS) analysis of volatile compounds was carried out, resulting in a higher classification capability showed by the coupling of E-nose and ANNs. Later on, Branciari et al. (2012) studied the consumer acceptability and the composition of Pecorino cheese obtained from ewes fed an extruded linseed-enriched diet. The results showed that the diet affected cheese composition after 60 days of ripening; in particular, linseed supplementation decreased protein and fat content while it increased the concentration of monounsaturated, polyunsaturated and long-chain fatty acids. A variation of the aromatic properties, however, might represent an obstacle for consumer acceptability even though there is an ever-higher concern for healthy products. In this regard, an E-nose was used to find any differences among the three types of cheese. The lack of clustering into the PCA plot, according to the different milk they were produced with, shows that the E-nose was not able to recognize any remarkable difference in the headspaces of the analysed samples. Results from this study suggest that the enriched cheese acceptability was similar to the control and that extruded linseed supplementation seems a possible strategy for improving the nutritional quality of lipids in Pecorino cheese without affecting its sensory properties. In addition, Sberveglieri (2016), employing E-nose and SPME-GC/MS technologies, was able to identify specific markers of Parmigiano Reggiano cheese aroma, which were then used to determine the percent of grated pulp and rind of commercially utilized products. "Parmigiano Reggiano" is one of the oldest traditional cheeses produced in Europe and still one of the most esteemed PDO Italian cheeses. For any food product, to gain the PDO title, the link between "authentic and unvarying local methods" and "tradition" is dramatically important, implying distinctive sensory characteristics strongly connected to the traditional production methods. For this reason, the grating operation must take place in the defined geographical area of the cheese production, and packaging must start immediately. As grated cheese has to be characterized by no more than 18% crust in its moisture, the typical fraud that can affect grated Parmigiano Reggiano cheese is an excess in crust. Therefore, the development of a rapid and non-destructive method for the determination of authenticity and ensure compliance with the regulations, is a priority. The author, using a thin films/nanowires MOS based gas sensor array, distinguished the percentage of grated Parmigiano Reggiano with good accuracy, even when very close to law limits. On the contrary, Renna et al. (2012) reported that the use of a MOS sensors-based E-nose did not provided helpful information regarding the seasonal variations in Asiago d'Allevo PDO cheese, producing an unsatisfactory classification rate of the collected samples. Further, an E-nose based on coated piezoelectric quartz crystals was used to distinguish cheese made from ewes' milk, and to distinguish cheese varieties. Two sensors coated with Nafion and Carbowax could certify half the ewes' cheese samples, exclude 32 cheeses made from cow's milk and classify half of the ewes' cheese samples as possibly authentic. Two other sensors, coated with polyvinylpyrrolidone and triethanolamine clearly distinguished between Flamengo, Brie, Gruyère and Mozzarella cheeses. Brie cheeses were further separated according to their origin, and Mozzarella grated cheese also appeared clearly separated from non-grated Mozzarella (Pais et al., 2012).

Finally, Lorenzen et al. (2013) developed a fast and non-destructive method for butter type differentiation using an E-nose equipped with 10 MOS sensors.

Fish

Fishery products are vital from a nutritional point of view, and the fishery industry is an important economic source related to the international trades for many coastal states and districts. Fish, however, is also one of the most vulnerable and perishable food items, and paying attention to its quality changes it is particularly important. Freshness is one of the utmost principal attributes of fish quality, which makes a major contribution to the quality of fish and fishery products. It is well-known that the internal and intrinsic characteristics (fragile muscle tissue and activity of endogenous protease), and inappropriate handling methods and storage conditions (time-temperature history, chilling or freezing, and refrigeration and preservation condition) can be prone to resulting in physical, chemical, biochemical and microbial changes and contaminations in fish, thus affecting its freshness quality (Cheng et al., 2015).

Conventionally, freshness assessment has been based on organoleptic tests by humans, including evaluating the appearance of skin and eyes, testing the flesh firmness and elasticity, and smelling the odour (Azeriee et al., 2009). The human array evaluation is an immediate assessment of quality and can be done with no damage to fish. On the other hand, it is not easy to keep evaluation results consistent among different assessments, because the evaluation score is subjective. Other methods, including microbiological measurement and chemical methods, are limited in the laboratory and not convenient on-site because they are time consuming and destructive detection (Huang et al., 2015).

Among different instrumental and sensory methods developed for fish freshness assessment, E-nose based methods are an attractive choice. A variety of smell sensors, coupled with different features extraction and data processing methods, have been employed for olfactory detection or freshness assessment of fishery products. Guohua et al. (2012), for example, used a self-developed, portable E-nose for monitoring the changes in the volatile composition of grass carp (*C. idellus*), generated by microbial propagation. Moreover, Guney and Atasoy (2013) evaluated fish freshness by means of E-nose, combining SVMs and k-NN (k-nearest neighbours) algorithms in a decision tree structure. Prawn and shrimp quality evaluation was also performed (Wei et al., 2015; Jiang et al., 2016). E-nose responses were processed with different multivariate statistical methods, leading to freshness predictive models with 94.29% and 91.67% accuracies, respectively. A new concept E-nose, based on colorimetric sensor arrays, transforming olfactory information into visual information by colour's change after pigments reacted with volatile compounds, was developed to determine fish quality (Huang et al., 2011). Compared with traditional E-noses, this technology has the advantages of high precision, wide range and it is not influenced by humidity (Huang et al., 2009).

Later on, an E-nose based on conducting polymer sensors, was used to detect the presence of malodorous off-flavours compounds in catfish meat fillets (Wilson et al., 2013). Total volatile basic amines (TVBAs) are among the most widely used chemical indicators of seafood quality. These include trimethylamine and dimethylamine produced by the autolytic enzymes of spoilage bacteria during frozen storage, and ammonia produced by the deamination of amino acids and nucleotide catabolites (Huss, 1995).

Furthermore, several authors reported the application of E-nose technology to discriminate different fish species. Guney and Atasoy (2015) distinguished between horse mackerel, anchovy and whiting, while Zhang et al. (2012) were able to recognize three different kinds of puffer fishes.

Since the wild fish production stagnated, particularly in the developing countries, aquaculture has turn out to be very important. Productivity of aquaculture farms rely on water quality; therefore, a proper monitoring is essential for any thriving operation (Adnan et al., 2014). Adnan et al. (2014) showed the suitability of an E-nose, combined with LDA, for the rapid classification of fresh, brackish and unclear water samples, usually used in aquaculture farming.

Gas sensors have also been used to study the aroma profile of different fish products, such as the popular fish sauce, manufactured from the fermentation of fish with salt (Nimsuk, 2014). Briefly, fish sauce is an amber-coloured translucent liquid, with a fishy flavour and salty taste, commonly used as a condiment in various cuisines, especially in Southeast Asia (Lopetcharat et al., 2001; Ijong and Ohta, 1995; Jiang et al., 2007; Zarei et al., 2012). Since Asian food has become popular worldwide, and fish sauce is essentially needed as a condiment in these oriental dishes, it is internationally traded nowadays (Nimsuk, 2014). Therefore, the monitoring of the quality of fish sauce products has become a necessity. An E-nose, together with ANNs, was employed to identify pure and mixed fish sauces, according to their total nitrogen content (Nimsuk, 2014). Furthermore, the effect of the fermentation by *Aspergillus oryzae* on the biochemical and sensory properties of fish sauce was investigated, revealing that the odour of the fish sauce was more balanced and favourable after fermentation (Sun et al., 2016).

Finally, E-nose technology was used to explore the relationship between different processing temperatures and bigeye tuna volatile composition (Sun et al., 2013). In particular, the E-nose, coupled with PCA and LDA, was able to discriminate the tuna samples into three distinct groups: untreated, 70-120°C and 150°C heat-treated. Based on the analysis of the E-nose, the untreated samples and samples heated to 100°C and 150°C were further analysed using SPME-GC/MS. Compounds that contribute the fresh flavour of bigeye tuna meats were found in the untreated sample group. However, as heating temperature increased, the relative amount of aldehydes, ketones and alcohols decreased rapidly, while hydrocarbons and heterocyclic compounds concentration increased.

FUTURE TRENDS IN ANIMAL SOURCE FOOD E-NOSE ANALYSIS

Data Fusion

Sometimes, because of the high complexity of animal source food aroma, the use of an E-nose alone could result insufficient to achieve the aimed purpose. For this reason, multi-sensor data fusion technique, combining the outputs of multiple instrumental sources, emerged as a powerful tool for improving the quality assessment and authentication of food products (Di Rosa et al., 2017). Generally, the fusion of data from complementary sensors, responding to different signature phenomena, may increase the probability of correct classification; therefore, characteristics that are not classified by one sensor may be apparent or measured by another (Banerjee et al., 2016).

Fusion processes have been categorized in a three-level model, distinguishing low, intermediate, and high level fusion (Elmenreich, 2002). In the low-level fusion, data from all sources are simply concatenated sample-wise into a single matrix that has many rows as samples and as many columns as signals obtained by the different instruments (Borras et al., 2015). In order to fuse raw sensor data, they must be commensurate, that is, observations of similar physical quantities must be considered (Hall and Llinas, 1997). Intermediate or mid-level fusion first extracts some relevant features from each data source separately, and then, concatenates them into a single array that is used for multivariate classification and regression. Hence, low and medium level data fusion, both join the data sources at data level. On the contrary, in high-level data fusion, a separate model is built for each available data set, and then, the model responses are combined to produce a final "fused" response. (Doeswijk et al., 2011). Therefore, low-level fusion is considered conceptually simple (Borras et al., 2015) and believed as the most accurate technique (Hall and Llinas, 1997), due to the fact that the originality of information from each sensor is maintained and used in further processes (Masnan et al., 2012). Some limitations are a high volume data and the possible predominance of one data source over the others. This is partially overcome by mid-level fusion, where feature extraction significantly reduces the data dimensionality (Borras et al., 2015). This approach appear very useful for non-commensurate type data, *i.e.*, if sensors are looking for different physical parameters (Banjeree et al., 2016). High-level fusion, finally, allows focusing on each individual techniques, but the final identity declaration is obtained from only a few values. One advantage is that every individual matrix is treated independently, and the results from inefficient techniques do not worsen the overall performance as much as in the other fusion levels.

So far, data fusion has been applied to a wide range of food and beverages, enhancing significantly the performance of classification or quality evaluation of the products under consideration (Borras et al., 2015). Huang et al. (2014) attempted the measurement of TVB-N content in pork meat, combining E-nose with NIR spectroscopy and CV. The intention of the authors was to take advantage of the three techniques, one performing a local measurement of inner chemical properties, the others performing a global assessment of colour, texture and smell, and to combine those types of measurements to improve the freshness evaluation. In this work, PCA was implemented on the characteristic variables from single sensors and fusion of two or three sensors data, then, the top PCs were extracted as the input of a BP-ANN model. According to the authors, the result obtained with the model integrating the three techniques showed its own superiority, when compared with single technique models.

Multi-sensor data fusion, involving E-nose measurements, have also been used to authenticate honey. Zakaria et al. (2011) found the E-nose and the E-tongue to be ineffective in the discrimination of different honey samples. However, by applying data fusion, the classification of honey with different floral origin, sugar syrups and adulterated samples was achieved. In this experiment, PCA and LDA were used to perform the low-level fusion, while PNN was employed to verify and validate the fusion of data. Later on, Maamor et al. (2014) successfully applied a combination of E-nose, E-tongue and FT-IR spectroscopy to distinguish between pure and adulterated honey samples. PCA was prior performed to reduce high dimensional features, then, several methods have been chosen for data analysis. However, it can be clearly concluded, from the results obtained, that k-NN method outperformed the other classification techniques. Subari et al. (2014) were able to estimate the concentration of honey purity in adulterated honey solution, integrating an E-nose with FT-IR spectroscopy. The information provided by the two instruments were fused together to produce a new set of data, which was employed as input for ANN learning. Finally, a comparison between data from single modality and fusion methods, to classify Tualang honey as pure or adulterated, was carried out, resulting in a better performance by combined E-nose and FT-IR spectroscopy (Subari et al., 2012). On the other hand, the E-nose resulted uneffective for the identification of Sicilian honeys botanical origin, both in single modality and adding E-tongue and computer vision data (Di Rosa et al., 2018).

The suitability of a multi-sensor system in the determination of brand and aging time of pasteurised milk was also investigated (Bougrini et al., 2014). In this experiment, a hybrid E-nose and a voltammetric E-tongue were fused at a mid-level of abstraction, outperforming the classification results obtained by the two instruments taken individually.

Moreover, Han et al. (2014) developed a new method to detect fish freshness by combining E-nose and E-tongue, in conjunction with chemometric methods. Collected data were analysed by PCA, and a different number of PCs was used for the construction of RBNN models. Thus, the performances of the different models were compared. The experimental results revealed that the best RBFNN model was acquired at seven PCs of E-nose data, with an optimal performance of 87.9% and 80.0% in the training set and prediction set respectively. At the same time, the best RBFNN model of E-tongue data analysis was at 5 PCs with an optimal performance of 86.3% in the training set and 81.8% in prediction set. Another RBFNN model was built with the combination of E-nose and E-tongue. The result shows that the discrimination rates improved to 94.0% and 93.9% in the training set and prediction set, respectively.

Anyway, despite the promising results obtained so far, data fusion strategies are still object of investigations. Borras et al. (2005) identified several future trends, including combination of low and mid-level data fusion, fusion of second and higher-order data, *i.e.*, hyperspectral images or data from hyphenated techniques, and the combination of first order-data from evolving systems, *i.e.*, on-line monitored systems.

Hybrid E-Noses, Bioelectronics Noses and "Electronic Mucosa"

Although the E-Nose technology offers various advantages over traditional analysis, such as the reduction in time and cost of analysis, and its environmentally friendly nature (Smyth and Cozzolino, 2013); the sensors still present a number of weak points. These includes the noise by major compounds, which may not be relevant to aroma (Rodriguez-Mendez et al., 2004); the sensor drift or poisoning (Loutfi et al., 2015), and the responses obtained by each sensor, which provide highly ambiguous qualitative or quantitative information, resulting in a sort of "fingerprint" of the sample (Ampuero and Bosset, 2003). Another aspect to consider is the humidity of the headspace above a food sample. With sensors that are reactive to the presence of water, this may be an interfering factor, as the water vapour concentration in the headspace is normally orders of magnitude higher than the concentration of the aroma compounds, whose presence one wishes to detect (Visser and Taylor, 1998).

Recent trends to overcome sensor shortcomings includes the development of hybrid E-noses, based on the combination of different sensor technologies. The most common utilization of hybrid systems involve MOS and MOSFET, however, commercial system are available that use MOS, MOSFET and QCM in various combinations. The major advantage of this type of electronic nose is that it incorporates the advantages inherent to the different transducer technologies, at the same time giving a choice for which chemical sensors can be used (Kiani et al., 2016).

However, even with technological advances and promising results, these sensors still cannot mimic the biological features of human counterparts. On the contrary, bioelectronic systems can provide signals triggered by the binding activities between selective receptors and odorous compounds. Recently, bioelectronic noses, using human olfactory receptors as primary recognition elements and nanomaterials as secondary transducers, have been reported (Yoon et al., 2009; Kim et al., 2008, Lee et al., 2012a; Jin et al., 2012; Park et al., 2012a; Lee et al., 2012b). Bioelectronic noses rapidly recognize target molecules and can be operated relatively simply. In addition, they can be easily integrated and multiplexed into small chips that enable the systems to be portable and, thus, suitable for on-site analysis (Son et al., 2017). Beyond human receptors, some efforts to design bioelectronic noses have used olfactory receptors of various mammals, such as dogs and mice, which perform better than humans in smell sensing (Park et al., 2012b; Goldsmith et al., 2011).

Nevertheless, electronic mucosa, which is an improvement of the E-nose technology, was developed. In this instrument, the sensor arrays have been combined with two retentive columns (mimicking two nasal paths) in order to add another dimension to the common E-nose, using temporal information from sensors. The retentive columns, coated with an absorbent material, acts like the mucous layer in the nasal cavity by delaying selectively some odour molecules, while letting other pass through with minimal delay. This provides differential temporal data, which are complementary to the conventional response (Ghasemi-Varnamkhasti and Aghbashlo, 2014).

CONCLUSION

In the last few decades, much research has been performed to mimic the olfactory perception of humans with artificial systems. Advances in sensor technology, microelectronics and artificial intelligence led to the development of devices such as the E-nose, capable of measuring and characterizing the complex odour of various products. Particularly, concerning animal source food quality control and authentication, this technology has been widely used in the recent years. The E-nose emerged as a rapid, precise, and accurate approach for the determination of volatile compounds, with minimal or no sample preparation and reagent consumption, being also suitable for real-time monitoring of several processes. Anyway, despite the promising results obtained so far, there is still room for improvement. The fusion of E-nose data with signals provided by other instrumental sources, as well as novel gas sensing technologies, are being studied nowadays. Therefore, keep exploring this area could results in successfully applications towards the next years.

REFERENCES

Adnan, K. N. A. K., Yusuf, N., Maamor, H. N., Rashid, F. N. A., Ismail, S. W. M., Thriumani, R., . . . Ahmad, M. N. (2014, August). *Water quality classification and monitoring using e-nose and e-tongue in aquaculture farming*. Paper presented at the 2nd International Conference on Electronic Design, Penang, Malaysia. doi:10.1109/ICED.2014.7015827

Ai, N., Liu, H., Wang, J., Zhang, X., Zhang, H., Chen, H., ... Sun, B. (2015). Triple-channel comparative analysis of volatile flavour composition in raw whole and skim milk via electronic nose, GC-MS and GC-O. *Analytical Methods*, *7*(10), 4278–4284. doi:10.1039/C4AY02751E

Ampuero, S., & Bee, G. (2006). The potential to detect boar tainted carcasses by using an electronic nose based on mass spectrometry. *Acta Veterinaria Scandinavica*, *48*(Suppl 1), 1. doi:10.1186/1751-0147-48-S1-P1

Ampuero, S., & Bosset, J. O. (2003). The electronic nose applied to dairy products: A review. *Sensors and Actuators. B, Chemical*, *94*(1), 1–12. doi:10.1016/S0925-4005(03)00321-6

Andresen, O. (2006). Boar taint related compounds: Androstenone/skatole/other substances. *Acta Veterinaria Scandinavica*, *48*(Suppl 1), S5. doi:10.1186/1751-0147-48-S1-S5

Azeriee, A. A., Hashim, H., Jarmin, R., & Ahmad, A. (2009, March). *A study on freshness of fish by using fish freshness meter*. Paper presented at the 5th International Colloquim o Signal Processing and Its Applications, Kuala Lumpur, Malaysia.

Balasubramanian, S., Amamcharla, J., & Shin, J.-E. (2016). Possible application of electronic nose systems for meat safety: An Overview. In M. L. Rodriguez-Mendez (Ed.), *Electronic Noses and Tongues in Food Science*. London, UK: Academic Press. doi:10.1016/B978-0-12-800243-8.00007-X

Banerjee, R., Tudu, B., Bandyopadhyay, R., & Bhattacharyya, N. (2016). A review on combined odor and taste sensor systems. *Journal of Food Engineering*, *190*, 10–21. doi:10.1016/j.jfoodeng.2016.06.001

Bertoncelj, J., Golob, T., Kropf, U., & Korosec, M. (2011). Characterisation of Slovenian honeys on the basis of sensory and physicochemical analysis with a chemometric approach. *International Journal of Food Science & Technology*, *46*(8), 1661–1671. doi:10.1111/j.1365-2621.2011.02664.x

Borras, E., Ferré, J., Boqué, R., Mestres, M., Acena, L., & Busto, O. (2015). Data fusion methodologies for food and beverage authentication and quality assessment - A review. *Analytica Chimica Acta*, *891*, 1–14. doi:10.1016/j.aca.2015.04.042 PMID:26388360

Bougrini, M., Tahri, K., Haddi, Z., El Bari, N., Llobet, E., Jaffrezic-Renault, N., & Bouchikhi, B. (2014). Aging time and brand determination of pasteurized milk using a multisensor e-nose combined with a voltammetric e-tongue. *Materials Science and Engineering C*, *45*, 348–358. doi:10.1016/j.msec.2014.09.030 PMID:25491839

Branciari, R., Valiani, A., Trabalza-Marinucci, M., Miraglia, D., Ranucci, D., Acuti, G., ... Mughetti, L. (2012). Consumer acceptability of ovine cheese from ewes fed extruded linseed-enriched diets. *Small Ruminant Research*, *106*, 43–48. doi:10.1016/j.smallrumres.2012.04.032

Brown, M. L., Holbrook, D. M., Hoerning, E. F., Legendre, M. G., & St. Angelo, A. J. (1986). Volatile indicators of deterioration in liquid egg products. *Poultry Science*, *65*(10), 1925–1933. doi:10.3382/ps.0651925

Capone, S., Epifani, M., Quaranta, F., Siciliano, P., Taurino, A., & Vasanelli, L. (2001). Monitoring of rancidity of milk by means of an electronic nose and a dynamic PCA analysis. *Sensors and Actuators. B, Chemical*, *78*(1-3), 174–179. doi:10.1016/S0925-4005(01)00809-7

Cevoli, C., Cerretani, L., Gori, A., Gallina Toschi, T., & Fabbri, A. (2011). Classification of Pecorino cheeses using electronic nose combined with artificial neural network and comparison with GC–MS analysis of volatile compounds. *Food Chemistry*, *129*(3), 1315–1319. doi:10.1016/j.foodchem.2011.05.126 PMID:25212373

Chang, C.-C., Yang, M.-H., Wen, H.-M., & Chern, J.-C. (2002). Estimation of total flavonoid content in propolis by two complementary colorimetric methods. *Journal of Food and Drug Analysis*, *10*(3), 178–182.

Cheng, H., Qin, Z. H., Guo, X. F., Hu, X. S., & Wu, J. H. (2013). Geographical origin identification of propolis using GC–MS and electronic nose combined with principal component analysis. *Food Research International*, *51*(2), 813–822. doi:10.1016/j.foodres.2013.01.053

Cheng, J., Sun, D.-W., Zeng, X.-A., & Liu, D. (2015). Recent advances in methods and techniques for freshness quality determination and evaluation of fish and fish fillets: A review. *Critical Reviews in Food Science and Nutrition*, *55*(7), 1012–1225. doi:10.1080/10408398.2013.769934 PMID:24915394

Cozzolino, D., Corbella, E., & Smyth, H. E. (2011). Quality control of honey using infrared spectroscopy - A review. *Applied Spectroscopy Reviews*, *46*(7), 523–538. doi:10.1080/05704928.2011.587857

Cubero, S., Aleixos, N., Moltó, E., Gómez-Sanchis, J., & Blasco, J. (2011). Advances in machine vision applications for automatic inspection and quality evaluation of fruits and vegetables. *Food and Bioprocess Technology*, *4*(4), 487–504. doi:10.1007/s11947-010-0411-8

d'Acampora Zellner, B., Dugo, P., Dugo, G., & Mondello, L. (2011). Milk and dairy foods. In L. M. L. Nollet & F. Toldrà (Eds.), *Sensory analysis of foods of animal origin*. Boca Raton, FL: CRC Press.

de Sousa, J. M. B., de Souza, E. L., Marques, G., Benassi, M. D. T., Gullon, B., Pintado, M. M., & Magnani, M. (2016). Sugar profile, physicochemical and sensory aspects of monofloral honeys produced by different stingless bee species in Brazilian semi-arid region. *Lebensmittel-Wissenschaft + Technologie*, *65*, 645–651. doi:10.1016/j.lwt.2015.08.058

Di Rosa, A. R., Leone, F., Cheli, F., & Chiofalo, V. (2017). Fusion of electronic nose, electronic tongue and computer vision for animal source food authentication and quality assessment – A review. *Journal of Food Engineering*, *210*, 62–75. doi:10.1016/j.jfoodeng.2017.04.024

Di Rosa, A. R., Leone, F., Scattareggia, C., & Chiofalo, V. (2018). Botanical origin identification of Sicilian honeys based on artificial senses and multi-sensor data fusion. *European Food Research and Technology*, *244*(1), 117–125. doi:10.1007/s00217-017-2945-8

Dias, L. G., Veloso, A. C. A., Sousa, M. E. B. C., Estevinho, L., Machado, A. A. S. C., & Peres, A. M. (2015). A novel approach for honey pollen profile assessment using an electronic tongue and chemometric tools. *Analytica Chimica Acta, 900*, 36–45. doi:10.1016/j.aca.2015.10.014 PMID:26572837

Doeswijk, T. G., Smilde, A. K., Hagemann, J. A., Westerhuis, J. A., & van Eeuwijk, F. A. (2011). On the increase of predictive performance with high-level data fusion. *Analytica Chimica Acta, 705*(1-2), 41–47. doi:10.1016/j.aca.2011.03.025 PMID:21962346

Elmenreich, W. (2002). *Sensor fusion in time-triggered systems*. Retrieved April 26, 2017, from https://mobile.aau.at/~welmenre/papers/elmenreich_Dissertation_sensorFusionInTimeTriggeredSystems.pdf

Gan, Z., Yang, Y., Li, J., Wen, X., Zhu, M., Jiang, Y., & Ni, Y. (2016). Using sensor and spectral analysis to classify botanical origin and determine adulteration of raw honey. *Journal of Food Engineering, 178*, 151–158. doi:10.1016/j.jfoodeng.2016.01.016

Gaspardo, B., Procida, G., Toso, B., & Stefanon, B. (2008). Determination of volatile compounds in San Daniele ham using headspace GC-MS. *Meat Science, 80*(2), 204–209. doi:10.1016/j.meatsci.2007.11.021 PMID:22063323

Ghasemi-Varnamkhasti, M., & Aghbashlo, M. (2014). Electronic nose and electronic mucosa as innovative instruments for real-time monitoring of food dryers. *Trends in Food Science & Technology, 38*(2), 158–166. doi:10.1016/j.tifs.2014.05.004

Ghasemi-Varnamkhasti, M., Mohtasebi, S. S., Rodriguez-Mendez, M. L., Lozano, J., Razavi, S. H., & Ahmadi, H. (2011). Potential application of electronic nose technology in brewery. *Trends in Food Science & Technology, 22*(4), 165–174. doi:10.1016/j.tifs.2010.12.005

Goldsmith, B. R., Mitala, J. J. Jr, Josue, J., Castro, A., Lerner, M. B., Bayburt, T. H., ... Johnson, C. (2011). Biomimetic chemical sensors using nanoelectronic readout of olfactory receptor proteins. *ACS Nano, 5*(7), 5408–5416. doi:10.1021/nn200489j PMID:21696137

Gomes, M. T. S. R. (2016). Electronic nose in dairy products. In M. L. Rodriguez-Mendez (Ed.), *Electronic noses and tongues in food science*. London, UK: Academic Press. doi:10.1016/B978-0-12-800243-8.00003-2

Gram, L., Ravn, L., Rasch, M., Bruhn, J. B., Christensen, A. B., & Givskov, M. (2002). Food spoilage - Interactions between food spoilage bacteria. *International Journal of Food Microbiology, 78*(1-2), 79–97. doi:10.1016/S0168-1605(02)00233-7 PMID:12222639

Guney, S., & Atasoy, A. (2013, July). *Fish freshness assessment by using electronic nose*. Paper presented at the 36th International Conference on Telecommunications and Signal Processing, Rome, Italy. doi:10.1109/TSP.2013.6614036

Guney, S., & Atasoy, A. (2015, July). *Discrimination of different fish species by E-nose*. Paper presented at the 38th International Conference on Telecommunications and Signal Processing, Prague, Czech Republic. doi:10.1109/TSP.2015.7296458

Gunn, M., Allen, P., Bonneau, M., Byrne, D. V., Cinotti, S., & Fredriksen, B. (2004). Welfare aspects of the castration of piglets. Scientific report on the scientific panel for animal health and welfare on a request from the Commission related to welfare aspects of the castration of piglets (Question No. EFSA-Q-2003-091). *EFSA Journal*, 91.

Guohua, H., Lvye, W., Yanhong, M., & Lingxia, Z. (2012). Study of grass carp (*Ctenopharyngodon idellus*) quality predictive model based on electronic nose. *Sensors and Actuators. B, Chemical*, *166-167*, 301–308. doi:10.1016/j.snb.2012.02.066

Hall, D. L., & Llinas, J. (1997). An introduction to multisensor data fusion. *Proceedings of the IEEE*, *85*(1), 6–23. doi:10.1109/5.554205

Han, F., Huang, X., Teye, E., Gu, F., & Gu, H. (2014). Nondestructive detection of fish freshness during its preservation by combining electronic nose and electronic tongue techniques in conjunction with chemometric analysis. *Analytical Methods*, *6*(2), 529–536. doi:10.1039/C3AY41579A

Haugen, J.-E. (2006). The use of chemical sensor array technology, the electronic nose, for detection of boar taint. *Acta Veterinaria Scandinavica*, *48*(Suppl 1), S15. doi:10.1186/1751-0147-48-S1-S15

Hinrichsen, L. L., & Pedersen, S. B. (1995). Relationship among flavour, volatile compounds, chemical changes, and microflora in Italian-type dry-cured ham during processing. *Journal of Agricultural and Food Chemistry*, *43*(11), 2932–2940. doi:10.1021/jf00059a030

Howe, P. E. (1950). Foods of animal origin. *Journal of the American Medical Association*, *143*(15), 1337–1342. doi:10.1001/jama.1950.82910500017008 PMID:15428269

Hu, G., Zheng, Y., Liu, Z., & Deng, Y. (2016). Effects of UV-C and single- and multiple-cycle high hydrostatic pressure treatments on flavour evolution of cow milk: Gas chromatography-mass spectrometry, electronic nose, and electronic tongue analysis. *International Journal of Food Properties*, *10*, 1–12.

Huang, L., Liu, H., Zhang, B., & Wu, D. (2015). Application of electronic nose with multivariate analysis and sensor selection for botanical identification and quality determination of honey. *Food and Bioprocess Technology*, *8*(2), 359–370. doi:10.1007/s11947-014-1407-6

Huang, L., Zhao, J., Chen, Q., & Zhang, Y. (2014). Nondestructive measurement of total volatile basic nitrogen (TVB-N) in pork meat by integrating near infrared spectroscopy, computer vision and electronic nose techniques. *Food Chemistry*, *145*, 228–236. doi:10.1016/j.foodchem.2013.06.073 PMID:24128472

Huang, X., He, Y., Chen, Z., & Hu, C. (2009). Colorimetric sensor for anion recognition based on the proton transfer signaling mechanism. *Chinese Journal of Chemistry*, *27*(8), 1526–1530. doi:10.1002/cjoc.200990257

Huang, X., Lv, R., Yao, L., Guan, C., Han, F., & Teye, E. (2015). Non-destructive evaluation of total volatile basic nitrogen (TVB-N) and K-values in fish using colorimetric sensor array. *Analytical Methods*, *7*(4), 1615–1621. doi:10.1039/C4AY02623C

Huang, X., Xin, J., & Zhao, J. (2011). A novel technique for rapid evaluation of fish freshness using colorimetric sensor array. *Journal of Food Engineering*, *105*(4), 632–637. doi:10.1016/j.jfoodeng.2011.03.034

Huss, H. H. (1995). *Quality and quality changes in fresh fish.* Retrieved April 26, 2017, from http://www.fao.org/docrep/v7180e/v7180e00.htm

Ijong, F. G., & Ohta, Y. (1995). Microflora and chemical assessment of an Indonesian traditional fermented fish sauce "Bakasang". *Journal of the Faculty of Applied Biological Science*, *34*, 95–100.

Jiang, J., Li, J., Zheng, F., Lin, H., & Guohua, H. (2016). Rapid freshness analysis of mantis shrimps (*Oratosquilla oratoria*) by using electronic nose. *Journal of Food Measurement and Characterization*, *10*(1), 48–55. doi:10.1007/s11694-015-9275-y

Jiang, J. J., Zeng, Q. X., Zhu, Z. W., & Zhang, L. Y. (2007). Chemical and sensory changes associated Yu-lu fermentation process - A traditional Chinese fish sauce. *Food Chemistry*, *104*(4), 1629–1634. doi:10.1016/j.foodchem.2007.03.024

Jin, H. J., Lee, S. H., Kim, T. H., Song, H. S., Park, T. H., & Hong, S. (2012). Nanovesicle-based bioelectronic nose platform mimicking human olfactory signal transduction. *Biosensors & Bioelectronics*, *35*(1), 335–341. doi:10.1016/j.bios.2012.03.012 PMID:22475887

Karoui, R., Kemps, B., Bamelis, F., De Ketelaere, B., Decuypere, E., & De Baerdemaeker, J. (2006). Methods to evaluate egg freshness in research and industry: A review. *European Food Research and Technology*, *222*(5-6), 727–732. doi:10.1007/s00217-005-0145-4

Khan, M. I., Jo, C., & Tariq, M. R. (2015). Meat flavor precursors and factors influencing flavour precursors - A systematic review. *Meat Science*, *110*, 278–284. doi:10.1016/j.meatsci.2015.08.002 PMID:26319308

Kiani, S., Minaei, S., & Ghasemi-Varnamkhasti, M. (2016). Fusion of artificial senses as a robust approach to food quality assessment. *Journal of Food Engineering*, *171*, 230–239. doi:10.1016/j.jfoodeng.2015.10.007

Kim, T. H., Lee, S. H., Lee, J., Song, H. S., Oh, E. H., Park, T. H., & Hong, S. (2008). Single-carbon-atomic-resolution detection of odorant molecules using a human olfactory receptor-based bioelectronic nose. *Advanced Materials*, *21*(1), 91–94. doi:10.1002/adma.200801435

Labreche, S., Bazzo, S., Cade, S., & Chanie, E. (2005). Shelf life determination by electronic nose: Application to milk. *Sensors and Actuators. B, Chemical*, *106*(1), 199–206. doi:10.1016/j.snb.2004.06.027

Larsen, C. S. (2003). Animal source foods and human health during evolution. *The Journal of Nutrition*, *133*(11), 3893–3897. PMID:14672287

Laureati, M., Buratti, S., Giovanelli, G., Corazzin, M., Lo Fiego, D. P., & Pagliarini, E. (2014). Characterization and differentiation of Italian Parma, San Daniele and Toscano dry-cured hams: A multidisciplinary approach. *Meat Science*, *96*(1), 288–294. doi:10.1016/j.meatsci.2013.07.014 PMID:23927917

Lee, S. H., Jin, H. J., Song, H. S., Hong, S., & Park, T. H. (2012b). Bioelectronic nose with high sensitivity and selectivity using chemically functionalized carbon nanotube combined with human olfactory receptor. *Journal of Biotechnology*, *157*(4), 467–472. doi:10.1016/j.jbiotec.2011.09.011 PMID:21945089

Lee, S. H., Kwon, O. S., Song, H. S., Park, S. J., Sung, J. H., Jang, J., & Park, T. H. (2012a). Mimicking the human smell sensing mechanism with an artificial nose platform. *Biomaterials*, *33*(6), 1722–1729. doi:10.1016/j.biomaterials.2011.11.044 PMID:22153868

Lippolis, V., Ferrara, M., Cervellieri, S., Damascelli, A., Epifani, F., Pascale, M., & Perrone, G. (2016). Rapid prediction of ochratoxin A-producing strains of *Penicillium* on dry-cured meat by MOS-based electronic nose. *International Journal of Food Microbiology*, *218*, 71–77. doi:10.1016/j.ijfoodmicro.2015.11.011 PMID:26619315

Liu, P., & Tu, K. (2012). Prediction of TVB-N content in eggs based on electronic nose. *Food Control*, *23*(1), 177–183. doi:10.1016/j.foodcont.2011.07.006

Lopetcharat, K., Choi, Y. J., Park, J. W., & Daeschel, M. A. (2001). Fish sauce products and manufacturing: A Review. *Food Reviews International*, *17*(1), 65–88. doi:10.1081/FRI-100000515

Lorenzen, P. C., Walte, H.-G., & Bosse, B. (2013). Development of a method for butter type differentiation by electronic nose technology. *Sensors and Actuators. B, Chemical*, *181*, 690–693. doi:10.1016/j.snb.2013.01.034

Loutfi, A., Coradeschi, S., Mani, G. K., Shankar, P., & Balaguru Rayappan, J. B. (2015). Electronic noses for food quality: A review. *Journal of Food Engineering*, *144*, 103–111. doi:10.1016/j.jfoodeng.2014.07.019

Maamor, H. N., Rashid, F. N. A., Zakaria, N. Z. I., Zakaria, A., Kamarudin, L. M., Jaafar, M. N., . . . Adnan, K. N. A. K. (2014, August). *Bio-inspired taste assessment of pure and adulterated honey using multi-sensing technique*. Paper presented at the 2nd International Conference on Electronic Design, Penang, Malaysia. doi:10.1109/ICED.2014.7015812

Masnan, M. J., Mahat, N. I., Zakaria, A., Shakaff, A. Y. M., Adom, A. H., & Sa'ad, F. S. A. (2012). Enhancing classification performance of multisensory data trough extraction and selection of features. *Procedia Chemistry*, *6*, 132–140. doi:10.1016/j.proche.2012.10.139

Millward, D. J. (1999). The nutritional value of plant-based diets in relation to human amino acid and protein requirements. *The Proceedings of the Nutrition Society*, *58*(02), 249–260. doi:10.1017/S0029665199000348 PMID:10466163

Mottram, D. S. (1998). Flavor formation in meat and meat a review. *Food Chemistry*, *62*(4), 415–424. doi:10.1016/S0308-8146(98)00076-4

Neumann, C., Harris, D. M., & Rogers, L. M. (2002). Contribution of animal source foods in improving diet quality and function in children in the developing world. *Nutrition Research (New York, N.Y.)*, *22*(1-2), 193–220. doi:10.1016/S0271-5317(01)00374-8

Nielsen, S. (Ed.). (2010). *Food Analysis*. Springer. doi:10.1007/978-1-4419-1478-1

Nimsuk, N. (2014, May). *Neural network based electronic nose for discrimination of fish sauces*. Paper presented at the 11th International Conference on Electrical Engineering/Electronics, Computer, Telecommunications and Information Technology, Nakhon Ratchasima, Thailand. doi:10.1109/ECTI-Con.2014.6839743

Nurjuliana, M., Che Man, Y. B., Mat Hashim, D., & Mohamed, A. K. S. (2011). Rapid identification of pork for halal authentication using the electronic nose and gas chromatography mass spectrometer with headspace analyzer. *Meat Science*, *88*(4), 638–644. doi:10.1016/j.meatsci.2011.02.022 PMID:21420795

Ocaño-Higuera, V. M., Maeda-Martinez, A. N., Marquez-Rios, E., Canizales-Rodriguez, D. F., Castillo-Yañez, F. J., Ruiz-Bustos, E., ... Plascencia-Jatomea, M. (2011). Freshness assessment of ray fish stored in ice by biochemical, chemical and physical methods. *Food Chemistry*, *125*(1), 49–54. doi:10.1016/j.foodchem.2010.08.034

Ojha, K. S., Tiwari, B. K., Kerry, J. P., & Cullen, P. J. (2016). Online meat quality and compositional assessment techniques. In E. J. Cummins & G. L. Lyng (Eds.), *Emerging Technologies in Meat Processing: Production, Processing and Technology* (pp. 375–390). Chichester, UK: John Wiley & Sons. doi:10.1002/9781118350676.ch14

Oroian, M., Amariei, S., Leahu, A., & Gutt, G. (2015). Multi-element composition of honey as a suitable tool for its authenticity analysis. *Polish Journal of Food and Nutrition Sciences*, *65*(29), 93–100.

Pais, V. F., Oliveira, J. A., & Gomes, M. T. (2012). An electronic nose based on coated piezoelectric quartz crystals to certify ewes' cheese and to discriminate between cheese varieties. *Sensors (Basel)*, *12*(2), 1422–1436. doi:10.3390/s120201422 PMID:22438717

Pan, D. D., Wu, Z., Peng, T., Zeng, X. Q., & Li, H. (2014). Volatile organic compounds profile during milk fermentation by *Lactobacillus pentosus* and correlations between volatiles flavor and carbohydrate metabolism. *Journal of Dairy Science*, *97*(2), 624–631. doi:10.3168/jds.2013-7131 PMID:24359834

Papadopoulou, O. S., Panagou, E. Z., Mohareb, F. R., & Nychas, G.-J. E. (2013). Sensory and microbiological quality assessment of beef fillets using a portable electronic nose in tandem with support vector machine analysis. *Food Research International*, *50*(1), 241–249. doi:10.1016/j.foodres.2012.10.020

Park, K., Lim, J. H., Jin, H. J., Namgung, S., Lee, S. H., Park, T. H., & Hong, S. (2012b). A bioelectronic sensor based on canine olfactory nanovesicle–carbon nanotube hybrid structures for the fast assessment of food quality. *Analyst (London)*, *137*(14), 3249–3254. doi:10.1039/c2an16274a PMID:22497005

Park, S. J., Kwon, O. S., Lee, S. H., Song, H. S., Park, T. H., & Jang, J. (2012a). Ultrasensitive flexible graphene based field-effect transistor (FET)-type bioelectronic nose. *Nano Letters*, *12*(10), 5082–5090. doi:10.1021/nl301714x PMID:22962838

Peris, M., & Escuder-Gilabert, L. (2013). On-line monitoring of food fermentation processes using electronic noses and electronic tongues: A review. *Analytica Chimica Acta*, *804*(4), 29–36. doi:10.1016/j.aca.2013.09.048 PMID:24267060

Perna, A., Simonetti, A., Intaglietta, I., Sofo, A., & Gambacorta, E. (2012). Metal content of southern Italy honey of different botanical origins and its correlation with polyphenol content and antioxidant activity. *International Journal of Food Science & Technology*, *47*(9), 1909–1917. doi:10.1111/j.1365-2621.2012.03050.x

Piana, M. L., Persano Oddo, L., Bentabol, A., Bruneau, E., Bogdanov, S., & Guyot Declerck, C. (2004). Sensory analysis applied to honey: State of the art. *Apidologie*, *35*(Suppl. 1), 26–37. doi:10.1051/apido:2004048

Renna, M., Cornale, P., Lussiana, C., Giordano, M., Belviso, S., Zeppa, G., & Battaglini, L. M. (2012). Efficacy of fatty acids and terpenoids and weakness of electronic nose response as tracers of Asiago d'Allevo PDO cheese produced in different seasons. *Dairy Science & Technology*, *92*(3), 203–218. doi:10.1007/s13594-012-0056-7

Rodriguez-Mendez, M. L., Arrieta, A. A., Parra, V., Bernal, A., Vegas, A., Villanueva, S., ... de Saja, J. A. (2004). Fusion of three sensory modalities for the multimodal characterization of red wines. *IEEE Sensors Journal*, *4*(3), 348–354. doi:10.1109/JSEN.2004.824236

Sberveglieri, V. (2016). Validation of Parmigiano Reggiano cheese aroma authenticity, categorized through the use of an array of semiconductors nanowire device. *Materials (Basel)*, *9*(2), 81. doi:10.3390/ma9020081 PMID:28787881

Schaller, E., Bosset, J. O., & Escher, F. (1998). "Electronic noses" and their application to food. *Lebensmittel-Wissenschaft + Technologie*, *31*(4), 305–316. doi:10.1006/fstl.1998.0376

Sinanoglou, V. J., Strati, I. F., & Miniadis-Meimaroglou, S. (2011). Lipid, fatty acids and carotenoid content of edible egg yolks from avian species: A comparative study. *Food Chemistry*, *124*(3), 971–977. doi:10.1016/j.foodchem.2010.07.037

Singh, T. K., Drake, M. A., & Cadwallader, K. R. (2003). Flavour of Cheddar cheese: A chemical and sensory perspective. *Comprehensive Reviews in Food Science and Food Safety*, *2*(4), 139–162. doi:10.1111/j.1541-4337.2003.tb00021.x

Smyth, H., & Cozzolino, D. (2013). Instrumental methods (spectroscopy, electronic nose, and tongue) as tools to predict taste and aroma in beverages: Advantages and limitations. *Chemical Reviews*, *113*(3), 1429–1440. doi:10.1021/cr300076c PMID:23256680

Son, M., Lee, J. Y., Ko, H. J., & Park, T. H. (2017). Bioelectronic nose: An emerging tool for odour standardization. *Trends in Biotechnology*, *35*(4), 301–307. doi:10.1016/j.tibtech.2016.12.007 PMID:28089199

Sorrenti, V., Di Giacomo, C., Acquaviva, R., Barbagallo, I., Bognanno, M., & Galvano, F. (2013). Toxicity of ochratoxin A and its modulation by andioxidants: A review. *Toxins*, *5*(10), 1742–1766. doi:10.3390/toxins5101742 PMID:24152986

Spiteri, M., Jamin, E., Thomas, F., Rebours, A., Lees, M., Rogers, K. M., & Rutledge, D. N. (2015). Fast and global authenticity screening of honey using ^{1H}-NMR profiling. *Food Chemistry*, *189*, 60–66. doi:10.1016/j.foodchem.2014.11.099 PMID:26190601

Subari, N., Saleh, J. M., & Shakaff, A. Y. M. (2014, September). *Fusion technique for honey purity estimation using artificial neural network*. Paper presented at the 2nd International Conference on Advances in Intelligent Systems in Bioinformatics, Chem-Informatics, Business Intelligence, Social Media and Cybernetics (IntelSys), Jakarta, Indonesia. doi:10.2495/Intelsys130071

Subari, N., Saleh, J. M., Shakaff, A. Y. M., & Zakaria, A. (2012). A hybrid sensing approach for pure and adulterated honey classification. *Sensors (Basel)*, *12*(10), 14022–14040. doi:10.3390/s121014022 PMID:23202033

Suman, M., Riani, G., & Dalcanale, E. (2007). MOS-based artificial olfacotry system for the assessment of egg products freshness. *Sensors and Actuators. B, Chemical*, *125*(1), 40–47. doi:10.1016/j.snb.2007.01.031

Sun, J., Wang, Q.-J., Huang, J., Hou, Y.-D., Chen, Y.-F., & Su, X.-R. (2013). Influence of heating temperature on the development of volatile compounds in bigeye tuna meat (*Thunnus obesus*) as assessed by E-nose and SPME-GC/MS. *International Food Research Journal*, *20*(6), 3077–3083.

Sun, J., Yu, X., Fang, B., Ma, L., Xue, C., Zhang, Z., & Mao, X. (2015). Effect of fermentation by *Aspergillus oryzae* on the biochemical and sensory properties of anchovy (*Engraulis japonicus*) fish sauce. *International Journal of Food Science & Technology*, *51*(1), 133–141. doi:10.1111/ijfs.12981

Surai, P. E., & Sparks, N. H. C. (2001). Designer eggs: From improvement of egg composition to functional food. *Trends in Food Science & Technology*, *12*(1), 7–16. doi:10.1016/S0924-2244(01)00048-6

Tian, X., Wang, J., & Cui, S. (2013). Analysis of pork adulteration in minced mutton using electronic nose of metal oxide sensors. *Journal of Food Engineering*, *119*(4), 744–749. doi:10.1016/j.jfoodeng.2013.07.004

Tiwari, K., Tudu, B., Bandyopadhyay, R., & Chatterjee, A. (2013). Identification of monofloral honey using voltammetric electronic tongue. *Journal of Food Engineering*, *117*(2), 205–210. doi:10.1016/j.jfoodeng.2013.02.023

Ulloa, P. A., Guerra, R., Cavaco, A. M., da Costa, A. M. R., Figuera, A. C., & Brigas, A. F. (2013). Determination of the botanical origin of honey by sensor fusion of impedance e-tongue and optical spectroscopy. *Computers and Electronics in Agriculture*, *94*, 1–11. doi:10.1016/j.compag.2013.03.001

Vestergaard, J. S., Haugen, J.-E., & Byrne, D. V. (2006). Application of an electronic nose for measurements of boar taint in entire male pigs. *Meat Science*, *74*(3), 564–577. doi:10.1016/j.meatsci.2006.05.005 PMID:22063061

Visser, F. R., & Taylor, M. (1998). Improved performance of the Aromascan A32S Electronic Nose and its potential for detecting aroma differences in dairy products. *Journal of Sensory Studies*, *13*(1), 95–120. doi:10.1111/j.1745-459X.1998.tb00077.x

Wang, J., Kliks, M. M., Jun, S., Jackson, M., & Li, Q. X. (2010). Rapid analysis of glucose, fructose, sucrose, and maltose in honeys from different geographic regions using Fourier transform infrared spectroscopy and multivariate analysis. *Journal of Food Science*, *75*(2), 208–214. doi:10.1111/j.1750-3841.2009.01504.x PMID:20492227

Wang, Q., Jin, G., Ma, M., Wang, N., Liu, C., & He, L. (2014). Discriminating eggs from different poultry species by fatty acids and volatiles profiling: Comparison of SPME-GC/MS, electronic nose, and principal component analysis method. *European Journal of Lipid Science and Technology*, *116*(8), 1044–1053. doi:10.1002/ejlt.201400016

Wardencki, W., Chmiel, T., Dymerski, T., & Biernacka, P. (2009). Instrumental techniques used for assessment of food quality. *Proceedings of ECOpole*, *3*(2), 273.

Warren, M. W., Larick, D. K., & Ball, H. R. (1995). Volatiles and sensory characteristics of cooked egg yolk, white and their combinations. *Journal of Food Science*, *60*(1), 79–84. doi:10.1111/j.1365-2621.1995.tb05611.x

Wei, L., Yuanyuan, H., Yanping, C., Jiaojiao, J., & Guohua, H. (2015). *Penaeus orientolis* prawn freshness rapid determination method based on electronic nose and non-linear stochastic resonance technique. *Bioengineered*, *6*(1), 45–52. doi:10.4161/21655979.2014.993294 PMID:25551520

Wilson, A. D., & Baietto, M. (2009). Applications and advances in electronic-nose technology. *Sensors (Basel)*, *9*(7), 5099–5148. doi:10.3390/s90705099 PMID:22346690

Wilson, A. D., Oberle, C. S., & Oberle, D. F. (2013). Detection of off-flavor in catfish using a conducting polymer electronic-nose technology. *Sensors (Basel)*, *13*(12), 15968–15984. doi:10.3390/s131215968 PMID:24287526

Yongwei, W., Wang, J., Zhou, B., & Lu, Q. (2009). Monitoring storage time and quality attribute of egg based on electronic nose. *Analytica Chimica Acta*, *650*(2), 183–188. doi:10.1016/j.aca.2009.07.049 PMID:19720190

Yoon, H., Lee, S. H., Kwon, O. S., Song, H. S., Oh, E. H., Park, T. H., & Jang, J. (2009). Polypyrrole nanotubes conjugated with human olfactory receptors: High-performance transducers for FET-type bioelectronic noses. *Angewandte Chemie*, *48*(15), 2755–2758. doi:10.1002/anie.200805171 PMID:19274689

Zakaria, A., Shakaff, A. Y., Masnan, M. J., Ahmad, M. N., Adom, A. H., Jaafar, M. N., ... Fikri, N. A. (2011). A biomimetic sensor for the classification of honeys of different floral origin and the detection of adulteration. *Sensors (Basel)*, *11*(8), 7799–7822. doi:10.3390/s110807799 PMID:22164046

Zarei, M., Najafzadeh, H., Eskandari, M. H., Pashmforoush, M., Enayati, A., Gharibi, D., & Fazlara, A. (2012). Chemical and microbial properties of mahyaveh, a traditional Iranian fish sauce. *Food Control*, *23*(2), 511–514. doi:10.1016/j.foodcont.2011.08.023

Zhang, M., Wang, X., Liu, Y., Xu, X., & Zhou, G. (2012). Species discrimination among three kinds of puffer fish using an electronic nose combined with olfactory sensory evaluation. *Sensors (Basel)*, *12*(9), 12562–12571. doi:10.3390/s120912562 PMID:23112731

KEY TERMS AND DEFINITIONS

Aquaculture: Also known as aquafarming, is the farming of fish, crustaceans, mollusks, aquatic plants, algae, and other aquatic organisms. Aquaculture involves cultivating freshwater and saltwater populations under controlled conditions, and can be contrasted with commercial fishing, which is the harvesting of wild fish.

Artificial Neural Networks: Computational models inspired by biological nervous systems whose goal is to solve problems in the same way that the human brain would. Artificial neural networks are composed of a large number of densely interconnected adaptive simple processing elements distributed in different layers. Backpropagation (BP) and radial basis function (RBF) networks are the most frequently used.

Chemometrics: The science of extracting information from chemical systems by data-driven means.

Computer Vision: The science that develops theoretical and algorithmic basis to automatically extract and analyze useful information about physical objects from images. Computer vision includes the capturing, processing, and analysis of two-dimensional images, with others noting that it aims to duplicate the effect of human vision by electronically perceiving and understanding an image.

Fingerprint: The unique response pattern, produced by the array of sensors, which reflects the aroma complexity of the sample under consideration.

Headspace: The gaseous constituents of a closed space above liquids or solids.

Odour-Active Molecules: Volatilized chemical compounds that human or other animals perceive by their sense of olfaction.

Pattern Recognition: A branch of machine learning that focuses on the recognition of regularities in data.

Tualang: Also known as *Koompassia excelsa* or Mengaris, is an emergent tropical rainforest tree species in the *Fabaceae* family. It is found in Indonesia, Malaysia, the Philippines, and Thailand. Tualangs grow branches above the canopy and have slippery trunks, which makes them attractive to giant honeybees, *Apis dorsata*, who hang their huge combs from the branches.

Chapter 9
Application of Electronic Nose to Ambient Air Quality Evaluation With Respect to Odour Nuisance in Vicinity of Municipal Landfills and Sewage Treatment

Jacek Gębicki
Gdańsk University of Technology, Poland

Paulina Kolasińska
Gdańsk University of Technology, Poland

Bartosz Szulczyński
Gdańsk University of Technology, Poland

Tomasz Dymerski
Gdańsk University of Technology, Poland

Hubert Byliński
Gdańsk University of Technology, Poland

Jacek Namieśnik
Gdańsk University of Technology, Poland

ABSTRACT

This chapter contains information about application of the electronic nose type instruments to evaluation of ambient air odour quality connected with such sectors of human activity as municipal landfills or sewage treatment plants. The authors present potential sources of emission from these sectors, characterize the chemical compounds responsible for presence of unpleasant odours, describe the influence of those compounds on human health and related discomfort. Legal aspects pertaining to admissible odour concentration levels in selected countries are also presented. The chapter describes instrumental and sensory methods utilized for determination of odour concentration, odour intensity, or hedonic quality. The chemical sensors potentially and currently employed in the electronic nose type devices are also characterized. Moreover, the future prospects of application of the electronic nose instruments to evaluation of ambient air with respect to odour nuisance are discussed.

DOI: 10.4018/978-1-5225-3862-2.ch009

INTRODUCTION

Progress in economic and industrial development significantly contributes to increased emission of pollutants to ambient air. Industrial plants, such as production plants, refineries, meat processing plants, breweries and distilleries, agricultural plants, metallurgical industry, and municipal plants: landfills, sewage treatment plants and solid waste processing plants generate significant amount of environmentally hazardous pollutants characterized by different physical and chemical properties. Such situation occurs despite widespread implementation of modern technologies in the aforementioned plants aimed at limitation of malodorous substances emission to the environment. Negative impact of these compounds can results from both intrinsic properties of individual chemicals as well as synergistic amplification of hazardous properties of various interacting chemical compounds (Byliński, Lewkowska, Gębicki, Dymerski, & Namieśnik, 2016). In some cases individual substances reveal small impact on particular elements of biotic and abiotic environment, nevertheless, their influence can significantly increase upon presence of other, interacting compounds. Among many different chemical compounds the odours (odorous compounds) play a significant role. The odours are volatile chemical organic and inorganic compounds sensed by animals and humans via olfactory receptors at very low concentrations and identified by brain as unpleasant sensations. The compounds characterized by unpleasant odour include:

- **Inorganic Compounds:** Hydrogen sulphide, ammonia, Sulphur dioxide, nitrogen oxide, hydrogen fluoride and hydrogen arsenide,
- **Organic Compounds:** Thiols, sulphides and disulphides, amines, carboxylic acids, aldehydes, ketones, aromatic hydrocarbons.

General definition identifies the odour as an individual chemical compound or a mixture of chemical compounds containing osmophoric groups in their structure. Table 1 presents the exemplary osmophoric groups and names of the chemical compounds containing these groups.

Reduction of emission of the odorous compounds becomes a priority for these fields of industry, which are responsible for their emission to ambient air. This reduction can consist in:

- Prevention of odorous compounds emission,
- Implementation of deodorization systems in already existing plants,
- Appropriate design and location of new-built facilities.

Irrespective of applied technology resulting in limitation of odours emission it is necessary to check (monitor) effectiveness of deodorization via measurement of odour intensity or odour concentration at inlet and outlet of an installation. One of the most popular methods of odour measurement, apart from olfactometric techniques, are instrumental techniques including electronic nose method. This technique, similarly to the olfactometric methods, utilizes holistic analysis without a need for identification of particular components contributing to a resultant odour. It belongs to a group of dynamically developing instrumental techniques and is more and more frequently employed to monitoring of unpleasant odours originating from different fields of human activity.

Table 1. Exemplary osmophoric groups

Osmophoric group	Chemical compounds
—OH	Alcohols
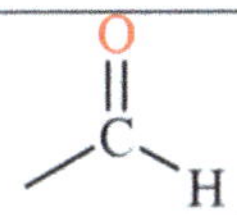	Aldehydes
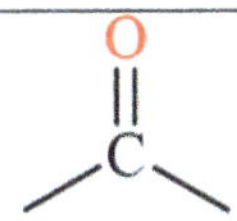	Ketones
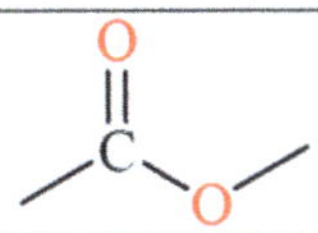	Esters
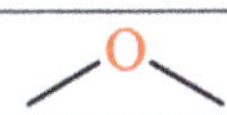	Ethers
—NH_2	Amines
	Sulfides
	Disulfides
—SH	Thiols
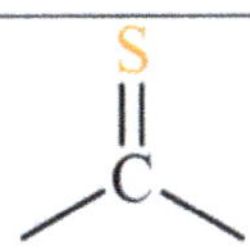	Thioketones
	Aromatic heterocyclic compounds

CHARACTERISTICS OF DEODORIZATION METHODS FOR GASES CONTAINING ODOROUS POLLUTANTS

Odour nuisance in a vicinity of residential areas and increasing social awareness impose a pressure on companies and industrial entities forcing them to undertake more and more efficient steps towards minimization (or elimination) of the effects of their activity connected with emission of malodorous substances. As far as the new-built facilities are concerned the prevention of malodorous substances emission should be taken into account already upon selection of technology and design of process devices. In case of the existing plants all available technological modifications should be employed.

Deodorization of the gases polluted with the odorous compounds can consist in:

- Removal of malodorous pollutants (often present at the trace levels),
- Transformation of malodorous pollutants into odourless chemical compounds or the compounds characterized by high threshold of odour sensing,

- Introduction of additives changing the character of odour or decreasing its intensity of hedonic quality (masking compounds).

Selection of the most effective deodorization method is difficult and dependent on many factors including: emission intensity, total content of pollutants, odour character of emitted gases. General characteristics of the most popular gas deodorization methods together with the characteristics of their advantages and disadvantages is shown in Table 2.

Most literature data and information concerning the number of complaints on odour quality of ambient air point to the following fields of human activity as the ones having the biggest impact:

- Municipal economy,
- Agriculture,
- Industry.

Table 3 presents a list of municipal and industrial facilities constituting the potential sources of odours (Gebicki, 2016).

The most onerous sectors of human activity with respect to odour emission are: municipal landfills and sewage treatment plants.

Table 2. Advantages and disadvantages of the most frequently used gas deodorization technologies and mean efficiency of removal of malodorous pollutants using different deodorization techniques

Deodorization Technologies	Disadvantages	Advantages	Mean Efficiency of Odour Removal
Absorption	• problem of absorbent regeneration or disposal remains • possible secondary emissions whose source is used absorbent • high costs of pumping • further stages of gas treatment are often required	• low operating costs • treatment of gases containing high odourant concentrations • possibility of recovery of adsorbed compounds	50 – 75%
Adsorption	• problem of adsorbent regeneration or disposal • necessary to regenerate a deposit using large amounts of gases (subsequent dilution of pollutants) • possible secondary emissions whose source is used adsorbent	possibility of recovery of adsorbed compounds	> 90%
Thermal neutralization	• a high content of inflammable pollutants is required • generation of secondary pollutants • high energy consumption of the process	• ensures high effectiveness of deodorization • waste-free process • simple design and operation of the installation	> 95%
Biological gas treatment	• treated gases must contain biodegradable components • problem with an excessive amount of biomass - installation overgrowth • for biofiltration - a large surface area of the installation	• low operating costs • high effectiveness when biological material is well selected • possible to treat gases with low odourant concentrations	90%

Table 3. List of exemplary economy sectors coupled with potential emission sources and types of emitted odours

Type of Object or Installation	Emitted Odorous Substances	Potential Source of Odour Emission
animal breeding: animal farms, poultry houses	organic acids, ammonia, phenols, sulphur compounds, toluene	improper management of waste and liquid manure, lack of hermetization and deodorization
chemical and petroleum industry, production of paints, lacquers, solvents, processing of crude oil	hydrocarbons, sulphur compounds, nitrogen compounds, aldehydes	water-waste management, failures of installations, transport and storage
dairies (depending on production magnitude)	ammonia, sulphur compounds	leaks in ammonia cooling installations, improper waste management (high tendency of dairy waste to fast fermentation)
municipal waste management objects: landfills, screening plants and composting plants	ammonia, sulphur compounds	anaerobic processes, lack of appropriate hermetization
processing of meat and fish	ammonia, nitrogen oxides, hydrogen sulphide, phenols, aldehydes	collection and storage of raw material, preparation of raw material, composting, smoking, production of meat-and-bone meal
sewage treatment plants: biological, mechanical	sulphur compounds, nitrogen compounds, aldehydes, ketones	anaerobic operation, preliminary operations, for example collection of sewage, drain points
animal breeding: animal farms, poultry houses	organic acids, ammonia, phenols, sulphur compounds, toluene	improper management of waste and liquid manure, lack of hermetization and deodorization

CHARACTERISTICS OF MALODOROUS POLLUTANTS EMISSION SOURCES IN MUNICIPAL LANDFILLS AND SEWAGE TREATMENT PLANTS

Landfill is a bioreactor, in which microbiological and biochemical transformations occur due to atmospheric factors and microorganisms activity (Figure 1). Initial decomposition of waste is an aerobic process as just after waste disposal there are air pockets inside the landfill. Soon after the decomposition onset all oxygen is consumed and an anaerobic process starts as fresh oxygen supply to the landfill is impossible. Anaerobic decay of organic waste can be simplified to two fundamental processes, although it must be emphasized that biochemistry of microbe processes is very complicated. In the first stage complex organic substances are biodegraded to simple organic substances including acetic acid (CH_3COOH), propionic acid (C_2H_5COOH), pyropropionic acid ($CH_3COCOOH$) or other simple organic acids and alcohols. Bacteria take part in the process and the chemical reactions provide them with the energy necessary for development; certain amount of organic waste is transformed into cellular and extracellular substances. The second stage of anaerobic decay is methanogenesis. Methanogenic bacteria consume the products of the first stage and produce oxygen as well as carbon dioxide. Oxygen, which forms carbon dioxide originates from the organic substrates or from sulphate ions. The methanogenic bacteria prefer neutral pH and they avoid acidic environment. Formation of acids in the first stage reduces pH and if this process continues, activity of the methanogenic bacteria can cease. Municipal landfills are the significant emission sources of methane (CH_4) and carbon dioxide (CO_2) and for these compounds exceeding of the threshold values defined in the Regulation (EC) No. 166/2006 of the European Parliament and Council of 18 January 2006 on establishment of a European Pollutant Release and Transfer Register is the most probable. Moreover, there could be emission of smaller amount of non-methane volatile organic compounds (NMVOC), dinitrogen monoxide (N_2O), carbon monoxide (CO) ammonia (NH_3), Sulphur oxides (SO_x) and nitrogen oxides (NO_x). Additionally, unloading and planting of waste as well as operation of the machines and vehicles on the landfill constitute sources of dust (PM) emission. It is obvious that odour of the products of aerobic degradation of organic compounds is less onerous than the odour of the products of anaerobic denitrification and desulphurization.

In the case of sewage treatment plants three basic sources of emission can be distinguished depending on the plant type (Figure 2):

- Sewage feed and its mechanical treatment,
- Biological treatment,
- Sludge processing.

Highly onerous anaerobic processes usually occur in the sewage feed and sludge processing zones. Composition and odour of gas mixture mainly depends on biodegradation conditions deciding about the type of bacteria participating in the process. The most important parameter is sewage aeration degree – it is necessary to prevent oxygen shortage and anaerobic bacteria development.

Emission of air pollutants from particular sources of the sewage treatment plants, which can have influence on odour nuisance over the adjacent areas, is presented in Figure 3. Emitted odorous compounds engulf: organic compounds of sulphur, indoles, skatoles, organic acids, aldehydes and ketones.

Figure 1. An example scheme for the construction of a municipal landfill

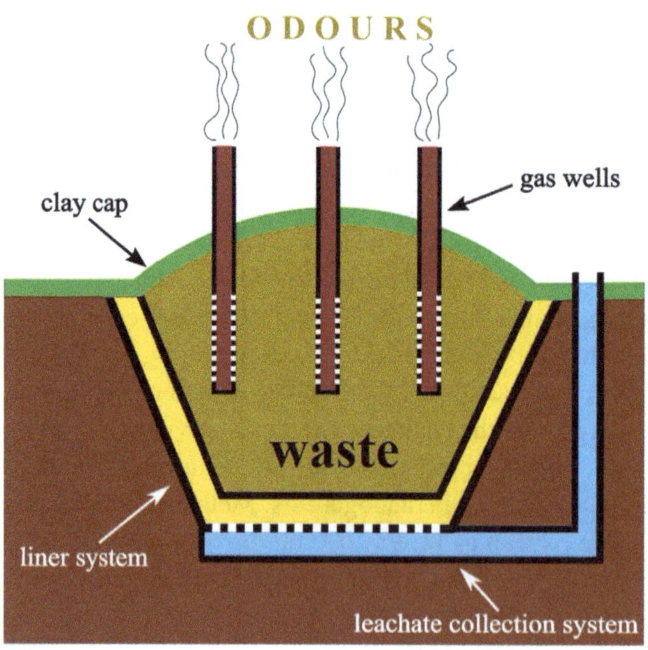

Figure 2. An example scheme for the construction of a wastewater treatment plant

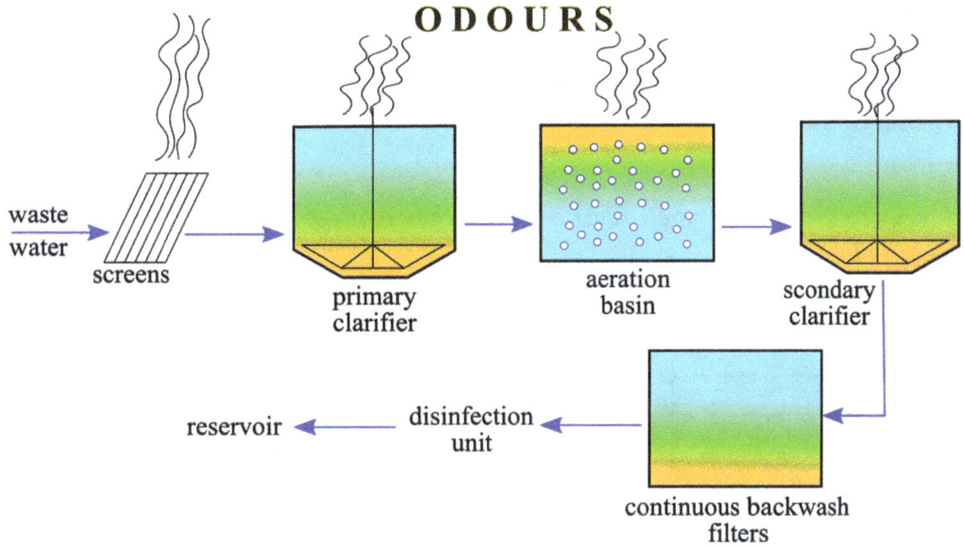

Figure 3. Average percentage distribution of odour emission sources from a sewage treatment plant

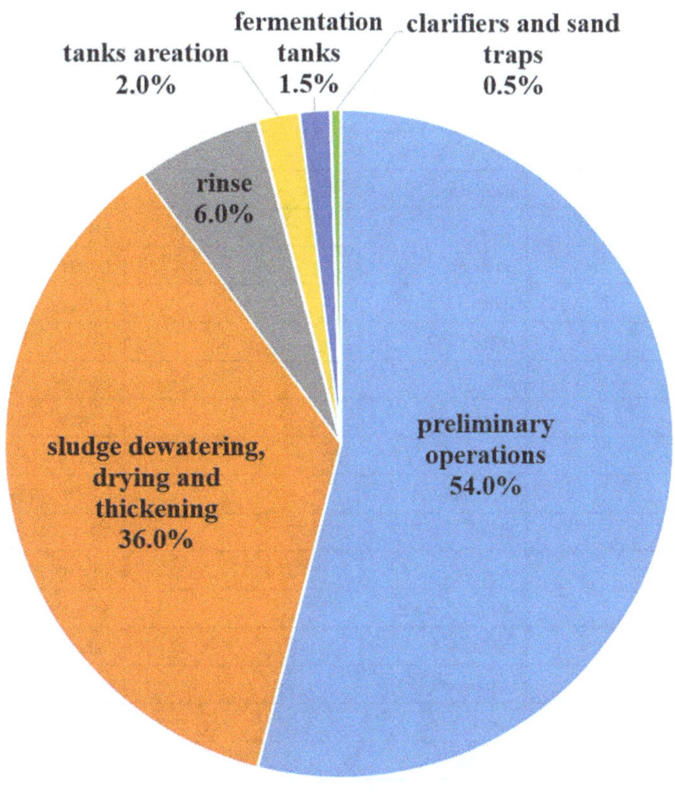

In Table 4 collects the information about odour sensing thresholds and smell sensations associated with selected chemical compounds emitted from municipal landfills and sewage treatment plants (Czerny et al., 2008; Fang et al., 2012; Gebicki, 2016; Hansen, Jonassen, & Feilberg, 2014; Lehtinen & Veijanen, 2011; Nagata, 2003; Talaiekhozani, Bagheri, Goli, & Talaei Khoozani, 2016; Wenjing et al., 2015; Zhu, Chen, Wang, Niu, & Xiao, 2017).

Table 4. Odours thresholds and smell sensations for chemical compounds emitted from landfills and sewage treatment plants

Compound	Kind of Fragrance	Odour Threshold [ppm]
acetaldehyde	fresh	6.0×10^{-3}–1.5×10^{-3}
acetone	pungent	4.0×10^{-1}–4.2×10^{1}
acetonitrile	ethereal	1.3×10^{1}
acrolein	pungent	3.6×10^{-3}
acrylonitrile	suffocating	1.6 - 8.8
ammonia	characteristic	1.0 – 1.7×10^{1}
benzaldehyde	almond	3.5×10^{-1}
benzene	characteristic	4.7×10^{-1} – 2.7
benzyl mercaptan	garlic	1.9×10^{-4}
butanone	pungent, sweet	1.0×10^{1}
butyl acetate	pear	5.8×10^{1}
butyric acid	sweet	1.9×10^{-4}
carbon disulfide	sweet, pleasant	8.1×10^{-3}–2.1×10^{-1}
carbonyl sulfide	unpleasant	5.5×10^{-2}
diacetyl	chloric	5.0×10^{-5}
diehylamine	ammonia	4.8×10^{-2}
dimethyl sulfide	rotten	1.0×10^{-4} – 0.5
dimethyl trisulfide	cabbage	1.9×10^{-3}
dimethylamine	fish	3.4×10^{-1}
ethanethiol	garlic	8.7×10^{-6}–1.9×10^{-4}
ethanol	alcoholic	5.2×10^{-1}
ethyl acetate	fruit, pleasant	8.7×10^{-1}–6.2×10^{3}
ethyl formate	fruit	2.7
ethylamine	ammonia	4.6×10^{-2}
ethylbenzene	aromatic, sweet	3.0×10^{-3}–1.7×10^{-1}
formaldehyde	suffocating	5.0×10^{-1}
furan	chloroform	9.9
haxanoic acid	sweet, cheesy	1.8
heptanal	pungent	1.3×10^{-3}–5.5×10^{-1}
heptanoid acid	sour	3.0
hexanal	mown grass	4.5×10^{-3}–6.8×10^{-3}
hydrogen sulfide	rotten eggs	4.1×10^{-4}–5.0×10^{-4}

continued on following page

Table 4. Continued

Compound	Kind of Fragrance	Odour Threshold [ppm]
indole	smell of feces	3.0×10^{-4}
isopropyl alcohol	alcoholic	1.0
isopropylamine	ammonia	2.5×10^{-2}
limonene	citrus	1.8×10^{-3}–3.8×10^{-2}
m-cresol	unpleasant, pungent	1.0×10^{-4}
methanethiol	rotten cabbage	7.0×10^{-5}–1.1×10^{-3}
methanol	alcoholic	3.3×10^{1}–1.0×10^{2}
methyl formate	pleasant	1.3×10^{2}
methyl methacrylate	fruit	2.1×10^{-1}
methylamine	fish	3.5×10^{-2}–4.7
m-xylene	pleasant	4.1×10^{-2}
naphthalene	characteristic	2.7×10^{-2}–3.8×10^{-2}
n-butane	solvent	3.8×10^{-2}
n-heptane alcohol	characteristic	4.8×10^{-3}
n-hexane alcohol	fruit	6.0×10^{-3}
n-octane alcohol	unpleasant	2.7×10^{-3}
n-pentane alcohol	fruit	1.0×10^{-1}
o-cresol	unpleasant, pungent	2.8×10^{-4}
octanal	citrus	0.1-1.2
octanoic acid	sweet, cheesy	1.9
o-xylene	pleasant	8.1×10^{-2}– 3.8×10^{-1}
p-cresol	unpleasant, pungent	5.4×10^{-5}
phenol	characteristic	5.6×10^{-3}
propanethiol	cabbage	1.5×10^{-5}
propionaldehyde	pungent	8.1×10^{1}
propionic acid	pungent, unpleasant	5.7×10^{-3}
propyl acetate	pear	2.4×10^{-1}
propyl foramte	rum-plum	9.6×10^{-1}
p-xylene	pleasant	5.8×10^{-2}–8.1×10^{-2}
pyridine	characteristic	6.3×10^{-2}
skatole	smell of feces	5.6×10^{-6}
styrene	unpleasant	3.5×10^{-2}–4.7×10^{-2}
sulphur dioxide	suffocating	9.0×10^{-3} – 2.7
toluene	pleasant, intense	2.1×10^{-2} – 2.8
triethylamine	fish	5.4×10^{-3}
trimethylamine	characteristic	3.2×10^{-5}
valeric acid	honey	3.7×10^{-5}
α-pinene	fir	2.9×10^{-3}–9.0×10^{-1}
β-pinene	pine	3.3×10^{-2}

INFLUENCE OF ODOROUS COMPOUNDS ON HUMAN HEALTH AND ENVIRONMENT

Negative impact of odours is most frequently a derivative of their subjective perception and their action consist in destructive influence on human psyche. This can result in appearance or intensification of somatic system malfunction. Hence, while speaking about odour nuisance one should discriminate between toxic effects and the effects caused by odour properties. Biological effects, resulting exclusively from odour nuisance, will trigger physiological changes at most, but without fatal implications, which can be caused by toxic substances.

Odours can induce a number of negative effects including (Nicell, 2009):

- Deterioration of natural environment quality,
- Damages to properties, fauna and flora,
- Discomfort,
- Deterioration of life safety and quality,
- Loss of normal ability to utilize real estate,
- Disturbances in normal business activity.

Odours can trigger a number of unwanted reactions in humans, from aggravation to documented health problems. In the societies exposed to odours emission the health problems connected with odour impact may not be evident immediately, however there can be morbidities or physical as well as mental handicap. Moreover, long-term exposure to odour can result in undesirable emotional reactions such as anxiety, feeling of discomfort, depression as well as physical symptoms including irritation, headache, respiratory track problems, nausea, vomiting . Exposure to unwanted odours can also lead to mental tension and caused such symptoms as insomnia, lack of appetite and irrational behaviour. High levels of odorants can be a source of acute toxicological symptoms and they can trigger the mechanisms connected with reaction of the organism to odour. These mechanisms can include intrinsic reluctance with respect to odour, heightening of fundamental conditions connected with odour perception (hypersensitivity), stress-induced diseases or even psychogenic diseases . Other identified odour influence of on human engulf: difficulties breathing, frustration, stress, increased tendency to cry, reduction of appetite and/or difficulties with consumption and preparation of meals, feeling of discomfort at nigh (difficulties with falling asleep, restless sleep) (Capelli, Sironi, Del Rosso, & Centola, 2009; Trincavelli, Coradeschi, & Loutfi, 2009). High complexity of odour nuisance problem is evidenced by numerous complaints of the residents from the regions located in direct vicinity of the main emitters of air pollutants. Every year the number of complaints on malodorous compounds presence in ambient air increases (for instance in Poland the number of complaints doubled during the period of 5 years). The number of complaints connected with odour nuisance in highly developed or developing countries also rises. The most important elements deciding about a degree of nuisance are: odour concentration, odour hedonic quality, frequency of unpleasant smell sensation occurrence and perceived odour intensity. Combination of those four properties of odour creates a complete image associated with presence of odour in ambient air. Each of these elements plays a significant role in perception of the entire phenomenon. An increase in odour concentration is connected with stronger smell sensations occurring upon presence in a vicinity of an emission source; odour hedonic quality influences on connotations related to perception of particular substance (most often a gas mixture of odorous compounds); frequency of unpleasant smell sensation occurrence determines frequency of increased odours emission situation; odour intensity described 'strength of perceived odour'.

LEGAL REGULATIONS AND LIMIT VALUES OF ODOUR CONCENTRATION IN SELECTED COUNTRIES

Due to odour impact of industrial, agricultural or municipal facilities and nuisance connected with odours emission the legislative bodies of highly developed countries face the problem of admissible emission of odorous pollutants within the frame of health and environmental protection programmes. So far suitable legal acts have been established in industry and agriculture as well as in other branches of economy in Germany, France, the Netherlands, the Great Britain, Japan, the USA and Canada. Introduction of universal measurement method in Europe according to the standard PN-EN 13725:2007 makes it necessary to define odour concentration and odour unit. Odour concentration is defined as the concentration, for which given person is able to sense single odorous substance – it is called individual (for particular person) detection threshold for given substance. Hence, the detection threshold of odour must be averaged value representative for a population (population detection threshold). The detection threshold is such concentration of the odorant, which causes smell sensation in half of the population (or representative group) exposed to odour impact. For individual substances the odour concentration will be equal to a quotient of odorant concentration and threshold concentration. In the case of odorous substances mixtures it is not possible to define odour concentration in a similar way because it is impossible to determine the detection threshold. Then the odour concentration cod can be defined as multiplicity of dilution of investigated sample (with clean, odourless air) necessary to achieve the detection threshold. In Table 5 presents the examples of admissible odour concentration accepted in different countries (Hung, Wang, & Shammas, 2012; Mahin, 2001; Naddeo, Zarra, Giuliani, & Belgiorno, 2012; Nicell, 2009; Ritvay & Kovács, 2006).

INSTRUMENTAL AND SENSORY METHODS UTILIZED FOR DETERMINATION OF ODOUR CONCENTRATION, ODOUR INTENSITY AND HEDONIC QUALITY

General classification of the techniques employed for ambient air quality evaluation with respect to odour distinguishes two fundamental approaches:

- Analytical (instrumental techniques),
- Sensory.

The sensory techniques, including most frequently applied dynamic olfactometry and field olfactometry, allow determination of summary concentration of the odorous substances present in an investigated sample, odour intensity and its hedonic quality. In case of these techniques human nose plays the role of 'measurement sensor'. Dynamic olfactometry is one of the most common solutions in the field of sensory techniques, which makes it possible to determine odour concentration expressed in European odour units per cubic meter (ou_E/m^3) (Figure 4). Concentration expressed in this way describes a degree of dilution of gaseous odorous sample with a stream of odourless air, which is necessary to achieve so-called detection

Table 5. Examples of acceptable odours values accepted in different countries

Country	Acceptable Odour Value (ou_E/m^3)	Averaging Time (min.)
Germany	1	60
France	5	60
Italy	1-5	60
Denmark	5-10	60
Netherlands	0.5-3.5	60
Hungary	3-5	60
Great Britain	5	1
Ireland	3-6	60
Norway	5-10	60
Australia	2-10	3-60
United States of America	1-20	< 5
China	5	5 s
Japan	< 5	no data
Korea	< 2	no data
New Zealand	2	60
Taiwan	50	no data
Tasmania	1	3

threshold. Detailed information about the methodology of determination of odour concentration with dynamic olfactometry technique are provided in the European standard EN 13725:2003: 'Air quality. Determination of odour concentration by dynamic olfactometry'. This document constitutes the only example of unified legal regulation concerning the methodology of air quality investigation with respect to odour properties. Nowadays the technique of dynamic olfactometry is commonly used for identification of odorous compounds emission sources and assessment of emission level.

However, in some cases this solution does not provide satisfactory results, for example in the case of low concentrations. Another limitation of dynamic olfactometry technique is a necessity of samples collection and their transport to laboratory, which increases the time of single analysis and has a negative impact on the results due to a risk of partial loss of analytes that can undergo adsorption processes inside a bag. It is difficult to determine odour concentration, especially in case of variability of emission over particular area. Then more convenient solution seems field olfactometry technique. Such approach allows characterization of properties of the odorants present in ambient air as a result of dilution of investigated air stream with odourless gas (typically air purified on carbon filters). The most popular device used in olfactometric measurements is a field olfactometer Nasal Ranger (St. Croix Sensory, USA), the scheme of which is presented in Figure 5.

This instrument is comprised of two replaceable filters made of active carbon, a valve adjusting flow rate of non-purified air (thus containing odorous compounds), replaceable mask for air inhalation via the olfactometer and a sensor allowing control over the flow rate of gas stream passing through the device, the value of which should be less than 20 dm^3/ml. The adjusting valve offers selection of one of six available levels of dilution of inhaled air: 60, 30, 15, 7, 4, 2. There is a possibility to take advantage of an

Figure 4. Example of a dynamic olfactometer

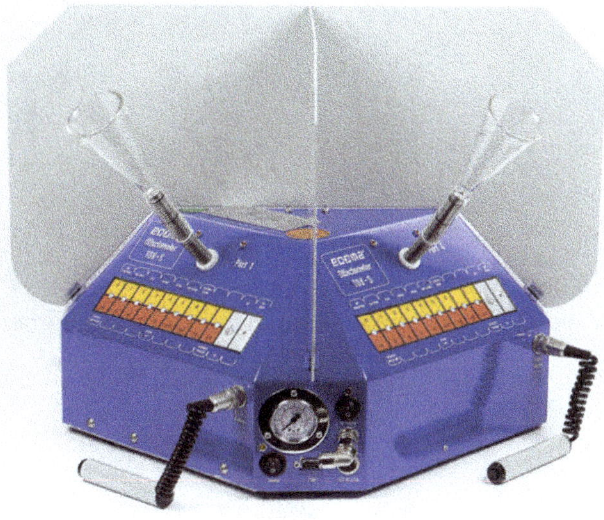

Figure 5. Scheme of Nasal Ranger field olfactometer

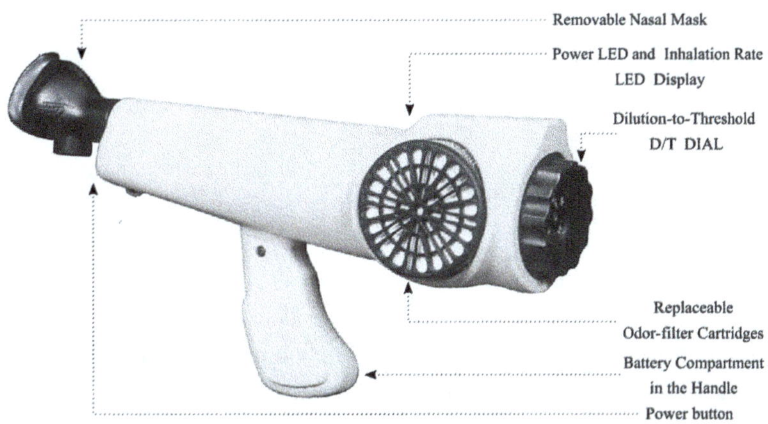

additional valve for higher concentration of odorants in air, which enables to obtain higher dilution. This valve provides the following values of dilution: 600, 500, 400, 300, 200, 100. Moreover, each available valve possesses 'blank' position (air passes only through the filters so the air inhaled by an assessor is purified and free from pollutants, thus also from odorous compounds). Accuracy and repeatability of the field olfactometer are ±10% and ±5%, respectively (Kośmider & Krajewska, 2007).

Application of the field olfactometers allows identification of the point sources of odours emission, which reveal the highest sensory nuisance over the regions neighbouring with the most important emitters of odorous air pollutants. The olfactometers can be used as both the measurement devices as well as the instruments for verification of the complaints about the odour nuisance over the areas in a vicinity of potential emitters of odorous compounds. In olfactometric measurements the assessors are expected to determine the ratio of purified air passed through the filters to non-purified air by-passing the filters

(D/T – Dilution to Threshold), which corresponds to the first moment when the odour of diluted air is sensed. Based on the D/T values it is possible to calculate individual odour concentration using the following formulas:

$$Z_{yes} = \left(D/T\right)_{yes} + 1$$

where:

Z_{yes} = dilution ratio, at which the odour became perceptible,
$(D/T)_{yes}$ = dilution ratio corresponding to the moment when the odour became perceptible for the first time,

$$Z_{no} = \left(D/T\right)_{no} + 1$$

where:

Z_{no} = dillution ratio, at which the odour was imperceptible,
$(D/T)_{no}$ = dillution ratio corresponding to the moment when the odour was imperceptible just before the dillution $(D/T)_{yes}$

$$Z_{ITE} = \sqrt{Z_{yes} \times Z_{no}}$$

where:

Z_{ITE} = individual odour threshold estimate.

Values of odour concentration are calculated as geometric mean from n-element set of all individual estimates for particular measurement point using the formula:

$$c_{od} = \sqrt[n]{\sum_{i=1}^{n} Z_{ITE,i}}$$

where:

c_{od} = value of odour concentration [ou/m^3],
n = number of all individual estimates.

Application of field olfactometry technique allows direct control over emission of odorous compounds released from different types of industrial and municipal facilities. Such investigations can contribute to verification of correct operation of production plants as far as amount of odours emitted during a single production cycles is concerned. Field olfactometry exhibits many advantages including relatively low cost of equipment (much cheaper than dynamic olfactometers), simple operation and possibility to determine odour intensity upon imission. The most significant drawbacks of this technique engulf lack of legal regulation, which would clearly define the conditions pertaining to assessors selection for the investigation, the way of measurements execution and the interpretation of obtained results. There are two dominant approaches in instrumental analysis of odorous samples. The first one consists in identification of volatile odorous compounds contained in the sample using gas chromatography (GC) couple with a suitable detector, the second one is based on 'holistic' analysis of an odorous mixture without separation into individual components, which employs the electronic nose instruments (Bai & Shi, 2007; Gallego, Perales, Roca, & Guardino, 2014; Harynuk & Górecki, 2004). The chromatographic techniques, including gas chromatography mass spectrometry (GC-MS), are one of the most common solutions utilized for identification and quantitative measurement of main ambient air pollutants, including odorous compounds. The information about chemical composition of the ambient air samples taking into account sampling and analysis procedures (analyte derivatization, conditions of separation process) enable preliminary estimation of emission level of atmospheric air pollutants over the investigated region. This information can be valuable at the stage of elaboration of efficient solutions for deodorization of odorous substances. Numerical values of concentration of particular chemical compounds present in the ambient air samples make it possible to determine odour activity value (OAV). This parameter is calculated as a sum of the ratios of concentration of particular chemical compounds present in ambient air to their individual odour detection threshold following the formula:

$$OAV = \sum_{i=1}^{n} \frac{c_i}{OT_i}$$

OAV = Odour Activity Value [ou/m^3],
c_i = concentration of compound i [mg/m^3],
OT_i = Odour Threshold value of compound i [mg/ou].

Determination of odour activity value allows visualization of odour nuisance degree over the investigated region via calculation of odour activity values for the ambient air samples collected from different locations in given time. Correlation of the odour activity values with chemical composition of the ambient air samples (types of compounds and their concentrations) makes it possible to identify these compounds, which have the biggest contribution to strength of perceived odour. Such approach is characterized by relatively high repeatability and reproducibility of results as compared to the olfactometric techniques. However, the main drawback of this approach is a lack of possibility of holistic characterization of many important properties of odorous mixtures – odour intensity and hedonic quality, which are important from the point of view of odour nuisance. Determination of OAV parameter seems to be good complement of the results obtained with the olfactometric techniques.

The electronic nose is defined as an analytical instrument intended for the rapid detection and distinction of mixtures of odorous substances as a result of imitating the principles of operation of the human sense of smell. In the case of electronic nose, the recognition of the composition of the gaseous mixture is possible, thanks to the stimulation of all sensors making up the electronic nose. This stimulation (signal sensor) occurs in each sensor to a varying degree. As a result of such stimulation, a characteristic collection of signals is created from single sensors, which is called the response vector. Analysis of the obtained response vector allows for classifying a given gaseous sample into the collection of previously known response vectors (during the electronic nose training process). The brains fulfils this role in a person as it compares nervous impulses obtained from the olfactory bulb to the database of odours that a person gets to know over their life. In the case of the electronic nose, the obtained response vector is assigned to the sets found in the computer database from the appropriate mathematical and statistical algorithms. Electronic noses are most often used for qualitative analysis tools; it is also possible to use electronic noses for quantitative analysis. To obtain information about quantitative nature, the following tools are used amongst other things:

- Multiple linear regression using selected features,

$$c_{od} = a_0 + \sum_{n=1}^{\infty}\left(a_n \mathrm{X}_{\mathrm{i,t}}\right)$$

where X is the value of the i-th signal at a given measuring moment during exposure of the sensor matrix to the tested gaseous mixtures, a_n - coefficients of the model,

- Multiple linear regression using extracted features,

$$c_{od} = a_0 + \sum_{n=1}^{\infty}\left(a_n \mathrm{PC}_{\mathrm{n}}\right)$$

where PC_n is the n-th main component selected as a result of transformation of the feature vector that is the development of the entire matrix of features,

- One-dimension of non-linear regression with the use of selected features,

$$c_{od} = ab^{X}$$

where *a* and *b* are coefficients of the model,

- One-dimension of non-linear regression with the use of extracted features,

$$c_{od} = ab^{PC1}$$

where PC1 is the first main component, *a* and *b* are coefficients of the model.

These are the simplest and the least required methods for solving the problem of determining the olfactory concentration using the electronic nose; however, the most frequent olfactory concentration is determined experimentally according to the following principle:

- Signals from sensors are normalized and averaged according to methods specified in Table 6.
- Normalized and averaged signal is calibrated according to the reference substance (n-butanol) at various levels of concentration,
- The correctness of calibration with the use of olfactometric techniques.

At present the e-nose instruments constitute only complementary solution with respect to sensory analysis of ambient air. However, in many cases subjectivity of the assessors participating in olfactory evaluations creates many doubts. The olfactometric techniques call for an experienced team of assessors, the smell sensitivity of who can significantly change upon numerous external factors. Special laboratory equipment utilized for olfactometric evaluations significantly elevates the cost of such analysis. These problems could be eliminated by application of the devices enabling fast and cheap analysis of odorous compounds. Looking at the potential and development pace of the e-nose instruments one can expect that their role in ambient air monitoring is going to increase continuously and current reference techniques for odour concentration determination will be supplemented with the electronic nose technique. Table 7 presents comparison of the olfactometric techniques (dynamic and field olfactometry) with the electronic nose technique with respect to selected aspects of discussed solutions.

CHARACTERISTICS OF CHEMICAL SENSORS USED FOR THE CONSTRUCTION OF THE ELECTRONIC NOSE

The level of odourant concentration in the air determines the selection of chemical sensors for their measurement. The application of single chemical sensors for air monitoring is limited due to very low odourant levels that usually occur at concentration levels of 0.1 ppb-0.1ppm. For emission measurements or measurements similar to emissions (i.e. at a close distance from the source of emissions), the level of odourant concentrations may exceed the value of 1 ppm v/v; then we can talk about the possibility of

Table 6. Basic methods of sensor signal normalization

Method	Formula	Comments
Differential	$X(t) = x_t - x_{0,t}$	Reduction of the drift signal with an additive character
Quotient	$X(t) = \frac{x_t}{x_{0,t}}$	Reduction of the drift signal with a multiplicative character
Fractional	$X(t) = \frac{x_t - x_{0,t}}{x_{0,t}}$	Reduction of the drift signal with a multiplicative and additive character
Logarithmic	$X(t) = \log \frac{x_t}{x_{0,t}}$	Reduction beneficial in the case of a large range of concentrations

where: X(t) - the value of the sensor signal at time *t* after the application of normalization, x_t - the value of the detection signal at time *t* before normalization, $x_{0,t}$ - the value of the sensor signal corresponding to the reference gas (clean gas).

Table 7. Comparison of electronic nose and olfactometry techniques

Properties	Olfactometry Techniques	Electronic Nose
legally regulated methodology	dynamic olfactometry – yes field olfactometry - no	no
element for odour measurement	human nose	sensor matrix
team of trained assessors	necessary for investigation	no need for training of the personnel participating in investigation
repeatability of obtained results	low	high
time of single analysis	short	short
mathematical algorithm	simple	complicated
possibility to conduct field measurements	yes (as an element of portable laboratory in case of dynamic olfactory)	yes
allows measurement of low concentrations of odorous compounds	dynamic olfactometry – no field olfactometry - yes	depending on applied sensor matrix
objectiveness of obtained results	low	high
limitation of the number of performed investigations	dependent on condition of the assessors (determining factor – smell adaptation)	no significant limitations
identification of odours	lack	a need to teach e-nose to identify and discriminate particular odours
a need of samples collection	dynamic olfactometry – yes (exception – mobile laboratories); field olfactometry - no	no need of samples collection in case of utilization of continuous monitoring devices
a need of samples preparation prior to analysis	lack	lack

practical use of chemical sensors in the electronic nose design. The selection of an appropriate chemical sensor/sensors for the construction of an electronic nose, of course, depends on the ingredients that are to be detected, the expected range of concentrations, whether the presence of other ingredients of the gaseous mixture should be identified (selectivity), which could influence the reading or the possibility of damage to the sensor. If a single sensor is used, requirements for these sensors are based on the best selective detection of a specific odourant, so that it can be detected and determined quantitatively at the lowest level of concentrations. Electronic noses, which are intended for distinguishing odourant mixtures with a complex composition, can contain chemical sensors in their design that are characterized by limited selectivity, which is, of course, a flaw for single sensors that should be characterized by high selectivity. For electronic noses, this feature of chemical sensors often becomes an advantage due to the integral element of the design of these devices - i.e. a reference recognition block, which is used for analysing signals from individual sensors. As a result, the universality of electronic nose devices increases. An additional advantage of electronic noses is the possibility of performing field analyses in an on-line system around odourant emission sources, as they do not exhibit olfactory adaptation, as opposed to the human nose. It is also possible to record short-term episodes of high odourant concentrations in a given area, which may not be recorded using dynamic or field olfactometry technique. Such a situation is related to the arrival time and the time when samples are collected by qualified personnel (for field olfactometry). Table 8 and Figure 6 presents the basic and most frequent chemical sensors used in the design of electronic nose devices. Their detection range — as well as advantages and disadvantages of each type of sensor — are described.

Table 9 summarizes basic information about types of sensors available on the market and their metrological parameters and intended for measuring odorous gases such as: hydrogen sulphide, ammonia, total volatile organic compounds. The table presents only chemical sensors that are characterized by the limit of quantification below 1 ppm v/v.

THE USE OF ELECTRONIC NOSE DEVICES FOR THE ASSESSMENT OF AIR QUALITY AROUND MUNICIPAL LANDFILLS AND SEWAGE TREATMENT PLANTS

A lot of methods for determining malodorous substances are considered to be too expensive to create commonly available measuring instruments or developed environmental monitoring systems. High hopes for the development of such systems are placed on electronic nose devices. Literature data show that odour analyses are performed more and more frequently with the use of electronic nose devices. Despite the low applicability of these devices for the monitoring of air pollution as compared with olfactometric techniques, their application is growing. The process of the appropriate selection of chemical sensors

Table 8. Comparison of chemical sensors used for the construction of electronic nose devices

Type of Sensor	Advantages	Disadvantages	Limit of Detection
Amperometric (EC)	high sensitivity, relatively good dynamic properties. Time constant of achieving 90% of response magnitude to a step change of measured gas concentration is typically shorter than 60 seconds, low energy consumption	medium selectivity, dependence of signal on operation temperature, relatively big dimensions	> 0.1 ppb
Semiconductor MOS-type	high sensitivity, low influence of temperature on sensor signal, low dependency of signal on humidity, short response time, fast regeneration, long lifetime, small, dimensions, low price	low selectivity, high operation temperature, high energy consumption, sensitivity to sulphur compounds and alcohols	> 0.1 ppm
Semiconductor CP-type	high sensitivity, short response time, high reproducibility, resistance to poisoning, relatively high selectivity, operation at ambient temperature	sensitivity to temperature and humidity changes, slow regeneration, short lifetime, signal drift, time-consuming process of controlling of this type of sensors	0.1 ppm
PID-type	limit of detection at ppb level, small size, mainly VOC detection	low selectivity, ionization only certain volatile chemical compounds, short lifetime of the lamp ionizing	1 ppb
Surface acoustic wave sensors (SAW)	low limit of detection, high sensitivity, high selectivity, good response time, diverse sensor coatings, small, inexpensive, sensitive to virtually all gases	complex circuitry, temperature sensitive, poor signal-to-noise ratio, specificity to analyte groups affected by polymeric film sensor coating	1 ppb
Quartz crystal microbalance sensor (QCM)	high selectivity, good precision, diverse range of sensor coatings, high sensitivity, short response time	complex circuitry, poor signal-to noise ratio, temperature sensitive, complicated production process precluding sensor reproducibility	> 0.1 ppm

Figure 6. Chemical sensors used for the construction of electronic nose devices

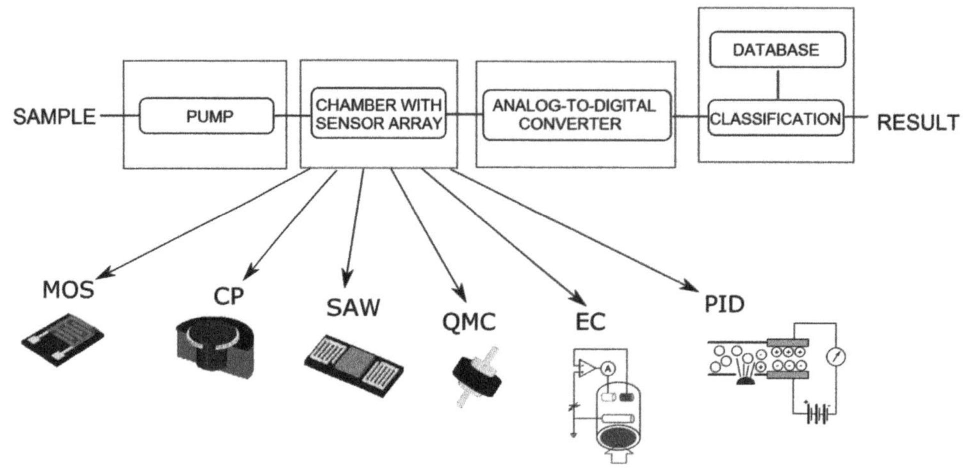

Table 9. Chemical sensors available on the market that are intended for determining odorous gases and are characterized by the limit of quantification below 1 ppm v/v

Gas	Manufacturer	Model	Sensor Type	LOD
H2S	Alphasense	H2S-B4	EC	40 ppb
H2S	Aeroqual	EHS / EHS2	MOS	40 ppb
NH3	Aeroqual	ENG	MOS	0.2 ppm
NH3	Alphasense	NH3-A1	EC	0.3 ppm
VOC	Aeroqual	VOC Sensor Head	MOS	0.1 ppm
VOC	Alphasense	PID-AH2	PID	1 ppb
VOC	Cambridge CMOS Sensors	CCS801	MOS	10 ppm
VOC	Environmental Sesnors CO	Formaldehyde Monitor Z-300	EC	0.1 ppm
VOC	ION Science	MiniPID	PID	5 ppb
VOC	Mocon Baseline Series	10.6, 10.0, 9.6 eV Photoionization Sensor	PID	1-250 ppb
VOC	Unietc SRL	SENS-IT (TF-MOS)	MOS	0.1 ppm
VOC	Unietc SRL	SENS-IT (EC)	EC	0.6 ppb

and the data analysis method allows for using electronic noses for tests of the quality of air polluted with odorous compounds of various origins. One of the main advantages of these devices is the possibility of skipping the first stage of preparing samples for the analysis and the possibility of their field use as a portable unit to monitor emission sources in real time. Table 10 contains information using exemplary applications of electronic nose devices for controlling and assessing air quality about odour emission sources such as landfills and sewage treatment plants. This table presents information about the type of electronic nose used, the chemical sensors it contains and the data analysis method used.

Table 10. Examples of the use of electronic nose devices for the assessment of air quality around municipal landfills and sewage treatment plants

Source of Malodours	Type of Sensors/ Name of Electronic Nose	Reference
Landfill	16 tin oxide	Micone & Guy, 2007
Landfil site	3xEOS835	Capelli, Sironi, Del Rosso, Centola, & Il Grande, 2008
Waste disposal, landfill areas	6 to 8 tin oxide sensors	Romain, Delva, & Nicolas, 2008
Wastewater treatment plant	FOX3000	Dewettinck, van Hege, & Verstraete, 2001
	eNOSE 5000	Bourgeois & Stuetz, 2002
	Neotronics Scientific Ltd. model D;	Onkal-Engin, Demir & Engin, 2005; Stuetz, Fenner, & Engin, 1999
	Pen-2	Nake, Dubreuil, Raynaud, & Talou, 2005; Littarru, 2007
	5× e-nose; 6 MOS	Nicolas, Cerisier & Delva, 2012
	EOS25, EOS28, EOS35	Capelli, Sironi, Centola, del Rosso, & Il Grande, 2008
Landfil	MOS	Gębicki, Dymerski, & Namieśnik, 2016

EXAMPLE OF THE USE OF ELECTRONIC NOSE FOR THE ASSESSMENT OF AIR QUALITY AROUND A MUNICIPAL LANDFILL

The study compares the discriminatory possibilities of two types of electronic noses in the distinction and classification of air samples collected around a municipal landfill (Gębicki, Dymerski & Namieśnik, 2017). One of the noses consists of four commercial solid-state sensors from the FIGARO company (TGS 832, TGS 2600, TGS 2602, TGS 2603), one photo-ionizing sensor of the PID type (PPB MiniPID) and two electrochemical sensors from the Figaro company (FECS44, FECS50) - self-assembled. The other one is based on the ultra-fast gas chromatography technology and it was bought from the Alpha MOS company. This device combines the advantages of gas chromatography with the advantages of electronic noses. Additionally, using Nasal Ranger field olfactometers, odour concentrations in the air were estimated at places where samples were collected for analysis using both types of electronic noses. Figure 7 presents a diagram of the place where the research is performed and samples are collected for the assessment of its quality as regards the odour. This research was conducted in the spring and winter period, and in the spring. The kNN classifier (where k=3) was used as a method of data analysis. On the basis of air sample classification around the municipal landfill, it was observed that the largest number of correctly classified samples came from checkpoints with the highest odour nuisance. The research conducted using a field olfactometer showed that the average odour concentrations at these points ranged from 14.5 to 32.2 [ou/m^3] depending on the season. The correctness of the classification of samples collected at those places was at the level of 100%, regardless of the type of electronic nose used for measurements or the season. At other measuring points, where odour concentrations were low (at the level of detection of a filed olfactometer where this level is 1.7 ou/m^3), the correctness of the classification was at a much lower level. The example of the use of electronic nose devices presented above show universality and a significant practical aspect in the use of electronic nose devices for the assessment of air quality around such sectors of human activity such as a municipal landfill.

Figure 7. Example of the use of electronic nose for the assessment of air quality around a municipal landfill

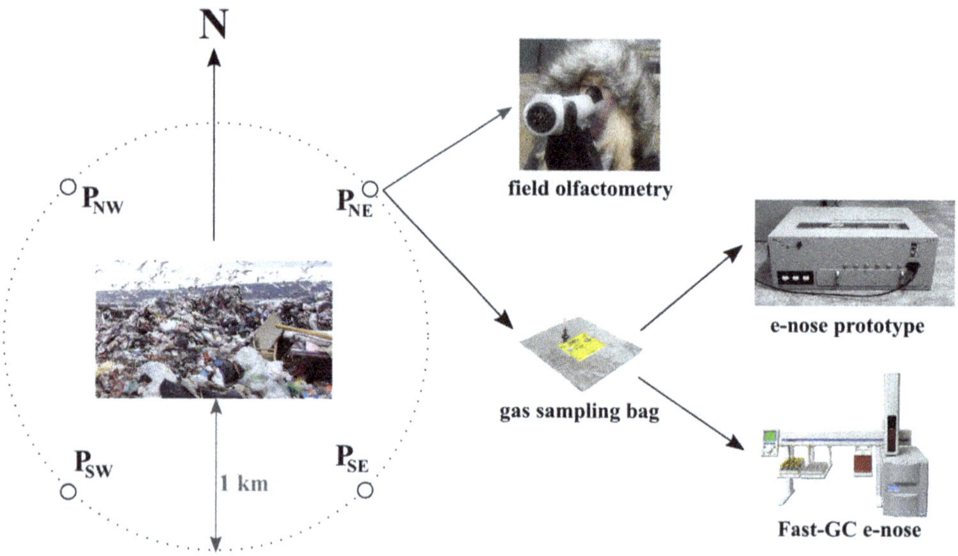

SUMMARY

Electronic nose devices are not commonly used for assessments of air quality with regard to odours due to numerous limitations:

- Variable sensitivity and drift of sensor signals in time,
- Variability of sensor signal depending on ambient atmospheric conditions,
- The influence of temperature and humidity on sensor signals,
- The lack or limitation of legal regulations allowing for the use of this type of device for routines measurements of air quality with regard to odours.

At present, electronic noses are only solutions that supplement sensory analysis methods as regards sensory assessment of atmospheric air. However, in many cases, the subjectivity of panellists who take part in olfactometric assessments may raise doubts. Moreover, the use of olfactometric techniques requires an experienced team whose olfactory sensitivity needs to be controlled on a regular basis, which makes olfactometric analyses labour- and time-consuming.

Potentialities and limitations of particular techniques suggest that the most advantageous solution seems to be combination of a few mutually supplementary approaches. In this way it is possible to obtain full picture of odour nuisance phenomenon, starting from odour generation in different industrial or municipal plants, through emission and imission processes, to evaluation of individual or population hazard due to exposure to malodorous compounds in particular area. Table 11 gathers selected examples of the papers, in which the minimum of 2 techniques were employed for evaluation of odorous properties of the gaseous mixtures originating from different fields of human activity.

Table 11. Examples of the ambient air investigation, in which the minimum of 2 measurement techniques were employed for characterization of odorous compounds

Techniques	Source of Emission	Results: Identified Compounds/Groups of Compounds/ Other Comments	Reference
DO & FO	WWTP	a discrepancy was revealed between the odour concentrations obtained via dynamic and field olfactometry	Barczak, Kulig, & Szyłak-Szydłowski, 2012
GC-MS & DO & E-NOS	Landfill	poor correlation between the odour concentrations obtained with olfactometric technique and the ones determined using concentration of particular compounds and detection threshold; significant potentialities associated with application of electronic nose instruments to identification of odorous compounds presence in air, for example due to installation failure	Capelli, Sironi, Del Rosso, Centola, & Il Grande, 2008
	Landfill	it was revealed that due to limitation of applied methods, only combination of many techniques allowed precise determination which odorous compounds have the biggest impact on environment	Romain, Delva, & Nicolas, 2008
GC-MS & DO	Landfill site	correlation was revealed between odour concentrations and total concentrations of volatile organic compounds in the investigated gas samples; it was shown that such groups of compounds as aldehydes, ketones and esters were the best indicators of elevated odour concentration;	Dincer, Odabasi, & Muezzinoglu, 2006
GC-MS & 8 E-NOS	Municipal solid waste	in most cases a good correlation was found between the results obtained with GC-MS and electronic nose;	Delgado-Rodriguez et al., 2012
GC-MS-O & DO & E-NOS	Waste management plant	demonstrated the potential of the techniques used to identify, distinguish and quantify odor concentrations of compounds released by the operation of waste management facilities	Giungato et al., 2016
DO & E-NOS	WWTP	complementary character of both presented techniques for efficient monitoring of odour concentration was revealed; it was found that both presented approaches provide objective, repeatable and reproducible results, typical for traditional chemical methods.	Littarru, 2007
	Municipal waste plant	it was found that sensor sets are an interesting solution for monitoring of odours.	Romain, Molitor, Adam, & Bietlot, 2016

abbreviations: DO – dynamic olfactometry; E-NOS – electronic nose; FO – field olfactometry; GC-MS – gas chromatography – mass spectrometry; GC-O – gas chromatography – olfactometry; WWTP – wastewater treatment plant.

Literature reports that are devoted to theoretical bases and principles of electronic nose operation and practical possibilities of their use are the basis for concluding that the development of odour analysis is connected with the use of electronic noses. It seems that many countries will soon extend reference methods in the analysis of malodorous substance to include the possibility of using the electronic nose; this mostly applies to countries where such legal regulations exist. While observing the potential and the rate of development, it can be supposed that the role of electronic noses in odour monitoring in the air will keep growing and this technique, apart from supplementing olfactometric techniques may also replace them in some aspects. For short-term odour nuisance (episodes), it will be possible to register them using electronic nose devices, which, in turn, is often not possible after qualified personnel arrives on site to perform field olfactometry measurements.

ACKNOWLEDGMENT

The research was supported by the Grant No. PBSII/B9/24/2013 from the National Centre for Research and Development.

REFERENCES

Bai, H., & Shi, G. (2007). Gas Sensors Based on Conducting Polymers. *Sensors (Basel)*, *7*(3), 267–307. doi:10.3390/s7030267

Barczak, R., Kulig, A., & Szyłak-Szydłowski, M. (2012). Olfactometric methods application for odour nuisance assessment of wastewater treatment facilities in Poland. *Chemical Engineering Transactions*, *30*, 187–192.

Bourgeois, W., & Stuetz, R. M. (2002). Use of a chemical sensor array for detecting pollutants in domestic wastewater. *Water Research*, *36*(18), 4505–4512. doi:10.1016/S0043-1354(02)00183-5 PMID:12418653

Byliński, H., Lewkowska, P., Gębicki, J., Dymerski, T., & Namieśnik, J. (2016). Identification of aromatic compounds in odours mixture by gas chromatography and field olfactometry techniques. *Proceedings of the 12th Modern Analytical Chemistry*, 108–112.

Capelli, L., Sironi, S., Del Rosso, R., & Centola, P. (2009). Predicting odour emissions from wastewater treatment plants by means of odour emission factors. *Water Research*, *43*(7), 1977–1985. doi:10.1016/j.watres.2009.01.022 PMID:19232669

Capelli, L., Sironi, S., Del Rosso, R., Centola, P., & Il Grande, M. (2008). A comparative and critical evaluation of odour assessment methods on a landfill site. *Atmospheric Environment*, *42*(30), 7050–7058. doi:10.1016/j.atmosenv.2008.06.009

Capelli, S., Sironi, P., Centola, R., del Rosso, R., & Il Grande, M. (2008). Electronic noses for the continuous monitoring of odours from a wastewater treatment plant at specific receptors: Focus on training methods. *Sensors and Actuators. B, Chemical*, *131*(1), 53–62. doi:10.1016/j.snb.2007.12.004

Czerny, M., Christlbauer, M., Christlbauer, M., Fischer, A., Granvogl, M., Hammer, M., ... Schieberle, P. (2008). Re-investigation on odour thresholds of key food aroma compounds and development of an aroma language based on odour qualities of defined aqueous odorant solutions. *European Food Research and Technology*, *228*(2), 265–273. doi:10.1007/s00217-008-0931-x

Delgado-Rodriguez, M., Ruiz-Montoya, M., Giraldez, I., Lopez, R., Madejon, E., & Diaz, M. J. (2012). Use of electronic nose and GC-MS in detection and monitoring some VOC. *Atmospheric Environment*, *51*, 278–285. doi:10.1016/j.atmosenv.2012.01.006

Dewettinck, T., van Hege, K., & Verstraete, W. (2001). The electronic nose as a rapid sensor for volatile compounds in treated domestic wastewater. *Water Research, 35*(10), 2475–2483. doi:10.1016/S0043-1354(00)00530-3 PMID:11394782

Dincer, F., Odabasi, M., & Muezzinoglu, A. (2006). Chemical characterization of odorous gases at a landfill site by gas chromatography-mass spectrometry. *Journal of Chromatography. A, 1122*(1-2), 222–229. doi:10.1016/j.chroma.2006.04.075 PMID:16697389

Fang, J. J., Yang, N., Cen, D. Y., Shao, L. M., & He, P. J. (2012). Odor compounds from different sources of landfill: Characterization and source identification. *Waste Management (New York, N.Y.), 32*(7), 1401–1410. doi:10.1016/j.wasman.2012.02.013 PMID:22480726

Gallego, E., Perales, J. F., Roca, F. J., & Guardino, X. (2014). Surface emission determination of volatile organic compounds (VOC) from a closed industrial waste landfill using a self-designed static flux chamber. *The Science of the Total Environment, 470-471*, 587–599. doi:10.1016/j.scitotenv.2013.09.105 PMID:24176707

Gebicki, J. (2016). Application of electrochemical sensors and sensor matrixes for measurement of odorous chemical compounds. *Trends in Analytical Chemistry, 77*, 1–13. doi:10.1016/j.trac.2015.10.005

Gębicki, J., Dymerski, T., & Namieśnik, J. (2016). Evaluation of Ambient Air Odour Quality in Vicinity of Municipal Landfill Using Electronic Nose Technique. *Chemical Engineering Transactions, 54*, 253–258.

Gębicki, J., Dymerski, T., & Namieśnik, J. (2017). Investigation of Air Quality beside a Municipal Landfill: The Fate of Malodour Compounds as a Model VOC. *Environments, 4*(1).

Giungato, P., de Gennaro, G., Barbieri, P., Briguglio, S., Amodio, M., de Gennaro, L., & Lasigna, S. (2016). Improving recognition of odors in a waste management plant by using electronic noses with different technologies, gas chromatography-mass spectrometry/olfactometry and dynamic olfactometry. *Journal of Cleaner Production, 133*, 1395–1402. doi:10.1016/j.jclepro.2016.05.148

Hansen, M. J., Jonassen, K. E. N., & Feilberg, A. (2014). Evaluation of Abatement Technologies for Pig Houses by Dynamic Olfactometry and On-site Mass Spectrometry. *Chemical Engineering Journal, 40*, 253–258.

Harynuk, J., & Górecki, T. (2004). Comprehensive two-dimensional gas chromatography in stop-flow mode. *Journal of Separation Science, 27*(5-6), 431–441. doi:10.1002/jssc.200301649 PMID:15335078

Hung, Y. T., Wang, L. K., & Shammas, N. K. (Eds.). (2012). Handbook of Environment and Waste Management: Air and Water Pollution Control. World Scientific Publishing Co. Pte. Ltd.

Kośmider, J., & Krajewska, B. (2007). Determining temporary odour concentration under field conditions - Comparison of methods. *Polish Journal of Environmental Studies, 16*, 215–225.

Lehtinen, J., & Veijanen, A. (2011). Odour monitoring by combined TD-GC-MS-sniff technique and dynamic olfactometry at the wastewater treatment plant of low H2S concentration. *Water, Air, and Soil Pollution*, *218*(1–4), 185–196. doi:10.1007/s11270-010-0634-3

Littarru, P. (2007). Environmental odours assessment from waste treatment plants: Dynamic olfactometry in combination with sensorial analysers "electronic noses,". *Waste Management (New York, N.Y.)*, *27*(2), 302–309. doi:10.1016/j.wasman.2006.03.011 PMID:16766175

Mahin, T. D. (2001). Comparison of different approaches used to regulate odours around the world. *Water Science and Technology*, *44*(9), 87–102. PMID:11762488

Micone, P. G., & Guy, C. (2007). Odour quantification by a sensor array: An application to landfill gas odours from two different municipal waste treatment works. *Sensors and Actuators. B, Chemical*, *120*(2), 628–637. doi:10.1016/j.snb.2006.03.026

Naddeo, V., Zarra, T., Giuliani, S., & Belgiorno, V. (2012). Odour Impact Assessment in Industrial Areas. *Chemical Engineering Transactions*, *30*, 85–90.

Nagata, Y. (2003). Measurement of Odor Threshold by Triangle Odor Bag Method. *Odor Measurement Review*, 118–127.

Nake, A., Dubreuil, B., Raynaud, C., & Talou, T. (2005). Outdoor *in situ* monitoring of volatile emissions from wastewater treatment plants with two portable technologies of electronic noses. *Sensors and Actuators. B, Chemical*, *106*(1), 36–39. doi:10.1016/j.snb.2004.05.034

Nicell, J. A. (2009). Assessment and regulation of odour impacts. *Atmospheric Environment*, *43*(1), 196–206. doi:10.1016/j.atmosenv.2008.09.033

Nicolas, J., Cerisier, C., & Delva, J. (2012). Potential of a Network of Electronic Noses to Assess in Real Time the Odour Annoyance in the Environment of a Compost Facility. *Chemical Engineering Transactions*, *30*, 133–138.

Onkal-Engin, G., Demir, I., & Engin, S. N. (2005). Determination of the relationship between sewage odour and BOD by neural networks. *Environmental Modelling & Software*, *20*(7), 843–850. doi:10.1016/j.envsoft.2004.04.012

Ritvay, D., & Kovács, D. (2006). The international and Hungarian practice of odor emission control (in Hungarian). National Environmental Conference, Siófok, 19-21 September 2006, Romain, A.C., Delva, J., & Nicolas J. (2008). Complementary approaches to measure environmental odours emitted by landfill areas. *Sensors and Actuators. B, Chemical*, *131*, 18–23.

Romain, A., Molitor, N., Adam, G., & Bietlot, E. (2016). Comparison " Low Cost Chemical Sensors – Analytical Instruments " for Odour Monitoring in a Municipal Waste Plant. *Chemical Engineering Transactions*, *54*, 211–216.

Stuetz, R. M., Fenner, R. A., & Engin, G. (1999). Assessment of odours from sewage treatment works by an electronic nose, H2S analysis and olfactometry. *Water Research*, *33*(2), 453–461. doi:10.1016/S0043-1354(98)00246-2

Talaiekhozani, A., Bagheri, M., Goli, A., & Talaei Khoozani, M. R. (2016). An overview of principles of odor production, emission, and control methods in wastewater collection and treatment systems. *Journal of Environmental Management*, *170*, 186–206. doi:10.1016/j.jenvman.2016.01.021 PMID:26829452

Trincavelli, M., Coradeschi, S., & Loutfi, A. (2009). Odour classification system for continuous monitoring applications. *Sensors and Actuators. B, Chemical*, *139*(2), 265–273. doi:10.1016/j.snb.2009.03.018

Wenjing, L., Zhenhan, D., Dong, L., Jimenez, L. M. C., Yanjun, L., Hanwen, G., & Hongtao, W. (2015). Characterization of odor emission on the working face of landfill and establishing of odorous compounds index. *Waste Management (New York, N.Y.)*, *42*, 74–81. doi:10.1016/j.wasman.2015.04.030 PMID:25997990

Zhu, J., Chen, F., Wang, L., Niu, Y., & Xiao, Z. (2017). Evaluation of the synergism among volatile compounds in Oolong tea infusion by odour threshold with sensory analysis and E-nose. *Food Chemistry*, *221*, 1484–1490. doi:10.1016/j.foodchem.2016.11.002 PMID:27979119

Chapter 10
Electronic Noses for Indoor Air Quality Assessment

Tomasz Majchrzak
Gdańsk University of Technology, Poland

Wojciech Karol Wojnowski
Gdańsk University of Technology, Poland

Tomasz Dymerski
Gdańsk University of Technology, Poland

Jacek Gębicki
Gdańsk University of Technology, Poland

Jacek Namieśnik
Gdańsk University of Technology, Poland

ABSTRACT

This chapter presents a proposal of the use of electronic noses in the monitoring of indoor air quality. The main focus is put on the detailed characteristics of today's indoor air quality control methods, the types of pollution in the air, and the development of electronic noses for air testing. Currently, scientists seek methodological and structural solutions that would enable real-time online indoor air control. It has been shown that using electronic noses in this situation is advantageous. In addition, potential uses of these devices are discussed, with particular focus on closed food processing spaces. The authors of the chapter argue that in the near future, the proposed solution could improve the quality of indoor air and thus the health of the users of the indoor environments, as well as the quality of the products prepared there.

DOI: 10.4018/978-1-5225-3862-2.ch010

INTRODUCTION

The composition of inhaled air has a direct impact on human health and well-being which is why it is crucial to examine the air quality in the environment. This applies to both outdoor and indoor air. Indoor air is a gas mixture in rooms, buildings and vehicles, which may be contaminated with compounds emitted by structural elements or room furnishings. Contamination can also come from outside the building, or from other rooms in the building. The composition of indoor air can change over time. It can be influenced by many factors such as the frequency of air exchange in the room, the temperature and the relative humidity of the air, or the manner of air exchange. The degree of air pollution in a room may also depend on the activities performed in it. According to various estimates, a person spends 70% – 90% of their life in closed spaces. These values refer to the time a person spends in houses, flats, public buildings, schools, offices, shops, factories, and means of transport. According to research conducted in the United States, on average people spend 88% of time in buildings, 7% in means of transport, and only 5% outside. In the gas mixture of a given room, there can be organic and inorganic compounds. These compounds can directly affect human health or be involved in chemical reactions that result in secondary pollution, which could be significantly more harmful than primary pollution. The topic of indoor air quality control is one of many fields of studies that have not been sufficiently researched/elaborated on. One of the priority needs for indoor air monitoring is to develop a tool to control the quality of air in real time. Sensory systems utilizing electronic nose technology are the perfect solution.

The authors of this chapter would like to review current and prospect solutions that use electronic noses to control indoor air quality. They believe that, in the first place, such a solution could be used in food processing plants, particularly in places where frying processes occur. The use of electronic noses in such places can not only improve the indoor quality, and thus the health of the people indoors, but also assure the quality of the products prepared there. This work also covers a whole range of aspects related to methods of indoor air quality control, pollution therein, as well as future directions for the development of electronic noses for indoor air testing.

BACKGROUND

History of IAQ Control

In 1962, a book entitled "Silent Spring" by Rachel Carson was published. Paradoxically, increasing pro-environmental awareness caused indoor air quality not to be of concern to environmental researchers. Interest in indoor air increased drastically after the 1960s. Threats connected with radioactivity or the presence of formaldehyde in the air, as well as the emergence of the term sick building syndrome (SBS) in the late 1970s have brought renewed interest in indoor air quality issues to agencies in the field of environmental quality assessment. On the other hand, it became clear that the condition of indoor air depends on many factors, including the state of the external environment. In the 1970s there was an energy crisis connected with the rise in fuel prices, which forced manufacturers to develop new insulation materials such as PVC, mineral wool or styrofoam. Consequently, this has led to excessive sealing of buildings and accumulation of harmful chemicals in enclosed spaces. In 1982, the World

Health Organization presented the definition of SBS. According to this definition, a sick building is a place where at least 20% of its inhabitants state that staying in it contributes to their ill health. These symptoms disappear when the users leave the building. SBS can also concern specific rooms. The most common symptoms include chronic headaches, skin sensitization, irritation of the mucous membranes of the nose, eyes and throat, trouble with concentration, chronic fatigue and respiratory, digestive, and nervous system problems. It has been noted that staying in a sick building results in decreased employee productivity. The most common SBS cases concern older buildings and rooms. This syndrome also affects modern, so-called 'airtight' buildings equipped with air conditioning and mechanical heating systems. A timeline of the development of IAQ control methods and electronic olfaction is depicted in Figure1.

Currently, reliable methods for assessing indoor air quality are being sought. The basic criterion for these methods is low price of a single measurement, short analysis time, adequate sensitivity and versatility of the proposed solution. At the present time, there is no one universal method that would meet the aforementioned requirements. Depending on needs, it is possible to use sensory analysis, gas analysers or instrumental analysis techniques. The most commonly used instrumental analysis technique is gas chromatography.

Figure 1. Development of IAQ analysis and electronic noses

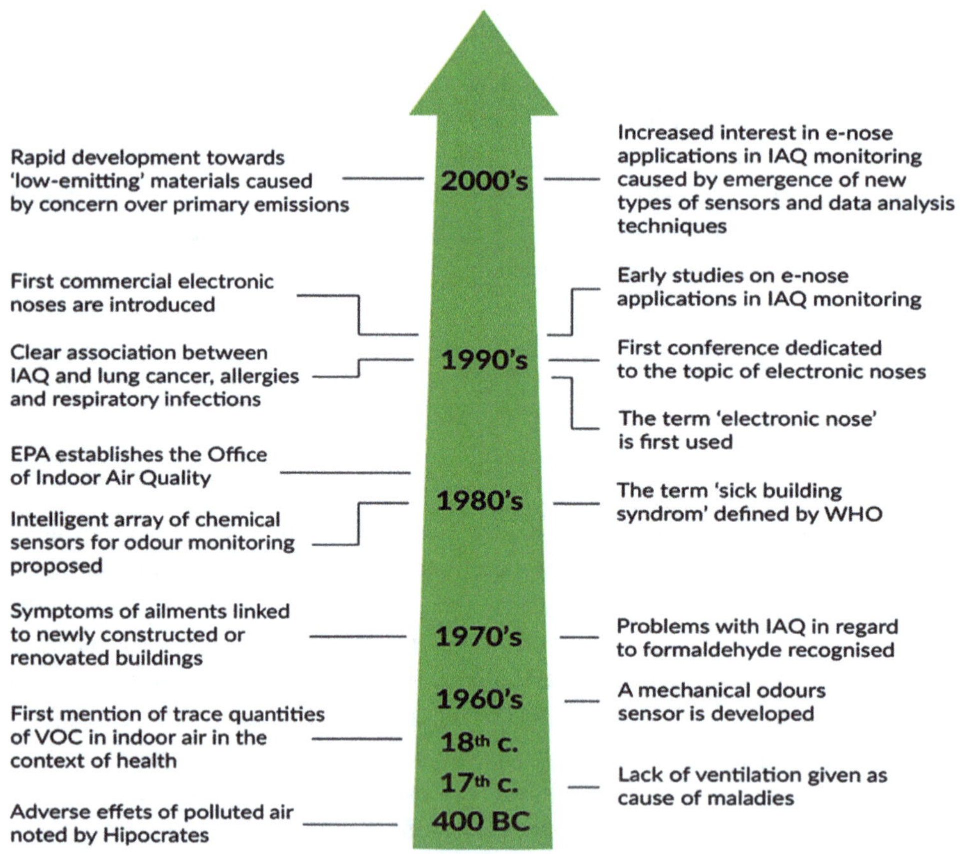

Sources of IA Contaminants

According to the information provided by US EPA, there are over 1000 different internal sources of pollution emission into the indoor air. Nearly 60 types of sources are responsible for the emission of carcinogens. The most frequently mentioned and most important sources of harmful substances include: furniture components, floor and wall coverings, thermal insulation, decorative or maintenance materials, but also emissions from heating, combustion of oil, coal, gas and tobacco, and emission from ventilation systems. It is worth noting that the composition of the air inside the building is also influenced by emission sources outside it.

The atmospheric air in the vicinity of a building may have an adverse effect on the quality of the indoor air. Airborne impurities enter a building through its ventilation system. The composition of the air entering the building depends on the area where the building is located. It could be an agricultural, industrial or urbanized area. The building can also be located near landfill sites, highways, sewage plants or airports. Indoor air quality can also be influenced by factors such as atmospheric conditions: wind strength and direction, temperature and insulation.

Selecting suitable materials for construction does not only affect the strength of a structure, its capacity or thermal insulation, but also the indoor air quality. Contamination sources include paints, varnishes and structural coatings and preservatives. For example, building materials such as concrete or bricks may be the source of radioactive radon emission. Wood materials (particle boards, plywood) can emit formaldehyde which is present in synthetic resins, into the indoor air. Indoor air pollutants which are associated with certain types of rooms are listed in Table 1.

As mentioned above, the quality of indoor air may be influenced by the emission of pollutants from materials and objects in the room or the migration of pollutants from the outdoor air. However, situations in which indoor air quality may be related to activities performed inside the building must be taken into account. Depending on their purpose, rooms can be divided into residential, office, or public utility (hospitals, museums, schools, or churches). Rooms can be used for sleeping, hygiene, health care,

Table 1. Indoor air pollutants occurring in different types of rooms

Type/Function of a Room	Indoor Air Pollutants
Offices	VOCs, BTEX, chlorinated hydrocarbons and halogenated organic compounds (Chao & Chan, 2001)
Museums	BTEX, naphthalene, benzoic acid, limonene,SO_2, NO_2, O_3, formic acid, acetic acid (Krupińska, Van Grieken, & De Wael, 2013), formaldehyde, styrene, benzaldehyde (Schieweck, Lohrengel, Siwinski, Genning, & Salthammer, 2005)
Libraries	VOCs, formaldehyde (Chao & Chan, 2001), SO_2, NO_2 (Andretta, Coppola, & Seccia, 2016)
Temples and churches	NO_2 (Worobiec et al., 2006), formaldehyde, BTEX (Zhang, Chen, Li, Yu, & Zhao, 2015)
Schools	VOCs, SVOCs, polychlorinated biphenyls (Herrick, Stewart, & Allen, 2016), D- limonene, formaldehyde, acetic aldehyde, benzene
Hospitals	VOCs, radon (Śmiełowska, Marć, & Zabiegała, 2017), N_2O, formaldehyde, glutaraldehyde, ethylene oxide (Zeiger, Gollapudi, & Spencer, 2005), BTEX (Dascalaki, Lagoudi, Balaras, & Gaglia, 2008), acetaldehyde, phthalates (Wang et al., 2015)
Elderly care centres	Nitrogen oxides, formaldehyde, VOCs (Mendes et al., 2015)
Underground parking	BTEX (Marć, Śmiełowska, & Zabiegała, 2016)

entertainment, as well as for preparing and consuming food. In this chapter, the authors focus on indoor air quality in food preparation and processing facilities.

Indoor Air in Food Processing and Distribution Areas

There are many types of food processing and distribution areas. Three of them should be highlighted in terms of potential indoor air contaminant sources:

IAQ in Greenhouses

In closed food production halls, especially greenhouses, the quality of indoor air may depend on the agricultural practices that are carried out there. Pollution emissions may depend on parameters such as temperature, humidity, the stage of plant development, treatments (weeding, fertilization) and the applied plant protection products. It has been shown that some pesticides may be present in indoor air and dust (Roinestad, Louis, & Rosen, 1993). Pesticides and their metabolites can cause various diseases in humans and be dangerous to nervous, digestive or reproductive systems.

IAQ in Warehouses

Food should be stored and transported under specified conditions. However, it is often the case that parameters such as high temperature or excessive humidity are permanently or periodically present in food storages. Under the influence of these parameters, and as a result of the presence of bacteria, food can become spoiled. The direct effect of food degradation is its lower quality, but also reduced quality of air inside the storage area. Spoilage organisms typically present in food are fungi and bacteria. For example, under aerobic conditions at low temperatures organisms such as *Pseudomonas* and *Bacillus* can develop. When access to oxygen is limited, yeast, *Enterobacteriaceae*, *Photobacterium*, clostridia, various fungi and lactic acid bacteria (LAB) can proliferate. The amount and type of microorganisms responsible for food spoilage depend primarily on temperature, oxygen level, pH and substrate (growth medium). Volatile compounds, including inorganic gases and microbial volatile organic compounds (MVOCs) are a byproduct of microbial nutrition. The inorganic compounds resulting from microbial activity are hydrogen sulphide, ammonia and carbon dioxide. The MVOC group includes organic sulfur compounds such as dimethyl disulphide, trimethylamine, certain esters, organic acids, diacetyl or biogenic amines. These compounds not only cause the unpleasant odor associated with food spoilage, but may also exhibit toxic and carcinogenic properties. It is also dangerous to inhale air where hyphae and bacteria are present. In addition, in some warehouses, silos or grain elevators, there is a risk of inhaling particles emitted from dusty materials – for example from flour. In such cases, workers' respiratory tracts are the most vulnerable, and symptoms such as dyspnea or cough can appear. It is possible for diseases such as pneumoconiosis to emerge. In order to improve safety, workers in this type of food warehouses should use personal protective equipment.

IAQ in Restaurants and Food Processing Plants

In restaurants, the quality of indoor air can be influenced primarily by the way food is prepared. Depending on the profile of the restaurant and the frequency of food preparation, there can be emission of chemical compounds which are harmful and generate unpleasant odor. The following parameters can determine the type of emitted compounds: type, quality and freshness of the food used, the type of food preparation process (cooking, frying, smoking) and the type and quality of the frying oil. Depending on the type of restaurant, the composition of the gases produced during cooking may vary. Comparing food processing procedures such as grilling, immersion cooking, steam cooking and classic cooking shows that the worst air quality can be observed in restaurants where food is prepared by grilling and frying. It has been noted that in this type of restaurants the highest concentration of formaldehyde, benzene, toluene and chloroform is present. In addition, PM_{10} and $PM_{2,5}$ particles, which are hazardous to human health, are emitted. It is worth noting that some of the determined chemical compounds, for example formaldehyde, are described as being emitted from building materials and furniture in a given space more often than from food preparation processes. However, it is clear that food preparation processes have a direct effect on the quality of indoor air and therefore on the comfort of the users – both employees and customers. Some compounds the presence of which in the air is associated with particular food processing and production areas are listed in Table 2.

INDOOR AIR CONTAMINANTS

World Health Organisation in its guidelines regarding chemicals that are commonly present in indoor air lists those which and can be potentially harmful to human health, e.g. benzene, carbon monoxide, formaldehyde, naphtalene, nitrogen dioxide, polycyclic aromatic hydrocarbons (PAHs), trichloroethylene and tetrachloroethylene. Apart from outdoor pollutants which make their way into indoor environment the main sources of these volatiles are building materials and furniture, and also the burning of fuel either for cooking or heating. Mentioned above are chemical compounds that are considered to pose risk to public health due to their properties or abundance. Some sources of indoor air contaminants are depicted in Figure 2.

Table 2. Chemical compounds occurring in indoor air associated with various food processing and production areas

Food Industry and Food Production Area	Compounds
Frying process	1-pentanol, hexenal, furfuryl alcohol, (E) -2- heptenal, 5-methylfuran, (E) -2- octenal, nonanal, (E) -2-nonenal, hexadecanoic acid and pyrazine (Takeoka, Perrino, & Buttery, 1996).
Meat and meat products production	Acetone, Carbon disulphide, Dimethyl disulphide, Dimethyl trisulphide, Ethanol, Ethyl acetate, Heptane, Hexanal, Hydrogen sulphide, Methyl benzoate, Methyl thioacetate, Toluene, 2,3-Butadienol, 2,5-Dimethylpyrazine, 2-Heptanone, Methanol, Methoxybenzene, 5-Methylpyrimidine, 2-Octanone, 2-Octenal, Phenol, Phenyl ethyl alcohol, Isoamyl acetate, methanethiol (Alexandrakis, Brunton, Downey, & Scannell, 2012; Senter, Arnold, & Chew, 2000)
Beverage production (distilleries)	Ethanol, octyl acetate, nonanal, limonene, linalool, 1-hexanol, benzaldehyde, ethyl cinnamate, γ-Decalactone, 2,4-decadienal, 2,6-nonadienal (Śliwińska, Wiśniewska, Dymerski, Wardencki, & Namieśnik, 2015)
Food storage (including spoilage)	Esters, trimethyl amine (TMA), NH_3, H_2S, dimethyl disulphide, 2-methoxy-phenol, trichloroanisol (Gram et al., 2002)
Mould and fungus in food industry	3-methylfuran, Dimethyl disulfide, Limonene, Pinene, 1-hexanol, 1-pentanol, 2-methyl-1-propanol, 2-methyl-1-butanol, 1-octen-3-ol, geosmin, 2-methyl-isoborneol, 3-octanone, 2-heptanone, 2-pentanone (Kuske, Romain, & Nicolas, 2005)

Figure 2. Some sources of indoor air contaminants

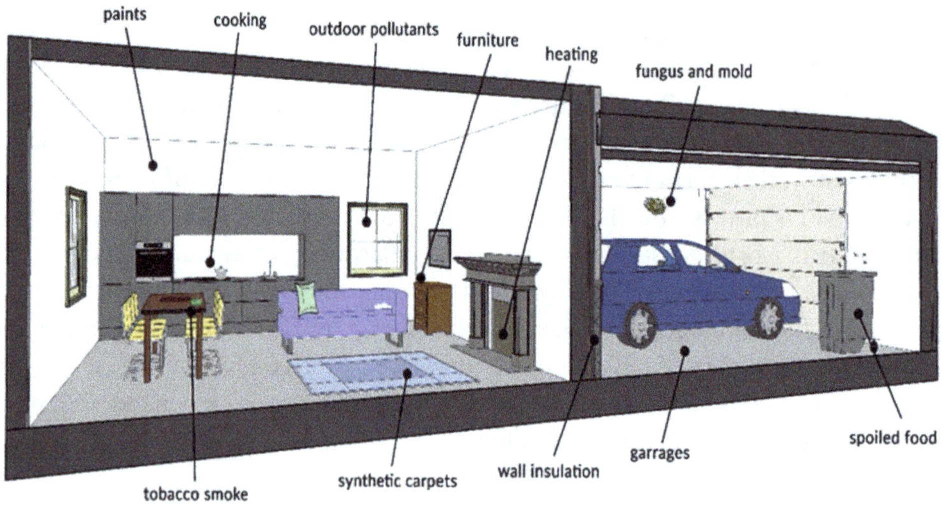

The determination of the above-mentioned compounds does not provide exhaustive information regarding indoor air quality. In reality, the constituents of air present at low concentration levels are numerous and their impact on human wellbeing is difficult to quantify, especially when one considers the effect not of a single gaseous substance, but of their complex mixture. Although construction and synthetic materials are the main sources of volatile compounds in indoor environments, emissions from processed and unprocessed food, potted plants, etc., should not be disregarded. For that reason, summary parameters, e.g. total volatile organic compounds (TVOC), were introduced, which, however, are not sufficient to determine air quality. An alternative approach is to measure the ambient air holistically, using an array of various sensors – electronic noses. This is especially important because of the growing popularity of the concept of so-called green buildings. Green buildings are buildings that are designed to promote efficient use of energy, water and materials, as well as sustainability. According to US EPA, green buildings are "the practice of creating structures and using processes that are environmentally responsible and resource-efficient throughout a building's life-cycle from siting to design, construction, operation, maintenance, renovation and deconstruction." (EN Standard 15251, 2007). It appears that staying in this type of buildings can have a negative effect on human health and well-being. Despite the fact that most studies have shown higher air quality in green buildings compared to that in traditional construction, the risk to human health cannot be ruled out for the former. It is believed that some "green" practices such as use of waste-based materials or economical ventilation of rooms may adversely affect indoor air quality. Using new construction materials or new ventilation solutions may lead to the emergence of new, hitherto unknown, air pollutants. Therefore, the authors of the chapter believe that the most appropriate way to monitor indoor air quality would be to use sensor systems. In order to develop a universally applicable solution, suitable both for "green" solutions and traditional construction, full air quality information should be used. One of the most important challenges would then be the extraction and statistical analysis of data, as well as development of a monitoring network in the building, which would provide reliable and reproducible results.

CLASSICAL IAQ ASSESMENT METHODS

Sensory Analysis

Sensory analysis is a measurement method that uses one or more senses such as sight, taste, smell or touch. Due to the direct nature of the measurement, sensory analysis is a commonly used tool for evaluating people's reactions to stimuli in the environment (Lewkowska, Dymerski, Gębicki, & Namieśnik, 2016). In order to understand in detail how the quality of internal air can be determined using this method, it is necessary to understand how the human sense of smell works. Olfactory receptors are located in an area of about 3 cm^3 in the nasal mucosa. The olfactory epithelium contains bipolar neurosensory cells, basal cells and sertoli cells. Bipolar neurosensory cells together with mitral cells form the olfactory bulb. Mitral cells are responsible for the supply and discharge of nerve impulses. They are connected with other neurons which are located in the hematoma. There, information concerning smell is recorded and processed.

Due to the fact that sensory analysis is based on the subjective assessment of the panelist, it is necessary to adhere to the appropriate measurement principles. Following the rules and using statistical methods can lead to reliable and reproducible results. Sensory analysis can be carried out in many ways. In terms of procedure, sensory analysis methods can be divided into four basic groups: differential, scaling, descriptive analysis and threshold determination. In differential methods, a group of trained panelists determines the similarities and differences between test samples and a reference sample. The intensity of the differences is not assessed, only a similarity or dissimilarity is indicated. Scaling methods are used to determine the scale of odour intensity differences. Numerical or graphical scales are often used to present results. Descriptive analysis is based on allocating the appropriate descriptors, that is the characteristics of the odour, to a specific odour. One of the most commonly used descriptive methods is Quantitative Descriptive Analysis (QDA), where panelists themselves create a list of descriptors they use to describe the tested sample. This measurement is therefore individual and simultaneous. Threshold determination techniques are used to detect or identify substances based on their odour detection thresholds.

Sensory analysis methods are used in indoor air assessment and can be used to identify the source of odorous compounds emissions in various rooms, such as offices, department stores or classrooms. In addition, sensory analysis may be complementary to gas chromatography tests. Olfactometry has also been used to assess indoor air quality in livestock buildings.

Total Volatile Organic Compounds (TVOCs) Measurement

In the past, the commonly used parameter for assessing indoor air quality was the concentration of carbon dioxide. Nowadays, this type of solution would not work, because there are numerous other chemical compounds in the air, mainly volatile organic compounds. The measurement of the concentration of these compounds is currently an indicator of indoor air quality. One of the most common parameters used to determine the concentrations of VOCs is the parameter termed Total Volatile Organic Compounds (TVOC). Because of the simplicity of measurement and relatively short analysis time, the TVOC determination is used in IAQ.

By determining TVOC values, comprehensive information on volatile organic compounds is obtained. This brings with it legitimate advantages, such as a holistic approach to the problem. The TVOC parameter describes the nature of a gas mixture quantitatively. It provides quantitative information even when the qualitative composition of the mixture changes. At the same time, it is one of the biggest limitations of this solution. Without knowing the composition of indoor air, we cannot determine the impact of chemicals on human health. Similarly, identifying the source of pollution can be difficult. It should be emphasized that only in an ideal scenario would TVOC represent the total concentration of volatile organic compounds present in an indoor air sample. In reality, the detectors such as FID, PID, or PAS are used to determine the TVOC parameter. In the first case, by using a flame ionization detector, a response signal is obtained through combustion of the substance in a hydrogen flame. Although the FID detector's response is very stable, not every volatile organic compound can give an analytical signal. In addition, two different VOCs with the same amount of carbons in the molecule, at the same concentration levels in the sample, can generate different signals. However, these detectors are most often used to determine the TVOC parameter. Using photoionization detectors (PIDs), it is possible to generate organic compound ions under the influence of UV radiation. For many VOCs, they are more sensitive than FIDs, but may be less stable, and compounds such as chlorinated VOCs are not ionized. A photo-acoustic (PAS) sensor can also be used to determine TVOCs. The signal is obtained as a result of both the change in vapor concentration of organic compounds caused by infrared radiation, and the temperature increase measurement. In this sensor, a change in acoustic wave frequency causes changes in the intensity of infrared radiation. These however are not as popular as older types detectors, since water vapor or methane concentration can lead to interference. It is worth emphasizing that usually these detectors are calibrated for one pure substance, such as n-hexane or toluene, thus the measurement may not produce sufficiently reliable results when measuring TVOC for a more complex matrix.

Passive and Active Sampling Coupled With Gas Chromatography

One of the major challenges in analyzing indoor air pollutants using gas chromatography is the method of sampling. When choosing a sampling technique, one should aim to obtain the most representative sample, enrich the analytes and reduce the matrix effect. Depending on the needs, the sampling method should allow the measurement of pollutants either for a long period of time (days, weeks, months), or for the moment in which the sample is being taken. The first solution involves continuous exposure to a given contaminant, while the second solution can be used for abnormal situations. Therefore, two main methods for sampling gas in closed spaces have evolved: the passive method and the active method. The principle of operation of both sampling methods is depicted in Figure 3.

Passive dosimetry is based on the extraction of analytes in the air without the use of forced airflow. This type of sampling is characterized by high susceptibility to air movements therefore its use in the assessment of outdoor air is very limited. However, because dosimeters are small devices that do not need to be plugged in and do not produce unpleasant sounds, they can be used indoors without disturbing the users of the room. Passive dosimetry is just as accurate as the active methods. The main drawback of this type of sampling is the relatively low sampling frequency. The result is most often presented as a time-weighted average (TWA) concentration. This type of approach is better suited for determining the long-term impact of harmful substances on the health of room users than the short-term concentrations measurement. For passive measurements, dosimeters are used, whose principle is based on diffusion or permeation phenomena. However, it should be remembered that the method of sampling depends mainly on the aim that is to be achieved. If we strive to determine the average concentration, passive dosimetry is more appropriate, but if we want to determine concentration levels over time, the active methods are better.

Figure 3. Development of IAQ analysis and electronic noses passive (a) and active (b) methods of sampling of volatile compounds present in the indoor environment

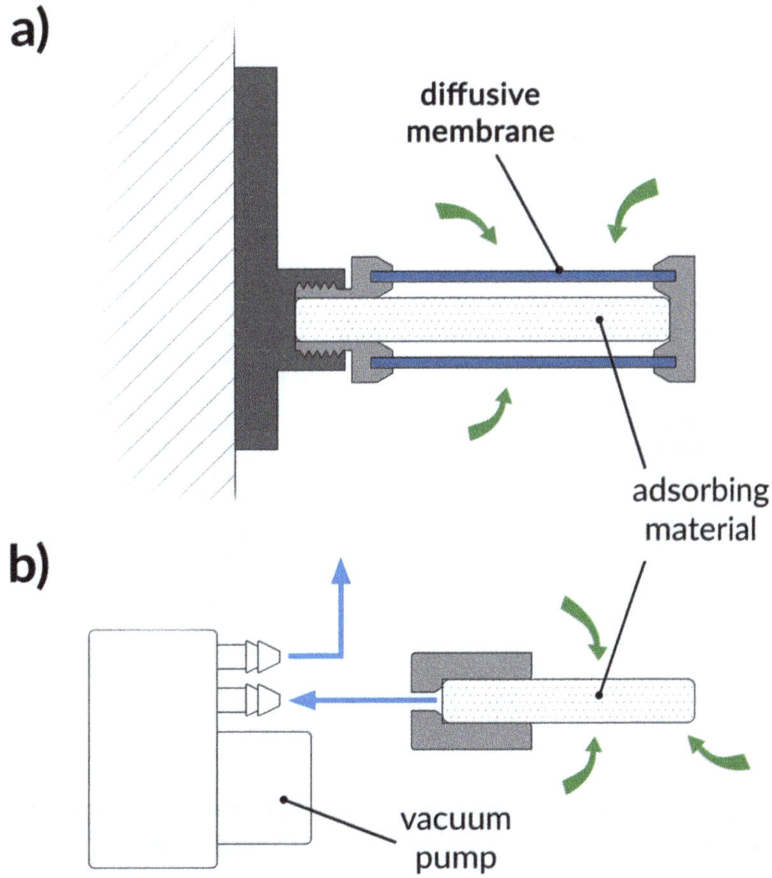

Active methods use forced air movement. Air is collected in sealed bags or containers, or passed through sorption tubes. Containers which will hold the air must be made of a possibly neutral material, such as Tedlar. However, a sample taken in this way is very diluted – most of the air in the container is inert. Therefore, a better solution is to pass the air through special sorption beds located in chemically inert tubes. The thus obtained analytes are either eluted with a solvent, or removed from the sorbent by thermal desorption. The most commonly used sorbents are 2,6-Diphenyl-p-phenyleneoxide (Tenax TA), molecular sieves (Carbosieve) or cross-linked polystyrene (Chromosorb). An important advantage of sorption tubes is their small size and storage possibility. By using thermal desorption, which is carried out directly in the chromatography system, the number of steps at the sample preparation stage is limited, which influences the duration of analysis as well as limits the possibility of errors. Such solutions can be used to determine pesticides and insecticides in indoor air, VOCs or polychlorinated biphenyls and fragrance allergens.

There are also other methods of sampling for GC analysis, including emission test chambers (ETCs) and field and laboratory emission cells (FLECs). In some cases it is worth using a natural sorbent present in most closed spaces, i.e. dust. For this purpose, a matrix solid-phase dispersion (MSPD) or solvent

Table 3. Advantages and disadvantages of various techniques used in IAQ analysis

Method	Main Advantages	Main Disadvantages
Sensory analysis	• Ease of use • Direct evaluation of quality • Relatively low cost • *In-situ* method	• Low sensitivity (detection threshold) • Lack of qualitative and quantitative information • The need to train a team of panellists • Sensory fatigue
TVOC	• Holistic analysis of indoor air • Ease of use • Measurement independent form qualitative change of air composition	• No quantitative identification of pollutants • Detectors calibrated to pure standards • High energy consumption • No information regarding toxicity
GC (passive sampling)	• Relatively small size • Low power consumption • No detrimental effect on the occupants of a room	• Susceptible to impact of air flow (drafts) • No possibility to register changes of concentration in time • Low sampling frequency • Poor enrichment of analytes
GC (active sampling)	• Good enrichment of analytes • High sampling frequency • Possibility to register changes of concentration in time (registering incidental emissions)	• Relatively expensive • Large sampling devices • Constant power consumption • Labour-intensive
Electronic nose	• Relatively small size • Capable of qualitative and quantitative analysis (depending on design) • Holistic evaluation of air quality • *In-situ* method • No sensory fatigue • Short time of a single analysis • Possibility to develop a maintenance-free solution	• Sensor drift • Moderate sensitivity • Need for statistical analysis of the response signal • Not suitable for screening of indoor air

extraction with solid-phase extraction (SPE) can be used to prepare a sample for analysis. Advantages and disadvantages of various analytical methods used in IAQ control are listed in Table 3.

ELETRONIC NOSES IN IAQ EVALUATION

Identification of Particular Compounds

One of the concepts of using electronic noses in indoor air monitoring is determining the concentration of particular harmful gaseous compounds. Unlike in the case of devices equipped with a single chemical sensor, such as a carbon monoxide sensor or a methane sensor, using a set of sensors allows multiple compounds to be analysed simultaneously. Additionally, using statistical data processing, the concentration of a chemical compound can be determined with greater accuracy and reliability. One of many examples of sensor systems used to determine the concentration of individual substances is a prototype built at Karlsruhe Forschungszentrum (Bender, Barié, Romoudis, Voigt, & Rapp, 2003). This device is equipped with 8 surface acoustic wave (SAW) detectors and is able to detect naphthalene, n-decane, styrene, xylene, ethylbenzene, n-octane, toluene, tetrachloroethene, benzene, n-heptane, tetrachloromethane, n-hexane, acetone, dichloromethane, ethanol and methanol at low- and sub-ppm concentrations. Before the analytes were introduced into the sensor chambers, they were enriched in a preconcentration unit - a sorption trap. The trap was made from a glass tube or PTFE (depending on the process conditions), where the sorption material was Tenax TA. The measuring system (without the preconcentration unit) is small in size – 6 cm x 3 cm. In addition to determining the concentration of the selected chemical compounds, a chemometric analysis was performed, in which PCA and LDA were used. The great advantage of this device is its sensitivity, which makes it possible to detect indoor air pollution below their perceptibility thresholds. In addition, the SAW sensors have a long life span, so they can be used for long-term, continuous indoor air monitoring.

A similar solution which used SAW sensors was presented at the 9th IEEE Conference on Nanotechnology in 2009 (Yao, 2009). A device composed of 7 piezoelectric sensors was used to determine the concentration of 5 volatile chemicals: ammonia, trimethylamine, methanol, ethanol and acetone. Two-way hierarchical cluster analysis was used to determine which sensors exhibited the highest sensitivity to the selected analytes. Under real conditions, it would be necessary to have several "electronic nose" devices placed throughout the building in order to comprehensively monitor indoor air quality. The devices, although identically constructed and equipped with the same set of sensors, can produce different output signals even if they are exposed to the same gas mixture. It is therefore necessary to properly calibrate the devices. For this purpose, global affine transformation (GAT) and Kennard-Stone sequential algorithm (KSS) can be used. The calibration was done for six devices: one served as "master" and the other was "slave". "Slave" devices were calibrated based on the information received from the "master" device. As a result of the on-line sensor calibration, each set of sensors worked with similar accuracy.

One of the hitherto proposed solutions has been the use of a matrix of 7 sensors with quartz crystal microbalance (QCM) to detect 23 different compounds belonging to alkanes, aromatic hydrocarbons, chlorocarbons, ketones and alcohols (Seyama, Sugimoto, & Miyagi, 2002). The sensors were covered with an organic film using plasma-deposition technology. The tests were conducted for two rooms: one with fresh air and the other – a smoking area. QCM sensors grant a significantly higher sensitivity than commercially available VOC analyzers, where MOS or PID sensors are most commonly used. Due to the sensitivity of detection, electronic noses with such sensors could be used in areas where very clean air is required, such as in cleanrooms, in some laboratories or in hospital rooms.

Holistic Approach

Due to the very complex composition of the gas mixture that is indoor air, it is difficult to determine the concentration of all pollutants, especially if there are chemical reactions leading to the formation of secondary compounds. In addition, there may be emerging contaminants in the indoor air, maximum permissible levels of which have not yet been determined. From a legal point of view, monitoring of the concentration of the chemical compounds specified in the regulations is justified. However, taking the health of the room users into account, holistic analysis of indoor air is also necessary. Then, such parameters as air quality indexes can be referred to. The obtained value does not provide information on the amount or type of the chemical compounds, but gives a simple signal which shows whether the air is of bad or good quality.

An important element in indoor air quality assessment is the monitoring of pollution in a building. However, it is important to remember that the quality of air in a room depends mostly on the materials used to build and furnish it. By controlling the quality of materials such as structural elements, thermal insulation, floor coverings or paints and varnishes used to finish surfaces, the state of the environment inside the room can be influenced. Currently, the use of paints and varnishes which contain volatile solvents or toxic protective substances is restricted in order to limit the emission of harmful or odorous chemical compounds. Therefore, acrylic paints or natural protection agents, such as linseed, are used. There are several solutions for using sensor systems to evaluate the quality of materials used in building a room. One example may be the use of an electronic nose with 32 sensors made from conductive polymers (Aromascan 32S) to identify indoor paints based on acrylic, alkyd, polyurethane and vinyl resins (Ramalho, 2000). The obtained results were compared with those obtained using gas chromatography and olfactometry. The measurements were carried out 3 days after the application of paints on the surface, which is a relatively short period. With longer exposure of the paint to room conditions such as forced airflow, lighting, or the presence of reactive inorganic compounds, it is possible that the impact of such a paint on the environment would be greater. Tests were carried out for eight pure compounds most often emitted from building materials: toluene, ethylbenzene, propylbenzene, styrene, phenylacetylene, ortho-xylene, meta-xylene, para-xylene, with the use of an electronic nose (Xu, He, Duan, Wu & Wang, 2016). These were some of the first uses of electronic noses in evaluating indoor air quality. Due to the fact that the experiment was not very advanced, pure chemical substances dissolved in synthetic air were analyzed and it cannot be determined whether this solution could in future be used for routine air quality monitoring.

One of the most commonly studied indoor air pollutants is formaldehyde. This is mainly due to the fact that it is a very volatile chemical that is suspected to be carcinogenic, and can lead to other health conditions, such as nervous system or immune system diseases. Classically used methods to identify formaldehyde in the indoor air are gas chromatography, FTIR and MS. By using these methods, low detection limits are possible, but the cost of analysis, the necessity of sampling, or relatively long analysis time, exclude such methods from routine, real-time measurements of formaldehyde concentration in the indoor air. The problem with chemical sensors used to detect volatile organic compounds is that they are not selective. The presence of compounds such as terpenes, which may come from natural sources (e.g. α-pinene) may impair the correct determination of formaldehyde concentration. The solution may be to use a sensor array and an electronic nose, and extract the results in such a way that information is obtained about the concentration of only one substance (Lv, Tang, Wei, Yu, & Huang, 2007). For this purpose, a miniature electronic nose system was used and the data was processed using artificial neural networks. Based on the results, formaldehyde can be detected at concentrations below ppm, even if there are impurities such as ethanol, toluene, acetone or α-pinene in the indoor air. However, it is important to remember that indoor air is a more complex matrix, thus it is important to continue this type of research in this area.

Another approach to assessing odour nuisance is the use of sensor systems in a composting hall (Nicolas, Romain, & Ledent, 2006). Nicolas et.al., similarly to previous studies, used olfactometry to compare results. The device they used was a prototype electronic nose equipped with 6 MOS sensors. It was proved that unsupervised methods (PCA) cannot be used to identify all sources of odour and it is necessary to use methods that utilize the learning stage. Due to the fact that during the storage of compost there is a variety of additional activities, it was attempted to take into account different sources of compound emissions that can affect the signal generated by the sensors. Four potential sources were identified: compost heap, vehicles and other machines in the composting facility, neutralizer added to the compost, and background. For this purpose, a discriminant function analysis (DFA) was used. Summarizing the study, it can be concluded that electronic noses can be used to detect sources and to determine the degree of odour nuisance in closed spaces.

Another specialized application is detecting fires using electronic noses. It is customary to use detectors equipped with single sensors for fire detection. They may, however, react to the presence of various chemical compounds, which did not originate from fires. This can lead to false alarms, which in the case of industrial plants, offices or public buildings, involves taking appropriate actions such as evacuation or shutdown. Persaud's research team dealt with separating and distinguishing the sensor response signal associated solely with the occurrence of a fire. They developed an electronic nose for this purpose in 2006 (Scorsone, Pisanelli, & Persaud, 2006), This device consisted of 8 sensors made of conducting polymers (CP) and the entire sensor array was the size of a coin. A specially prepared test kit was used in the study, consisting of a reaction chamber and a sampling system. Conditions present during a fire were simulated in the reaction chamber. Smoke was directed to a sampling system in which the gas stream was separated. Smoke samples were directed to a commercial carbon monoxide detector – a MOS sensor, electronic nose, optical sensor and the Fourier transform infrared spectrometer (FTIR). Furthermore, by placing an SPME fibre in the sampling system, it was possible to analyse the smoke composition using GC-MS. Based on the SPME-GC-MS analysis, it was possible to select markers for the test fires: smouldering cotton, wood, paper, cigarette smoke and flaming polyurethane. The electronic nose was initially used to determine the sensitivity to individual fire markers, and then the ability to classify the selected fires was tested using PCA. Using the electronic nose, it was possible to identify a fire event and a non-fire event. It is especially important to distinguish cigarette smoke from fire smoke, as the most common cause of false alarms is the presence of cigarette smoke in indoor air.

In addition, as mentioned above, indoor air quality also applies to means of transport such as cars, trains, aircraft, ships, etc. Quality control of particular vehicle equipment, such as vehicle seats, can significantly improve the quality of use of these vehicles. An example of the use of electronic noses for the characterization of volatile organic compounds emitted from car seat foams is research conducted by Morvan *et al.* (Morvan, Talou, & Beziau, 2003). Polyurethane foams, which are commonly used as filling in car seats, were used as samples. They differed in origin (3 different manufacturers), colour, moulding (warm or cold) and place of application (different parts of the seat). The NST3320 commercial electronic nose (Applied Sensor AG, Germany) with 11 MOS sensors and 11 MOSFET sensors was used for the characterization of volatile fraction. The obtained results were juxtaposed and compared with ATD-GC-MS and sensory analysis. It was shown that similar information can be obtained using an electronic nose and using a sensory panel. Researchers believe that information obtained using a sensor system provides less quantitative information than sensory analysis. In addition, an unsatisfactory correlation between the identified compounds and the sensor response signals was obtained. These studies may be proof that electronic nose systems cannot completely replace advanced instrumental analysis or the work of panellists in sensory analysis. They may, however, be supplementary to research, or be used for preliminary screening of a sample. It is estimated that concentrations of pollutants such as carbon monoxide, hydrocarbons, VOCs and nitrogen oxides in vehicles are significantly higher than the standards set by Occupational Safety and Health Administration (OSHA) and World Health Organization (WHO). This is primarily due to the materials used for the interior of vehicles, the internal temperature and the air exchange system. Air quality in vehicles can be influenced by parameters such as temperature, humidity and chemical composition of the air. By using a sensor system equipped with temperature sensors, humidity sensors, a CO2 sensor and a VOC (Indoor Air Quality Sensor - IAQ sensor) classes of air quality in vehicles can be distinguished (Szczurek & Maciejewska, 2015). It has been shown that the quality of air that the driver and passengers breathe changes with time spent in the car.

Identification of Microbal Volatile Organic Compounds (MVOCs)

The dampness in a room, as well as appropriately high temperature and growth medium may result in the emergence of germs indoors. Due to bacteria and fungi residing on walls, furniture, carpeting, etc., harmful compounds, such as mycotoxins and microbial volatile organic compounds (MVOCs) may be emitted. What is very dangerous is the appearance of fungi indoors. Some of them may be visible, for instance sporocarps of certain fungi, but the vast majority of fungi are not easily recognizable. In most cases, scientists focus on air pollution caused by emissions from surfaces of furniture or building materials. Volatile pollutants, which develop due to the activity of living organisms, are often forgotten. Microbial volatile organic compounds (MVOCs) include mostly alcohols, terpenes, ketones, furans and sulfur compounds. Part of them is not harmful to human health, however they can act as markers of fungi occurrence indoors. Such chemical compounds may be identified with a variety of instrumental techniques. Sensor systems enable us to carry out real-time monitoring. Existing research proves the usefulness of electronic noses in identifying certain fungi strains. The majority concerns the analysis of foods such as cereal and cereal products. There are, however, reports in the literature, in which mold detection indoors was the subject of research. Using an electronic nose equipped with 15 MOS sensors and employing chemometrics such as LDA, KNN or LS enabled to detect and classify five fungi which can commonly be found in indoor environments. Persuad's team used a system with 32 sensors to detect dry rot wood decay fungi. These types of fungi occur mostly in old, antique indoor environments with wooden construction elements. Using microextraction in a stationary phase as a method of retrieving and enriching MVOCs proved to be an interesting solution. Another application of an electronic nose was using commercial electronic noses MOSES II and KAMINA to assess the quality of indoor air in which MVOCs occur. In artificial conditions, on agar media, three types of mold were cultured: *Aspergillus niger*, *A. versicolor* and *Cladosporium sphaerospermum*. Analysis of the main components enabled to detect the mold, however the precise assessment of the strain of fungi was difficult. What is important, detecting MVOCs in indoor air may in practice facilitate an early detection of mold indoors, thereby allowing to take remedial measures.

Application in Frying Processes

Some volatiles in the indoor environment can be of fungal or bacterial origin, being emitted either by mold or microorganisms responsible for decomposition of food products. Volatiles from fungi and mold developing on food products such as cereals can be detected using electronic olfaction. Apart from extreme instances of spoilage, electronic noses can be successfully used to detect expired food products, either in retail stores or at home. For example, a system for automatic detection of spoiled fruit in a domestic environment based on an array of SAW sensors was tested successfully. A similar study was also carried out in regard to dairy products, namely milk and yogurt. A holistic approach was also taken when detecting volatile indicators of fish and meat spoilage in a supermarket, where there are numerous interferences from various sold food products. Still, it was possible to detect spoilage of particular products against the background noise of numerous different volatile compounds.

There are numerous methods of processing food. One of the crucial food processes, due to the frequency of occurrence in households, restaurants, as well as big food processing facilities, is frying. As a result of thermal processing, harmful volatile compounds may be generated. They can enter the human body through the respiratory system. Reactions which occur during frying are oxidation, hydrolysis, cyclization, polimerization, and Maillard reactions. As a result of frying, harmful substances may be created: aldehydes, ketones, chain hydrocarbons and aromatic hydrocarbons, lactones, carboxylic acids, and esters. With the increase in concentration of aldehydes, the food safety decreases. Aldehydes such as acrolein or olefin-aldehydes are hazardous to human health. In addition, oxidated aldehydes from α,β-unsaturated aldehydes (OαβUAs), which prove to be carcinogenic, can also cause diabetes or arteriosclerosis and may trigger Alzheimer's disease and Parkinson's disease. Another compound important due to its harmful properties is acrylamide, which is generated during a series of reactions called Maillard reactions. Acrylamide has serious neurotoxic effects, it can cause genetic and cancerous changes. Other products of the Maillard reaction are polycyclic aromatic amines (PAAs), which prove to have mutagenic and carcinogenic properties. Similar compounds may be produced during baking bread or pastry. In the air in such indoor environments, substances can be emitted from sources such as furniture, equipment or a ventilation system, as well as products of thermal degradation of food. It is important to emphasize that workers in such rooms rarely employ means of personal protection such as protective masks.

The quality of indoor air in places of food preparation depends mostly on the quality of food products, for instance the quality of frying oil. It is hence crucial to indicate the quality of the used frying oil. This is when electronic noses may be employed. The process of oil spoilage, which is caused by oil oxidation, is called rancidification. Using electronic noses enables to tell whether or not the oil is fit for consumption, as well as to assess its thermostability. Rancidification can also occur during storing oil, even if it is kept in a tightly closed container. On opening the container, the products of oil oxidization are emitted into the indoor air. Research using electronic noses was also carried out in this field, for example in canola oil (Mildner-Szkudlarz, Jeleń, & Zawirska-Wojtasiak, 2008; Shen et al. 2001), corn oil and soybean oil (Shen et al., 2001).

FUTURE RESEARCH DIRECTIONS

Wireless E-Nose Networks

Devices such as an electronic nose may be equipped with a Wi-Fi module. As a consequence, the device can be significantly smaller in size, because the data collecting and analyzing system may be omitted from its construction. Storing data and processing it is hence done in the central device, which collects data from numerous devices that make up a monitoring network. Moreover, powering the electronic nose from an interior source enables to place the device anywhere in the building and measure the air quality. A good example of a device employing the Wi-Fi module is an electronic nose equipped with eight MOS sensors (Wongchoosuk, Khunarak, Lutz, & Kerdcharoen, 2012). This device was placed in two different indoor environments, namely in an office and in a kitchen, and then measurements were taken for seven hours in each of these places. Some possible placements of electronic nose devices in indoor environment are depicted in Figure 4. The authors of this experiment focused mainly on defining the concentration of carbon monoxide in both of these environments. The advantage of this solution was the small size of the developed device (15x10 cm). What is more, employing the statistical data processing with a PCA method enabled to distinguish the air coming from the kitchen from the air coming from the office. Yet another example of an electronic nose equipped with a Wi-Fi module is a device composed of three MOS sensors. This device was developed at the University of Science and Technology Beijing (Xu et al., 2016). Because of the fact that MOS sensors are extremely susceptible to temperature and humidity, the device was equipped with an additional humidity and temperature sensor. To stream data, the team employed a commercially available Wi-Fi module Xbee. The research was carried out in controlled conditions, in a custom-made chamber. A model gas used in this experiment was formaldehyde, because it can be emitted from the surface of furniture or the equipment of indoor environments. To statistically process data, the team employed machine learning algorithms such as artificial neural networks with back propagation (BP ANN) and with radial basis function (RBF ANN) and support vector regression. The device's responses for different temperature and humidity conditions were assessed. The obtained results were subjected to validation. The use of this type of device, however, was limited to detecting formaldehyde in indoor air. The authors predict that these types of devices can in the nearest future be used for holistic assessment of indoor air quality. Another technical solution is Indoor Air Quality Monitoring System, presented in 2017 (Saad et al., 2017). It is a network used for monitoring air inside a building. It consists of units of measurement which form a sensor module cloud (SMC). Each sensor module consists of sensors used for measuring such pollutants as nitrogen dioxide, carbon dioxide, ozone, carbon monoxide, oxygen, volatile organic compounds and suspended matter PM10. In addition, this device was equipped with temperature and pressure sensors. These modules stream data using wireless network to a hub, where the signal is processed with probabilistic neural network and multilayer perception classifier. The obtained classification accuracy was satisfactory (99.98%) for five different samples of indoor air: ambient air, chemical presence, fragrance presence, foods and beverages, and human activity.

Electronic Noses as an Element of Heating, Ventilation, and Air Conditioning Systems

One of the biggest challenges that creators of electronic noses face is to reduce electrical energy use. The necessity for powering the system of chemical sensors, the support software and the pneumatic system, significantly limits the possibilities of using these types of devices for constant monitoring of indoor air quality. A solution to this problem may be to develop of an electronic nose without pneumatic system (valves, pumps, filters). This kind of device, proposed by Zampolli et al., was used to identify such gases as NO2 or CO (Zampolli et al., 2004). The device was equipped with four MOS sensors, driven by an ad hoc electronic circuit. Such a device is an example of a new class of low-cost e-noses, which can be used as integral elements of such devices as air conditioners or ventilation systems in buildings.

Figure 4. Possible placement of electronic noses in indoor environment: In ventilation ducts (a), on a wall next to potential sources of emission (b) or as a hand-held portable device (c)

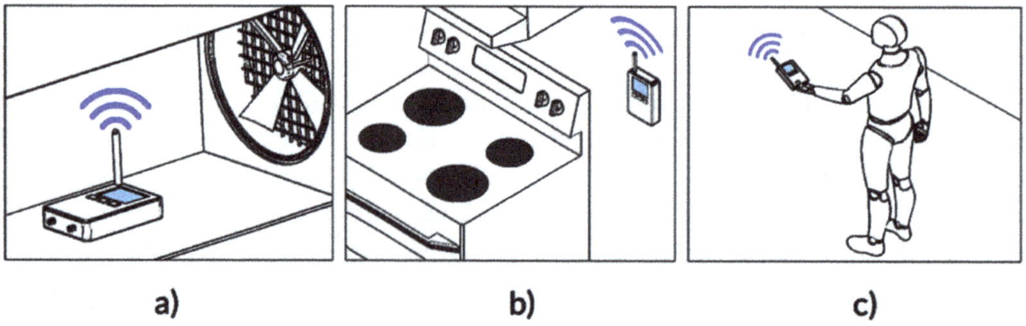

This solution can be used not only for monitoring the condition of air indoors, but also to improve the effectiveness of HVAC systems, which can lead to lowering overall power consumption. Miniaturized electronic noses integrated with the ventilation systems may be utilized in the newest building concept, namely smart buildings.

Electronic Noses in Smart Buildings

In pursuit of developing the indoor space in buildings in a user-friendly manner, scientists have devised a new concept called smart buildings. The concept is based on the idea that such elements as ventilation, lighting or anti-burglar devices could be automatically operated. The system is based on a network of sensors, including gas or movement sensors, which collect data from the surroundings (see Figure 5). With this information, the system can control such things as blinds in windows or air conditioning. It can also switch on and off some devices, for example television sets. Smart houses work like neural networks, which learn and remember preferences of the user of the building. Air quality is controlled by chemical sensors, which send response signals that may be used to regulate the workings of ventilation or air conditioning systems. Currently, there is no solution that takes electronic noses into account. The authors of this chapter are certain that employing e-noses would fit into the concept of smart houses perfectly. Using electronic noses as main elements of control systems in smart buildings would be the newest development direction for this type of device in indoor air quality assessment.

CONCLUSION

The development of electronic noses is closely linked with the development of sensor technology. Over the last 30 years, electronic noses have found many uses in various areas of industry.

Figure 5. Various possible components of a 'smart building' network

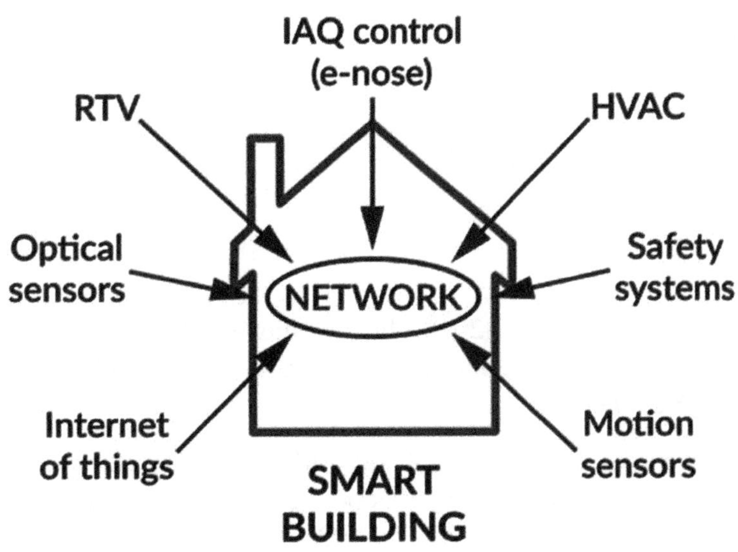

In most cases, however, highly skilled specialists are needed to understand the chemistry of the measurement process and statistical data analysis methods. Nevertheless, the authors of this chapter believe that in the near future it will be possible to utilize them in everyday life. These devices can be customized for a specific application. The measurement process can be automated and the final measurement result can be presented in a manner understandable to most potential users, for example "the air is clean/ safe". While using electronic noses to control indoor air quality, there are numerous ways to integrate these devices into the monitoring network. Their biggest advantage is that the results are obtained in real time and in digital form. This gives the possibility to create a wireless monitoring network with multiple electronic nose units, and integrate it into the overall room management system, which is the core element of smart homes. In addition, using the existing indoor systems, such as heating, ventilating or air conditioning systems, some elements of the pneumatic system in electronic noses can be omitted, thus reducing energy consumption.

The use of these devices for indoor air measurements seems to be more desirable also due to the fact that sensors used in electronic noses are not selective. It is thus possible to identify individual chemicals present in indoor air, or multicomponent gas mixtures, if the research is carried out in accordance with the holistic concept of measurement. This is the research approach that is more useful when different chemical processes occur in a closed space, often in a similar, but not identical way. A standard example may be the use of electronic noses to control air quality in food processing facilities (restaurant kitchens, large food processing facilities), and in particular in indoor environments where the frying process is carried out, as it may cause the production of carcinogenic chemicals. The use of electronic noses to control indoor air quality is a promising solution that can be used for indoor air monitoring, also because their chemical sensors are more suitable for emission rather than emission measurements.

ACKNOWLEDGMENT

This work was financially supported by the Grant No. PBSII/B9/24/2013 obtained from the National Centre for Research and Development of Poland and Grant no. 2015/19/B/ST4/02722 received from the National Science Center of Poland.

REFERENCES

Alexandrakis, D., Brunton, N. P., Downey, G., & Scannell, A. G. M. (2012). Identification of Spoilage Marker Metabolites in Irish Chicken Breast Muscle Using HPLC, GC–MS Coupled with SPME and Traditional Chemical Techniques. *Food and Bioprocess Technology*, *5*(5), 1917–1923. doi:10.1007/ s11947-010-0500-8

Andretta, M., Coppola, F., & Seccia, L. (2016). Investigation on the interaction between the outdoor environment and the indoor microclimate of a historical library. *Journal of Cultural Heritage*, *17*, 75–86. doi:10.1016/j.culher.2015.07.002

Bender, F., Barié, N., Romoudis, G., Voigt, A., & Rapp, M. (2003). Development of a preconcentration unit for a SAW sensor micro array and its use for indoor air quality monitoring. *Sensors and Actuators. B, Chemical*, *93*(1-3), 135–141. doi:10.1016/S0925-4005(03)00239-9

Chao, C. Y., & Chan, G. Y. (2001). Quantification of indoor VOCs in twenty mechanically ventilated buildings in Hong Kong. *Atmospheric Environment*, *35*(34), 5895–5913. doi:10.1016/S1352-2310(01)00410-1

Dascalaki, E. G., Lagoudi, A., Balaras, C. A., & Gaglia, A. G. (2008). Air quality in hospital operating rooms. *Building and Environment*, *43*(11), 1945–1952. doi:10.1016/j.buildenv.2007.11.015

Gram, L., Ravn, L., Rasch, M., Bruhn, J. B., Christensen, A. B., & Givskov, M. (2002). Food spoilage—interactions between food spoilage bacteria. *International Journal of Food Microbiology*, *78*(1-2), 79–97. doi:10.1016/S0168-1605(02)00233-7 PMID:12222639

Herrick, R. F., Stewart, J. H., & Allen, J. G. (2016). Review of PCBs in US schools: A brief history, an estimate of the number of impacted schools, and an approach for evaluating indoor air samples. *Environmental Science and Pollution Research International*, *23*(3), 1975–1985. doi:10.1007/s11356-015-4574-8 PMID:25940477

Krupińska, B., Van Grieken, R., & De Wael, K. (2013). Air quality monitoring in a museum for preventive conservation: Results of a three-year study in the Plantin-Moretus Museum in Antwerp, Belgium. *Microchemical Journal*, *110*, 350–360. doi:10.1016/j.microc.2013.05.006

Kuske, M., Romain, A.-C., & Nicolas, J. (2005). Microbial volatile organic compounds as indicators of fungi. Can an electronic nose detect fungi in indoor environments? *Building and Environment*, *40*(6), 824–831. doi:10.1016/j.buildenv.2004.08.012

Lewkowska, P., Dymerski, T., Gębicki, J., & Namieśnik, J. (2016). The Use of Sensory Analysis Techniques to Assess the Quality of Indoor Air. *Critical Reviews in Analytical Chemistry*, *47*(1), 37–50. doi:10.1080/10408347.2016.1176888 PMID:27105173

Lv, P., Tang, Z., Wei, G., Yu, J., & Huang, Z. (2007). Recognizing indoor formaldehyde in binary gas mixtures with a micro gas sensor array and a neural network. *Measurement Science & Technology*, *18*(9), 2997–3004. doi:10.1088/0957-0233/18/9/034

Marć, M., Śmiełowska, M., & Zabiegała, B. (2016). Concentrations of monoaromatic hydrocarbons in the air of the underground car park and individual garages attached to residential buildings. *Science of the Total Environment Journal*, *573*, 767–777. doi:10.1016/j.scitotenv.2016.08.173 PMID:27591527

Mendes, A., Bonassi, S., Aguiar, L., Pereira, C., Neves, P., Silva, S., ... Teixeira, J. P. (2015). Indoor air quality and thermal comfort in elderly care centers. *Urban Climate*, *14*, 486–501. doi:10.1016/j.uclim.2014.07.005

Mildner-Szkudlarz, S., Jeleń, H. H., & Zawirska-Wojtasiak, R. (2008). The use of electronic and human nose for monitoring rapeseed oil autoxidation. *European Journal of Lipid Science and Technology*, *110*(1), 61–72. doi:10.1002/ejlt.200700009

Morvan, M., Talou, T., & Beziau, J.-F. (2003). MOS–MOSFET gas sensors array measurements versus sensory and chemical characterisation of VOC's emissions from car seat foams. *Sensors and Actuators. B, Chemical*, *95*(1-3), 212–223. doi:10.1016/S0925-4005(03)00425-8

Nicolas, J., Romain, A.-C., & Ledent, C. (2006). The electronic nose as a warning device of the odour emergence in a compost hall. *Sensors and Actuators. B, Chemical, 116*(1-2), 95–99. doi:10.1016/j.snb.2005.11.085

Ramalho, O. (2000). Correspondences between olfactometry, analytical and electronic nose data for 10 indoor paints. *Analusis, 28*(3), 207–215. doi:10.1051/analusis:2000280207

Roinestad, K. S., Louis, J. B., & Rosen, J. D. (1993). Determination of pesticides in indoor air and dust. *Journal of AOAC International, 76*(5), 1121–1126. PMID:8241815

Saad, S. M., Shakaff, A. Y. M., Saad, A. R. M., Yusof, A. M., Andrew, A. M., Zakaria, A., & Adom, A. H. (2017). Analysis of feature selection with Probabilistic Neural Network (PNN) to classify sources influencing indoor air quality. *AIP Conference Proceedings*, 1808.

Schieweck, A., Lohrengel, B., Siwinski, N., Genning, C., & Salthammer, T. (2005). Organic and inorganic pollutants in storage rooms of the Lower Saxony State Museum Hanover, Germany. *Atmospheric Environment, 39*(33), 6098–6108. doi:10.1016/j.atmosenv.2005.06.047

Scorsone, E., Pisanelli, A. M., & Persaud, K. C. (2006). Development of an electronic nose for fire detection. *Sensors and Actuators. B, Chemical, 116*(1-2), 55–61. doi:10.1016/j.snb.2005.12.059

Senter, S. D., Arnold, J. W., & Chew, V. (2000). APC values and volatile compounds formed in commercially processed, raw chicken parts during storage at 4 and 13 C and under simulated temperature abuse conditions. *Journal of the Science of Food and Agriculture, 80*(10), 1559–1564. doi:10.1002/1097-0010(200008)80:10<1559::AID-JSFA686>3.0.CO;2-8

Seyama, M., Sugimoto, I., & Miyagi, T. (2002). Application of an array sensor based on plasma-deposited organic film coated quartz crystal resonators to monitoring indoor volatile compounds. *IEEE Sensors Journal, 2*(5), 422–427. doi:10.1109/JSEN.2002.804576

Shen, N., Moizuddin, S., Wilson, L., Duvick, S. A., White, P. J., & Pollak, L. M. (2001). Relationship of electronic nose analyses and sensory evaluation of vegetable oils during storage. *Journal of the American Oil Chemists' Society, 78*(9), 937–940. doi:10.1007/s11746-001-0367-z

Śliwińska, M., Wiśniewska, P., Dymerski, T., Wardencki, W., & Namieśnik, J. (2015). The flavour of fruit spirits and fruit liqueurs: A review. *Flavour and Fragrance Journal, 30*(3), 197–207. doi:10.1002/ffj.3237

Śmiełowska, M., Marć, M., & Zabiegała, B. (2017). Indoor air quality in public utility environments—a review. *Environmental Science and Pollution Research International*, 1–11. PMID:28236201

Szczurek, A., & Maciejewska, M. (2015). Classification of air quality inside car cabin using sensor system. Sensornets.

Takeoka, G., Perrino, C., & Buttery, R. (1996). Volatile Constituents of Used Frying Oils. *Journal of Agricultural and Food Chemistry, 44*(3), 654–660. doi:10.1021/jf950430m

Wongchoosuk, C., Khunarak, C., Lutz, M., & Kerdcharoen, T. (2012). WiFi electronic nose for indoor air monitoring. *9th International Conference on Electrical Engineering/Electronics, Computer, Telecommunications and Information Technology*, 1–4. doi:10.1109/ECTICon.2012.6254166

Worobiec, A., Samek, L., Spolnik, Z., Kontozova, V., Stefaniak, E., & Van Grieken, R. (2006). Study of the winter and summer changes of the air composition in the church of Szalowa, Poland, related to conservation. *Mikrochimica Acta*, *156*(3–4), 253–261. doi:10.1007/s00604-006-0619-5

Xu, L., He, J., Duan, S., Wu, X., & Wang, Q. (2016). Comparison of machine learning algorithms for concentration detection and prediction of formaldehyde based on electronic nose. *Sensor Review*, *36*(2), 207–216. doi:10.1108/SR-07-2015-0104

Xu, L., Yu, X., Liu, L., & Zhang, R. (2016). A novel method for qualitative analysis of edible oil oxidation using an electronic nose. *Food Chemistry*, *202*, 229–235. doi:10.1016/j.foodchem.2016.01.144 PMID:26920289

Yao, D. (2009). A gas sensing system for indoor air quality control and polluted environmental monitoring. In *9th IEEE Conference on Nanotechnology 2009* (Vol. 8, pp. 806–811). IEEE.

Zampolli, S., Elmi, I., Ahmed, F., Passini, M., Cardinali, G. C., Nicoletti, S., & Dori, L. (2004). An electronic nose based on solid state sensor arrays for low-cost indoor air quality monitoring applications. *Sensors and Actuators. B, Chemical*, *101*(1), 39–46. doi:10.1016/j.snb.2004.02.024

Zeiger, E., Gollapudi, B., & Spencer, P. (2005). Genetic toxicity and carcinogenicity studies of glutaraldehyde—a review. *Mutation Research/Reviews in Mutation Research*, *589*(2), 136–151. doi:10.1016/j.mrrev.2005.01.001 PMID:15795166

Zhang, J., Chen, W., Li, J., Yu, S., & Zhao, W. (2015). VOCs and Particulate Pollution due to Incense Burning in Temples, China. *Procedia Engineering*, *121*, 992–1000. doi:10.1016/j.proeng.2015.09.067

Chapter 11
Integrating Odor Sensing in Robotics

Alauddin Yousif Al-Omary
University of Bahrain, Bahrain

ABSTRACT

In this chapter, the benefit of equipping the robot with odor sensors is investigated. The chapter addresses the types of tasks the mobile robots can accomplish with the help of olfactory sensing capabilities, the technical challenges in mobile robot olfaction, the status of mobile robot olfaction. The chapter also addresses simple and complex electronic olfaction sensors used in mobile robotics, the challenge of using chemical sensors, the use of many types of algorithms for robot olfaction, and the future research directions in the field of mobile robot olfaction.

INTRODUCTION

Among the challenges facing robotics researchers is developing intelligent autonomous robot that can take decision by his own. This kind of robot should rely on complex sensors to gather the necessary information about the environment it navigates and should also use intelligent algorithms that enable the robot to take the necessary actions when needed. The robot consists mainly of many parts: the sensors, the controller, effectors and actuators. Sensors are used to provide the robot with necessary information about the environment in which the robot is navigating and discovering. Things that can be sensed includes light and sound (presence, color, intensity, content (mod), direction), sound (presence, frequency, intensity, content (mod), direction), heat (temperature, wavelength, magnitude, direction), chemicals (presence, concentration, identity, etc.), object proximity (presence/absence, distance, bearing, color, etc.), physical orientation/attitude/position, magnitude, pitch, roll, yaw, coordinates, etc. Commonly used sensors includes ultrasonic sensor, infrared sensor, touch sensor, sonic sensor, laser sensor, photoelectric light sensors, magnetic sensor and recently chemical sensor. Each one of these sensor measure different environmental phenomena and provide the robot with different information. Sensors that mimic the animal world (particularly the human senses) are finding their use in robotics like sensors for taste and smell (chemical sensors). Other part of robot is the controller which receive signals from sensors,

DOI: 10.4018/978-1-5225-3862-2.ch011

analyze these signals and take the appropriate action. The controller can be based on deliberative or reactive model. The deliberative model is representation dependent, need complex sensors, uses high level intelligence and therefore has slower response and variable latency. On the other side, the reactive model is representation free, uses low level intelligence and therefore it performs simple computation and has real time response. The effectors and actuators terms are often used interchangeably to mean: "whatever makes the robot take an action" but they aren't the same thing. An effector is any device that interact with the environment and is controlled by the robot controller. It can range from legs and wheels to arms and fingers that is used to grip objects. An actuator is the actual mechanism that enables the effector to execute an action. It can include electric motors (AC or DC motor, servo motor and stepper motor), hydraulic cylinders, and pneumatic cylinders etc. Robot can be autonomous or tele-operated. The autonomous robot can navigate and move without human intervention while tele-operated robot is controlled by human operator. Tele-operated operator can be wired controlled or wireless controlled. Wireless controlled can be achieved using infrared, radio frequency, bluetooth, WiFi or cellular/ GPRS signal. Robot also can work on land, air or underwater. Robot that works on land is called unmanned ground vehicle (UGV). Robot that works on air is called Unmanned Aerial Vehicle (UAV) or drone and finally robot that works underwater is called Unmanned Underwater Vehicle (UUV).

Robots are used in many applications and mainly in two application areas, i.e., industrial and domestic. Industrial robots were developed to perform many manufacturing tasks ranging from simple product assembly to complex manufacturing jobs. However, most industrial robots have no intelligence as they cannot hear, see, smell or feel and should be programmed precisely to do specific task. If assembly parts are not presented to the robot in a precise and repetitive fashion, the robot cannot perform its task properly. Many new applications have been emerged recently where there is a need to detect and analyze volatile chemicals. The role of the robot in these new applications need to make robot more autonomous. The autonomous robot can navigate and move in a dynamic environment that have not been specifically designed for it and are subject to sudden change therefore autonomous mobile robot should be able to work in dynamic environment and can make decisions independently and adaptively. Complex robotic tasks can include product transferring in manufacturing factory, rubbish cleaning in office, trash collection, bomb searching on battle field, searching for gas leak in pipes or in refineries, searching for drugs in custom, etc. Autonomous robot navigation requires the robot to have more intelligence and equipped with more capabilities such as e-nose or odor sensors. E-nose can recognize alcohol, blood, urine and sweet. Once it has collected data about a certain smell, the nose will send it to the robot which runs algorithms on the information, and calculates where it is coming from and how to get there. Electronic nose can be very useful for robot when it need to carry out some chemically related tasks, such as cleaning, finding gas leaks and in search-and-rescue operations. It was believed for long time that human noses have receptors that respond based on the shapes of odor molecules. But recently a conflicting and more exotic theory was presented by researchers in Greece (Randy Rieland 2013). The theory concluded that we can also sense smells through quantum physics. In this case the vibration of odor molecules. IBM, in its annual "5 in 5″ forecast–a list of technologies it believes will hit the mainstream in five years–focused exclusively on the development of the five human senses in machines. To mimic smelling, tiny sensors would be integrated into smartphones or other mobile devices and, as a breathalyzer can determine alcohol levels, they would gather data from the smell of your breath by detecting chemicals that humans wouldn't perceive and send it to a computer in your doctor's office. The thinking is that eventually this would be a core component of home health care–the ability to "smell" diseases remotely, such as liver or kidney ailments, asthma or diabetes. This new technology will help robot in tackling new applications that were not possible in the past.

BACKGROUND

In the last decade, researchers talked about the topic of biologically inspired robotics and developed evolutionary robotics (ER) which use a methodology based on artificial evolution to obtain self-adaptation and self-learning capabilities. Other investigations are being conducted on the use of robots in the imitation of life. One of the new promising way to increase the robot capabilities and intelligence is to equip the robot with olfaction capabilities that can be integrated with other sensor like vision, laser scanner and enable the robot to work faster. It can also increase the precision of robot motion by having additional sensing for validating measurements from a different position. Robot intelligence can also be increased by having more exploration capabilities. Thus, the robot can efficiently use the onboard computation that are available recently due to the advancement of in the area of microelectronics and VLSI.

Although one can find robots that use many types of sensors, sensor that mimic human sense like smell and taste, so far, are among the sensors that have limited use in robotics. This is due to the reason that in human motion we did not uses these senses and because the chemical sensors that mimic these senses still need more work and time to become similar to their biological counterparts. Despite these fact, many researches have tackled this area and experimented many possible applications in this field. Piotr Batog and Andrzej Wołczowski (Piotre & et al 2012) presented the development of the odor markers detection system for mobile robot navigation along a chemical path. The researcher develops a method of probe sampling in order to reduce the time parameters of gas sensors. They also introduced odor markers detection system for mobile robot navigation (Piotre & et al 2012). In their system they guide the robot navigation using Short Living Navigational Markers (SLNM), such as odor or heat. The lines marked in this way disappear after the robot completes its work. Siti Nurmaini, Bambang Tutuko and Aulia Rahman Thoharsin (Siti Nurmaini et al 2013) introduced Intelligent Mobile Olfaction of Swarm Robots. They present intelligent mobile olfaction design and experimental results of intelligent swarm robots to detect a gas/odour source in an indoor environment by using multi agent based on hybrid algorithm. Olivier (Olivier Rochel and et al 2001) introduced a neuromorphic robot using stereo olfaction and a sniffing system based on nonselective sensors. The system was developed for tracking a specific gas. They show that a biologically inspired approach using artificial spiking neurons can be used to estimate the gas-concentration gradient and recognize a particular gas independent of its intensity. The system has been implemented on an autonomous robot and is tested on real gases. Javier (Javier Gonzalez-Jimenez & et al 2011) introduced an improved multi-chamber electronic nose olfaction sensor for Mobile Robotics. They use four identical sets of Tin Oxide sensors accommodated in separate chambers which alternate between sensing and recovery states, providing, as a whole, a device capable of sensing changes in chemical concentrations faster than single Tin oxide sensor. They tested the performance of the multi chamber electronic nose in mobile robotic olfaction through several experiments involving rapid sensing of gas concentration and mobile robot gas mapping. Kai Song and et al (Kai Song & et al 2011) introduced olfaction and hearing based mobile robot navigation for Odor/Sound Source Search. They design an odor/sound tracking multi-robot system by integrating smell, hearing and touch. The olfactory robot can track the chemical odor plume within distance of 2 meter using information fusion from gas sensors and airflow sensors. They use two hearing robots to localize the sound source using time delay estimation

(TDE) and the geometrical position of microphone array. The two hearing robots can quickly localize and track the olfactory robot in 2 min. The researchers pointed out that the devised multi-robot system can achieve target search with a considerable success ratio and high stability.

TYPES OF ODOR SENSORS SUITABLE FOR ROBOTICS APPLICATIONS

Out of the numerous techniques available for detecting chemicals, only a few have been used and found to be suitable for robotics applications (R. Russell 2001, R. Andrew Russell 2006). To detect and analyze volatile chemicals, there exit many sensory systems. They range from those simple one used to measure alcohol in the breath (like tin oxide sensors) and those used to analyze complex mixtures of chemicals (like mass spectrometers and gas chromatographs). Researchers and industry are developing from time to time a new and improved chemical sensor. However, as have been mentioned before, these sensors need to be developed more to be suitable for robotic applications.

CHEMICAL SENSORS NEEDED ATTRIBUTES

Robotic applications need chemical sensors to include the following attributes (R. Russell 2001):

1. **Low Power Consumption:** All chemical sensors are active sensor and they need power to work. This power will be fed to these sensors from the robot's power supply (onboard batteries) and this put restriction on the amount of power needed to run the sensors.
2. **High Sensitivity:** The sensitivity of t the chemical sensor is crucial for robotic applications and increasing the sensitivity of such sensor will open the door for many more applications.
3. **Rapid Response:** Many autonomous robot systems operate in real-time and in order to provide real time control of such system there should be a limit in sensor response time (between milliseconds to a small number of seconds). Fast sensing allow sensor to get more information from the odor signal (Atema 1996).
4. **Robustness:** The sensor will be equipped on the autonomous robot which might collide or abused. Therefore, the sensor must be able to withstand harsh environment conditions. The sensor will also be used in an unstructured and unconstrained environment and therefore should stand the chemicals it will encounter in its environment that might cause damage to it.
5. **Compact Size:** Small size is preferable and should be related to the robot dimensions and space which are usually limited.
6. **The Sensor Cost and Availability:** The sensor cost should be reasonable and should be available commercially or at least can be easily constructed from common available materials.

CHEMICAL SENSORS USED IN ROBOTICS

The research in new chemical sensing technologies was motivated by the need of controlling industrial processes and the environment monitoring. In common applications, the main gas categories to be monitored are:

1. Oxygen, for the control of combustion processes.
2. Flammable gases in order to protect against fire or explosion.
3. Toxic gases for environmental monitoring. Many sensors that can be used to detect these gases have been tested on robotic platforms. The following sensors are employed widely in robotics research:
   a. The commercial Tin Oxide gas sensor which is mostly used because of their reasonable price and availability. The size of these sensors is small, and they are easy to interface. Tin Oxide sensors are manufactured using thick-film technology and have relatively low power consumption. Some researchers have developed their own Tin oxide sensors (Siegel 1990). The response of these sensors has been optimized for solvents, halocarbons, and a range of combustible gasses with claimed sensitivities of 1 ppm. One of the major disadvantages of Tin oxide technology is the long recovery period needed after each gas exposure which make it critical for electronic gas sensing. This restricts its usage in applications where the gas concentrations may change rapidly. In mobile robotic olfaction, the high delay in detecting the gas forces the robot to move at a very low speed, almost incompatible with any practical robot operation. To overcome this limitation, researchers proposed many methods to increase the sensing speed of the Tin Oxide sensor as will be discussed later.
   b. The Quartz microbalance sensors which is easy to fabricate (Deveza &Russell 1993), (Russell et al. 1995) and have been used widely in the fields of biology, material science, environmental monitoring, and electrochemistry (R. Russell 1999). Quartz microbalance sensors use a quartz crystal to evaluate odor molecules. A special chemical coating that have a specific attraction for the target odorant molecules is used. If the air has molecules of the odor is drawn over the crystal, some of the molecules become temporarily attached to the coating. This results in increasing the effective crystal mass and lowers its resonant frequency. The industry developed a Quartz microbalance based multi analyte detection system, composed of six chips sensor module coated with synthetic polypeptides together with conducting polymers. It can detect acetic acid, butyric acid, ammonia, dimethyl amine, benzene, chlorobenzene, and their mixtures
   c. Conducting polymer sensors (Stella et al. 1995) which is a cheap and safe method of making polypyrrole. It is developed using electrochemical polymerization of pyrrole between a pair of electrodes. The conducting polymers has advantages and disadvantages. There are many advantages of using sensing layers from conducing polymers. It can used in room-temperature operations and has high sensitivity and short response time. It has also facile property adjustment, and easy device fabrication. The disadvantages of using conducting polymers are long-time instability and irreversibility and low selectivity. A single sensor cannot distinguish different analyses and the response is easily influenced by the presence of other gases.
   d. Live moth antennae (Kuwana et al. 1996) which is used by animal to successfully locate their mate over long distances (hundreds of meters). It is used by researchers to help robot navigation. The advantage of this type of sensors is their high sensitivity. The sensitivity of the antennae to some odors reach parts per billion range (S. Schutz & et al 2000) while the best current artificial nose technology is able to detect compounds in the parts per million range. The disadvantage of the moth antennae sensor is their short lifetime problems.

Table 1 compares between Tin Oxcide, Quartz, conducting polymer and live moth antennae sensors in term of sensitivity and response time (R Russell 2001).

Table 1. Comparison between different sensors

Odor Sensor	Sensitivity	Response Time
Tin Oxide	1 to 10 ppm	30 ms to 10 s
The Quartz microbalance	1ppm	0.05 s
Conducting polymer	0.1-100 ppm	24 s
Live moth antennae	300 molecules/s	< 1 s

1. **IR Spectroscopy:** Which uses the phenomena that all gases interfere and absorb light at specific wavelengths of the infrared spectrum because of their natural molecular vibration. This property can be used to detect the concentration of different gases in the environment. There are monochromatic systems tuned with narrow-band interference filters or laser light sources for a specific gas (like $C0_2$) and there are spectroscopic systems able to determine the concentration of several gases at once. These sensors feature slow response, good linearity, low cross-sensitivity to other gases and fairly good accuracy, but they need frequent recalibration, are bulky and very expensive.
2. **Optical Fiber Sensors:** Optical fiber gas sensors are composed by a fiber bundle with its ends coated with a gas-sensitive fluorescent polymer. When light comes down the fiber, it excites the polymer, which emits at a longer wavelength to the detection system. The amount of returning fluorescent light is related with the concentration of the chemical species of interest at the fiber tips. Tufts University uses an array of 19 fibers coated with Nile Red dye. To differentiate among various chemical gases, they correlate the amount of returned intensity and the fluorescence lifetime in each fiber (Dickinson J, White J, Kauer J, Walt D 1996). This sensor can be used by drone to test the concentration of specific gas constration.

APPLICATIONS

The robot equipped by odor sensing capabilities can be used in many applications. These applications range from:

NAVIGATION ALONG A CHEMICAL PATH

One of the interesting application of using odor sensing is to help the robot in navigation. In industrial transport, the robot is used to carry material from place to another place. The navigation task consists of planning the path of movement and then controlling its realization (Gideon Kowadlo & R. Andrew Russell, 2008). To help the robot navigation, the environment is marked using painted line or using cables powered by high frequency alternating current. The robot can easily follow the observable path marked in the environment, but the disadvantage of this method is its low flexibility since changing the route need to remove the old lines and paint a new one (Wisnu Jatmiko & et al 2005). In order to add flexibility for robot navigation Short Living Navigational Markers (SLNM), such as odor or heat are used (Piotr Batoga & Andrzej Wołczowski 2012). The lines marked in this way disappear after the robot completes its work similar to the pheromones paths created by insects (A. Russell 1995). The problem of this method is that chemical sensors, mostly Tin Oxide, are relatively slow if compared with non-chemical sensors used for mobile robot navigation like the light sensors. In order to overcome this problem, researchers introduced many methods that accelerate the chemical senor speed. Javier Gonzalez-Jimenez, Javier G. Monroy and Jose Luis Blanco (Javier et al 2011) proposed e-nose, called Multi-Chamber Electronic Nose (MCE-nose) which uses 6 identical sets of Tin oxide sensors accommodated in separate chambers that alternate between sensing and recovery states, providing, as a whole, a device capable of sensing changes in chemical concentrations faster. Other researches use vision (Hiroshi Ishida & et al 2004) or sound (Kai & et al 2011) to assist the chemical sensor and the robot to achieve faster navigation. Hiroshi Ishida & et al (Hiroshi Ishida & et al 2004) uses vision so when the presence of gas is detected, the robot turns to the direction of the airflow that carries the gas, and uses vision sensor to look for any suspicious object. Then with the help of the two sensors, path planning is performed using a potential field method. Mexican-hat shaped potential field is formed around the visually detected object. Consequently, the robot is guided along the perimeter of the object to sniff out the location of the gas leak. Kai & et al (Kai & et al 2011) integrates smell, hearing and touch to design an odor/sound tracking multi-robot system. The olfactory robot can track the chemical odor plume within distance of 2 meter using information fusion from gas sensors and airflow sensors. They use two hearing robots to localize the sound source using time delay estimation (TDE) and the geometrical position of microphone array. The two hearing robots can quickly localize and track the olfactory robot in 2 min. The researchers pointed out that the devised multi-robot system can achieve target search with a considerable success ratio and high stability.

Other researches did many experimental studies that intends to find the best chemical volatiles to mark odourific trails with mobile robots. Sousa (Sousa P. Et al 2008) presented the characteristics of some chosen chemicals and analyzed the influence of the substrate permeability on the ability to detect those chemicals along the time. They propose algorithm to search and track chemical trails with a mobile robot is proposed. They validate the algorithm by simulations.

MONITORING OF HAZARDOUS INDUSTRIAL SITES

The mobile robot equipped by olfactory capability can be used to monitor hazardous site to detect gas leak and dangerous substances along predefined path. The monitoring site could be a factory or a mine. To increase the level of human operators, robot can be sued to monitor things like gas leakage location, estimation of the gas-concentration gradient and gas recognition independent of the intensity. Cosimo Distante (Cosimo & et al 2009) used mobile robot to detect hazardous fluid leakages in chemical warehouses and industrial sites. The proposed robot performs automated monitoring of hazardous industrial sites using odour-sensing capability. The odour sensor used is an array of tin oxide gas sensors mounted within a nylon chamber so-called nose cap. The researchers claimed that the robot was able to detect multiple odour sources and differentiate between sources very close to one another. The limitation of their work is that their mobile robot has been specifically designed to detect odour along a well-known path in a defined structured environment like that expected in industrial field. The problem of detection of leakages outside the search path is not addressed in their work. On the other hand, the international atomic energy agency published a report (Safety Report 1991) entitled" Safety aspects of nuclear power plant automation and robotics". The report emphasis on the use of robotic technology to enhance the safety of existing and future commercial nuclear power plants. Also, drone can be used to detect gas leak for example Örebro University has deployed its "Gasbot" mobile robot into several landfills to search for methane leaks. Gasbot has a particularly awesome method for hunting down these fissures. The robot shoots laser beams several meters ahead of itself, scouting for the gas's precise wavelength. When the beam bounces back, Gasbot can tell if how much of it was absorbed by nearby plumes of methane, to an accuracy of 5 parts per million. It's not "smell" in the familiar biological sense, but it clearly demonstrates that robots can root out dangerous leaks far more efficiently and safely than humans.

THE DETECTION OF ABANDONED LAND MINES

In war zone, a lot of land mines are left, and these cause a danger to civilian. Detecting and cleaning these mines is costly, dangerous and difficult task. The detection of mine is seen as target application for odor-sensing robots. In order to use the robot for mine detection it is needed to equip the robot with of a chemical-sensing system that can detect the vapor of TNT given off by the mine. The sensors that are developed so far are not sufficiently sensitive or selective for this kind of application. However, it is reported that a company involved in defense technology, named "Bofors", is developing a highly sensitive detector for TNT vapor. This device is reported to gather and concentrate the vapor before presenting it to a crystal coated in antibodies sensitive to TNT (Brink 1996). In the presence of TNT, the surface of the crystal is detached by TNT antibodies. This increase of resonant the crystal frequency, which indicates the presence of TNT. A sensitivity of better than 10 picograms is claimed for this system. The development of a very sensitive and specific sensor contribute to the mine sweeping industry and will be the significant step toward the creation of effective robot systems for a number of odor-sensing applications.

DETECTION OF DRUGS, WEAPONS, EXPLOSIVE, AND OTHER ILLEGAL MATERIALS

Traditional cargo-screening methods rely on a many separate method. For detecting drug sniffer dogs are used. External scanners are employed for detecting explosives. To detect human presence carbon dioxide probes and heartbeat monitors are used. Researchers have developed a 'cargo-screening ferret' for use at seaports and airports to detect drugs, weapons, explosives and illegal immigrants (R. Russell & et al 2009). A new type of 30cm-long robot that can help custom service is developed. This robot that can detect drugs, weapons, explosives and illegal immigrants concealed in cargo containers. The device was funded by the Engineering and Physical Sciences Research Council (EPSRC) and implemented at the University of Sheffield and (R. Russell & et al 2009). A sensitive sensor developed by the EPSRC-funded are used in this robot. These sensors are small enough to be carried on the robot and can detect the specific 'fingerprint' of illegal substances at much lower concentrations which enable the robot can detect tiny particles of different substances. The ferret will attach itself magnetically to the top inside a steel freight container, then automatically move around and seek out contraband, sending a steady stream of information back to its controller. The 'cargo-screening ferret offer major advantage is that the system uses sensors that detect tiny traces of carbon dioxide which indicate the presence of humans concealed in the containers avoiding the use of X-rays which is prohibited due to the harm the radiation that could do. It also reduces the need for customs and security officials to enter or unpack containers thus saving time and protect officers from danger or possible contamination by harmful substances.

SEARCH FOR VICTIMS OF AVALANCHES AND EARTHQUAKES

Robots can do the job of sniffer dogs. Dogs require extensive training and care and their living time is short. They also suffer from tiredness. Robotic systems could offer improvements in all of these areas. There cannot operate in environments that includes high concentrations of poison gas and nuclear radiation. Robots could be withstanding these conditions.

Russel (Russel & et al 2000) proposed a mobile robot called RAT robot that can be used in rescue operation as a replacement for sniffer dogs. RAT robot can be used in searching earthquake damaged buildings for survivors. It also can be used for robots which call for assistance when robot's batteries fail by releasing a chemical signal.

The proposed RAT robot was equipped with ultrasonic sensors, an airflow sensor, whisker bump sensors, and a chemical sensor. The RAT robot can sense and control the location of an odour source in a maze of narrow corridors (Russel & et al 2000). Recently, the UAV or drone are used for rescue operation. Drone can be flown much lower to the ground compared to helicopters, and can be outfitted with wide-angle lenses to get a better view of huge fields. They can cover huge areas of land in a fraction of the time that a search and rescue team can, which saves money and could end up being the difference between life and death. Equipping drone with e-nose enabled the drone to search through precarious terrain, following bodily scents to recover casualties similar to the search done by a rescue dog. It could be used in situations where an animal's life might be at risk, or perhaps in areas that are constricted or volatile in the wake of a natural disaster.

TASTE SENSING

Chemicals that have molecular weight greater than 300 do not produce detectable plumes because it has very low volatility. Such chemicals cannot be detected by plumes of volatile chemicals released into the atmosphere and can be detected only by taste. Russell (Russel & et al 2006) researcher describes a cleaning robot called TASTI equipped with conductivity sensor for detecting chemical deposits on the ground. A conductivity sensor is a simple method of detecting chemicals in solution and this form the basis of the robotic tongue used in this robot. Conductivity sensors can detect ionic compounds on the ground. It cannot discriminate different ions, but they do give an indication of total ion concentration (Russel & et al 2006) >The researcher used three search algorithms to test the cleaning robot in locating patches of chemical deposited on the ground using its chemical sensing tongue.

COLLABORATIVE APPLICATION

- **Collaboration Can Be Defined as:** *"The action of working with someone to produce or create something"*. In robotics, collaborative application enables many robots to work together to fulfill a specific task. Cooperative robotics Volatile chemicals can be used as temporary marks to coordinate a set of autonomous robots executing a common task. For example, if several cleaning robots clean a huge area, they can use odor information to detect places recently cleaned by other robots because the ammonia odor will be stronger in these places. In a similar way, security robots can mark its path with a volatile trail in order to detect paths recently patrolled. This can be done using chemical to pass information between these robots. Chemicals that pass information between members of the same species are called pheromones.

For robotic, versions of pheromones can be used to achieve many actions such as (Russell 2001):

- **Call for Assistance:** A robot with discharged battery could release a chemical as a call for help. Any nearby robot can give help by pushing the discharged battery robot to reach a central charging point.
- **Passing Information Between Robots About the Environment:** Useful information about the environment can be passed among group of robots using *pheromones* such as a warning of hazardous conditions ahead so that robot can avoid the hazardous place.
- **Helping to Organize Groups:** In conditions where electromagnetic communication is unusable or inadvisable, broadcast chemical signals could be used to modify the actions of large groups of robots.

Robotic swarms aims to study the autonomous behavior of large numbers of simple robots in order to find collaborative application and technological solutions to common complex tasks. Robotics researchers give attention to this topic and are investigation how to provide a robust, scalable approach for coordinating actions of large numbers of small scale robots. This can help in achieving large scale results in surveillance, reconnaissance, hazard detection, path finding, payload conveyance, and small-scale actuation. Robotics and artificial intelligence researchers at the University of Lincoln, UK (Swarm Robotics 2015) introduced effective and low-cost system which replicates in robots the pheromone-based communication of insect swarms. They created a robot called Colias (Honeybee' rorbots 2014) as a result of collaboration with Tsinghua University in China. Colias has small size (4cm diameter) and fast motion (35cm/s) that enable it to be used in fast-paced swarm scenarios over large areas - effectively replicating a honeybee swarm.

Ryusuke Fujisawa, Hikaru Imamura, Takashi Hashimoto and Fumitoshi Matsuno (Ryusuke Fujisawa & et al 2008) introduced a swarm robot that Communicate using Pheromone Field. In their research their goal was to use pheromone to enable swarm robots to solve real tasks in a real world. In their study, they pay attention to object transportation by the swarm. Which include searching objects and attracting another agent. Such function is needed when agents search a particular object from vast space, and when they find the object and cannot transport it by themselves. They also tackle the Dynamic interference (pushing, grasping, etc.) between objects and agents. Such function is needed when an object is too heavy for them to move and agents in a swarm should cooperate with each other to move the heavy object.

Purnamadjaja and Russell (Purnamadjaja 2004, 2005) implemented an algorithm in a swarm of robots that cooperated in pushing a "dead" robot. The swarm did not cooperate in locating the odor source, but rather in the necrophoric behavior.

SECURITY ROBOT

A robot equipped with a set of environmental monitoring sensors like temperature, humidity, smoke, flame and MOS gas sensors for detection of explosive and toxic gases. With these sensors the robot can perform a set of preventive tasks that have already made possible the detection of toxic gases, gas leaks, burning equipment, etc. Many companies are making these kind of security robots and are used commercially.

AGRICULTURE

The application of robotics in agriculture has been the target of several research groups recently. The objectives is the automation of tasks like measurement and control of environmental conditions, plant inspection, and spraying of pesticides, fungicides and other chemical products over the plants. These tasks if automated can have significant economic and health impacts, avoiding the workers exposure to insecticides and other dangerous chemical products (Mandow A, et al 1996). Electronic noses can be used with farming robots and will equip it with very useful capabilities and they will provide the necessary sensorial feedback for some of the most common farming tasks. For example, the robot can analyze in real-time the volatile compounds released by the soil and fertilize it with the estimated needs. This way the fertilizer is not wasted, and the soil does not become contaminated with too much nitrate. The electronic nose can analyze the environmental conditions to prevent diseases and actuate before the plants become ill. Some diseases release volatile compounds. In these cases, if the plants are already ill, the robot can report the situation in order to prevent the spread of the disease. A harvesting robot can use aroma information to selectively gather ripe fruits or adult flowers. If the robot knows a map of the plants around the farm, it can use odor information as a rough localization method: near a rose flowerbed, it should smell rose aroma.

CLEANING

A cleaning robot with an electronic nose can detect the ammonia odor of an already cleaned floor. This way the robot does not waste time cleaning the floor again.

ALGORITHMS USED BY ROBOTS FOR SEARCHING AN ODOR

The odor sensor when detecting the presence of chemical substance send a signal to the robot controller. The controller analyzes the received signal and take the appropriate action by sending appropriate signal to the effectors. The controller can be based on deliberative or reactive model. The deliberative model is representation dependent, need complex sensors, uses high level intelligence and therefore has slower response and variable latency. On the other side, the reactive model is representation free, uses low level intelligence and therefore it performs simple computation and has real time response. While it is possible to use the debilitative and reactive model, however, most of the reported work involving odor-mediated navigation has used relatively simple reactive control algorithms (Russell 2001). It is also worth mentioning that many of these algorithms, imitate the behavior observed in insects and other simple creatures like ants. The algorithms can be simple, complex or hybrid algorithm as summarized below:

TRAIL-FOLLOWING ALGORITHMS

This algorithm imitates ants employ chemical trails to guide workers between a source of food and their nest. The same method is used for an autonomous robot. In order to find its way around it follow a guide path. The algorithm is implemented using one sensor and two sensors.

TRAIL FOLLOWING WITH A SINGLE SENSOR

Russell, Thiel, and Mackay-Sim (Russell & et al 1997) introduced a robot using one sensor follows the left-hand edge of a chemical trail. The algorithm consists of three stages (Russell & et al 1997):

- Start rotating clockwise until the edge of the trail is detected.
- Rotate a fixed angle away from the edge of the trail (Counterclockwise).
- Move forward a fixed distance.
- Repeat.

TRAIL FOLLOWING WITH TWO SENSORS

Other researcher implements the algorithm using two sensors. Stella et al. (Stella et al. 1995) equipped their robot vehicle with a pair of polymer sensors and used them to follow a 15 cm wide felt strip soaked in alcohol. These forward-facing sensors were spaced 10 cm apart and positioned 1 cm above the ground. The control algorithm consists of (Stella et al. 1995):

- Starting the vehicle with both sensors over the felt strip.
- As the robot proceeds, it might deviate sufficiently that one sensor loses contact with the strip.
- The resulting fall in sensor response triggers a signal to turn the robot to bring the sensor back onto the tape.

Using this algorithm, the robot was able to follow the tape with a reported tracking speed of 60 mm/s. One drawback of this technique is the very wide chemical marking that it requires.

LOCATING THE SOURCES OF CHEMICAL PLUMES

Most of experiments involving odor-sensing robots have been directed toward finding the source of a volatile chemical. Using the combined odor and wind direction sensor described in Section 3.2, Ishida et al. (Ishida & et al 1994) implemented the following two algorithms for locating the source of an odor plume.

STEP-BY-STEP ALGORITHM

In this algorithm,

- Robot selects a heading in the upwind direction as measured by the anemometric sensors and toward the center of the odor plume as indicated by the gradient reading from the gas sensors.
- Having determined a suitable course, the robot moves 2 cm in this direction.
- The robot trajectory then consists of a repeated sequence of these sense-move operations.

The disadvantage of this algorithm it works with high concentration level of chemical gradient. The turbulence in the airflow causes the chemical plume to fluctuate and become difficult to get a reliable indication of chemical gradient and if the sensed chemical concentration falls below a threshold level, the robot is stopped, and it waits until the sensor registers a higher concentration again.

ZIGZAG ALGORITHM

In this algorithm:

- The robot is programmed to run diagonally across the full width of the odor plume.
- On reaching the far edge of the plume, the robot changes direction to cut across the plume in the opposite direction.
- On each sweep, the diagonal path is angled upwind so that the robot makes progress toward the odor source.
- If the robot loses contact with the plume, it is programmed to back up until it senses the chemical again

ODOR SOURCE LOCALIZATION USING PARTICLE SWARM OPTIMIZATION (PSO)

Particle Swarm Optimization (PSO) is a simple but powerful search technique that applies the concept of social interaction to problem solving. The concept of particle swarm was originated to simulate simplified social system. However, it was found that particle swarm model can be used as an optimize [20]. In PSO, a swarm of n individuals communicate either directly or indirectly with one another search directions (gradients). W. Jatmiko and et al [19] used PSO for real-world odor source localization. In their experiment, they claimed that PSO can localized the source in 360 seconds or bellow.

HYBRID ALGORITHM (FUZZY AND PSO)

Some researchers used Simple form of cooperation between Intervals Type-2 Fuzzy Logic and Particle Swarm Optimization (IT2FL-PSO). This hybrid algorithm is implemented in the olfaction strategies to detect a gas/odour source in an indoor environment. Siti Nurmaini and et al. (Siti Nurmaini and et al 2013) uses hybrid algorithm combination of particle swarm optimization and interval type-2 fuzzy logic algorithm for navigation and robotic search applications in multi robot environment. All the robots execute the same algorithm. Depending on the local information gathered, the algorithm focuses on how to compute, effectively and consistently, the goal position. The positions of two neighboring robots is taken as input, and the goal position is taken as output. The input information may directly come from sensors of the robot and the direction of motion can be implemented with the actuators of the robot. The researcher used simulation to show that a mobile robot is capable to locate the source of a simulated chemical leak in an environment while detecting and avoiding obstacles along its path. The researchers claimed that by using proposed IT2FLPSO algorithm, the swarm robots could successfully use olfaction system to achieve gas/odour source in unknown environment.

FUTURE RESEARCH DIRECTIONS

Equipping robot with chemical sensors will certainly increase his ability and intelligence. However, the chemical sensors needed for robot applications have different requirements than those used in industrial applications. In industrial applications, the most important parameter of chemical sensors is the gas concentration measurement accuracy but in mobile robotics applications the most important parameters are the sensitivity to tested chemical, the selection threshold, and the time of reaction to changes of gas concentration (Russell 1999), (Russell 2001). However, these parameters also may have different importance depending on the target application for example, in mobile robot navigation, detailed knowledge about gas concentration and emission rate is not really necessary. The only purpose is the detection of chemical path placement as precisely and quickly as possible. While in other application like gas leak detection and drugs, weapons and explosive detection sensor sensitivity is more important than time needed for detection. In order to make the use the best use of chemical sensors in robotic applications, many revolutionary methods for enhancing the current sensors and producing new sensors based on new technologies are needed to meet the requirement of different robot applications. Some of the trends can be summarized below:

1. Searching for other options to modify the gas sensor to achieve higher sensitivity and speed of detection and this can be achieved by changing the sensory systems organization using hybrid or multi sensors working together.
2. Search for other volatile substances which ensure certain properties of chemical trace.
3. Developing sensors that are able to distinguish between various chemicals, so that the robot could decide which goal to pursue, (which path to move).
4. Developing of a sensor system capable of evaluation of the concentration gradient in three-dimensional space (Nakamoto Ishida 1997), (Ishida & at al 1999).
5. Developing sensors dedicated for emergency and monitoring systems that can provide accurate and fast localization of sources of gas leak (Sutton R & et al 2007), (Waphare S & et al 2009).

In future, there will be applications that work on complex environments or underground, underwater, enclosed spaces, and interconnected tunnels. It is also going to move from 2D to 3D. It is likely that many of these methods will be enhanced and integrated. 3D localization will be focus of researchers and more attention will be given to obstacles detection and avoidance and how to deal with unstructured environments.

In the future, robots will be part of IoT network and have a general awareness of their surroundings. Therefore, new feeling sensors are required for these specific functions. For example, a mail delivery robot should be equipped with a sense of smell to be able to detect such things as smoke, gas, and rotting food. The robot could then act to these odors.

Other challenge is the smart system integration. This need developing and manufacturing smart objects and systems that closely integrate different types of sensors (like Micro Electro Mechanical Sensor (MEMS), odor sensor, light and sound sensors, etc..), actuators, controller, embedded memory, wireless communication and power supply in one small smart device. Other issues like low power consumption, capability to operate under various and harsh conditions and interoperability within systems of systems are needed and this will require technology breakthroughs.

In future there will be a need for multi-robot olfaction applications that uses Plug and Play (PnP) unit such as a communication interface, obstacle avoidance guidance control, etc... The PnP feature is necessary to easily and fast integrate robot units with minimum interfacing effort.

CONCLUSION

Robotic odor-sensing technology is emerging and in only short period of time many researchers have been investigating this topic potential in different applications from simple designs with limited sensing capabilities to multiphase algorithms that exploit a range of sensors. Chemical sensors used in robotic applications are limited and their sensitivity and selectivity are very poor compared with biological systems. In order to enhance the chemical sensors sensitivity and selectivity, researchers are using array of chemical sensors or augmenting them using other sensors like light and hearing sensors. Thus, we got a chemical/sensor combination with sufficiently good performance that can be used to explore the possibilities for robotic odor sensing. In this chapter many topics were addressed like the benefit of equipping the robot with odor sensors, the types of tasks the mobile robots can accomplish with the help of olfactory sensing capabilities, the technical challenges in mobile robot olfaction, the current status of mobile robot olfaction. The chapter also addressed simple and complex Electronic Olfaction sensors used in Mobile Robotics, the challenge of gas sensing and the use of many types of algorithms for robot olfaction.

REFERENCES

Andrew Russel, R. (1995). Laying and sensing odor markings as a strategy for assisting mobile robot navigation task. *IEEE Robotics & Automation Magazine*, *2*(3), 3–9. doi:10.1109/100.414920

Andrew Russel, R. (1999). Odour Detection by Mobile Robots. *World Scientific Series in Robotics and Intelligent System*, *22*, 197-217.

Andrew Russell, R. (2001). Survey of Robotic Applications for Odor-Sensing Technology. *The International Journal of Robotics Research*, *20*(2), 144–162. doi:10.1177/02783640122067318

Andrew Russell, R. (2006). Taste and Search in a Robotics Context. *Proc. Australasian Conf. on Robotics and Automation*, 1–6.

Andrew Russell, R., Thiel, D., & Mackay-Sim, A. (1997). Recruiting swarm robots using coded odour trails. *Proceedings of the International Conference on Field and Service Robotics*, 472–476.

Atema, J. (1996). Eddy chemotaxis and odor landscapes: Exploration of nature with animal sensors. *The Biological Bulletin*, *191*(1), 129–138. doi:10.2307/1543074

Batoga, P., & Wołczowski, A. (2012). Odor markers detection system for mobile robot navigation. *Proc. Eurosensors XXVI*, 1442-1445. doi:10.1016/j.proeng.2012.09.429

Brink, S. A. (1996). Bofors schnauzer - a biosensor for detection of explosives. *The EUREL International Conference on The Detection of Abandoned Land Mines Edinburgh*, *431*, 33-36. doi:10.1049/cp:19961074

Deveza, R., Russell, R. A., & Thiel, D. (1993). Robot navigation by smell: Problems and some solutions. *Proceedings of the Robots for Competitive Industries Conference*, 458–466.

Dickinson, J., White, J., Kauer, J., & Walt, D. (1996). A chemical-detecting system based on a cross reactive optical sensor array. *Nature*, *382*(6593), 697–700. doi:10.1038/382697a0 PMID:8751439

Distante, C., Indiveri, G., & Reina, G. (2009). An application of mobile robotics for olfactory monitoring of hazardous industrial sites. *Industrial Robot: An International Journal*, *36*(1), 51–59. doi:10.1108/01439910910924675

Gonzalez-Jimenez, Monroy & Blanco. (2011). The Multi-Chamber Electronic Nose— An Improved Olfaction Sensor for Mobile Robotics. *Sensors (Basel)*, *11*(6), 6145–6164. PMID:22163947

González-Jimenez, J., Monroy, J. G., & Blanco, J. L. (2013). Robots that can smell: motivation and problems. *The International Symposium on Olfaction and Electronic Nose (ISOEN)*.

Honeybee' robots. (2014). Retrieved from https://phys.org/news/2014-09-honeybee-robots-replicate-swarm-behaviour.html#jCp

Ishida, H., Kobayashi, A., Nakamoto, T., & Moriizumi, T. (1999). Three-Dimensional Odor Compass. *IEEE Transactions on Robotics and Automation*, *15*(2), 328–339. doi:10.1109/70.760346

Ishida, H., Ushiku, T., & Toyama, S. (2004). Mobile Robot Path Planning Using Vision and Olfaction to Search for a Gas Source. *International Conference on Intelligent Robots and Systems*, 313-318.

Jatmiko, W. (2005). Robotic applications for odor-sensing technology: Progress and challenge. *WSEAS Transactions on Systems*, *4*(7), 1134–1141.

Jatmiko, W. (2011). Robots implementation for odor source localization using PSO algorithm. *WSEAS Transactions on Circuits and Systems*, *10*(4), 115–125.

Kowadlo, G., & Russell, R. A. (2008). Robot Odor Localization: A Taxonomy and Survey. *The International Journal of Robotics Research*, *27*(8), 869–893. doi:10.1177/0278364908095118

Mandow, A., de Gabriel, J. G., Martlnez, J., Mufioz, V., Ollero, A., & Garcla-Cerezo, A. (1996). The Autonomous Mobile Robot AURORA for Greenhouse Operation. *IEEE Robotics & Automation Magazine*, *3*(4), 18–28. doi:10.1109/100.556479

Nakamoto, T., Ishida, H., & Moriizumi, T. (1997). Active odor sensing system. *Proceedings of the IEEE International Symposium on Industrial Electronics*, 128-133.

Nurmaini, S., Tutuko, B., & Thoharsin, A. R. (2013). Intelligent Mobile Olfaction of Swarm Robots. *International Journal of Robotics and Automation*, *2*(4), 189–198.

Politechnika Wrocławska. (2012). Semiconductor gas sensors for mobile robot navigation. *Electrical Review*, *88*(11b), 281-284.

Purnamadjaja, A., & Russell, R. (2004). Pheromone communication: implementation of necrophoric bee behaviour in a robot swarm. *Proceedings of the 2004 IEEE Conference on Robotics, Automation and Mechatronics,* 2, 638– 643.

Purnamadjaja, A. H., & Russell, R. A. (2005). Pheromone communication in a robot swarm: Necrophoric bee behaviour and its replication. *Robotica*, *23*(06), 731–742. doi:10.1017/S0263574704001225

Randy, R. (2013). *These Machines Will Be Able to Detect Smells Your Own Nose Cannot*. Retrieved from https://www.smithsonianmag.com/innovation/these-machines-will-be-able-to-detect-smells-your-own-nose-cannot-8790482/

Report, E. (2009). Robotic Ferret Will Detect Hidden Drugs and Weapons. *Engineering and Physical Sciences Research Council, Science Daily*. Retrieved from www.sciencedaily.com/releases/2009/06/090612115531.htm

RoboticsS. (2015). Retrieved from https://phys.org/news/2015-11-scientists-breakthrough-pheromone-based-swarm-robots.html#jCp

Rochel, O., & Martinez, D. (2002). Stereo-olfaction with a sniffing neuromorphic robot using spiking neurons. *16th European Conference on Solid State Transducers - EUROSENSORS.*

Russell, R. A., Kleeman, L., & Kennedy, S. (2000). Using Volatile Chemicals to Help Locate Targets in Complex Environments. *Australian Conference on Robotics and Automation*, 87-91.

Ryusuke Fujisawa. (2008). Communication Using Pheromone Field for Multiple Robots. *2008 IEEE/RSJ International Conference on Intelligent Robots and Systems*, 1391-1396.

Safety Report. (1991). Safety aspects of nuclear power plant automation and robotics. *The International Atomic Energy Agency*. Retrieved from http://www.iaea.org/inis/collection/NCLCollectionStore/_Public/24/007/24007130.pdf

Schutz, S., Schoning, M. J., Schroth, P., Malkoc, U., Weissbecker, B., Kordos, P., ... Hummel, H. E. (2000). An insect-based BioFET as a bioelectronic nose. *Sensors and Actuators. B, Chemical*, *65*(1-3), 291–295. doi:10.1016/S0925-4005(99)00325-1

Siegel, M. W. (1990). Olfaction metal oxide semiconductor gas sensors and neural networks. In T. C. Henderson (Ed.), *Traditional and Non-Traditional Robotic Sensors, NATO ASI Series* (Vol. F63, pp. 143–157). Berlin: Springer-Verlag. doi:10.1007/978-3-642-75984-0_11

Song, K., Liu, Q., & Wang, Q. (2011). Olfaction and Hearing Based Mobile Robot Navigation for Odor/ Sound Source Search. *Sensors (Basel)*, *11*(2), 2129–2154. doi:10.3390/s110202129 PMID:22319401

Sousa, P., Marques, L., & De Almeida, A. (2008). Toward chemical-trail following robots. *Seventh International Conference on Machine Learning and Applications (ICMLA08)*, 489–494. doi:10.1109/ ICMLA.2008.133

Stella, E., Musio, F., Vasanelli, L., & Distante, A. (1995). Goal-oriented mobile robot navigation using an odour sensor. *Proceedings of the Intelligent Vehicles Symposium*, 147–151. doi:10.1109/IVS.1995.528273

Sutton, R., Naeem, W., & Chudley, J. (2007). Chemical plume tracing and odour source localization by autonomous vehicles. *Journal of Navigation*, *60*(2), 173–190. doi:10.1017/S0373463307004183

Waphare, S., Botre, B., & Sadistap, S. (2009). Mobile odor tracking robot based on embedded technology. *International Conference on Emerging Trends in Electronic and Photonic Devices Systems*, 108-111.

ADDITIONAL READING

Barca, J. C., Lee, E. E., & Sekercioglu, A. (2013). Flexible Morphogenesis based Formation Control for Multi-Robot systems. *International Journal of Robotics and Automation*, *2*(1), 26–34.

Brattoli, M., De Gennaro, G., De Pinto, V., Demarinis Loiotile, A., Lovascio, S., & Penza, M. (2011). 1 Gianluigi de Gennaro, Valentina de Pinto, Annamaria Demarinis Loiotile, Sara Lovascio, and Michele Penza, " Odour Detection Methods: Olfactometry and Chemical Sensors. *Sensors (Basel)*, *11*(5), 5290–5322. doi:10.3390/s110505290 PMID:22163901

Marques, L., Nunes, U., & de Almeida, A. T. (2006). Particle swarm-based olfactory guided search. *Autonomous Robots*, *20*(3), 277–287. doi:10.1007/s10514-006-7567-0

Mendel, J., & John, R. (2002). Type-2 fuzzy sets made simple. IEEE Trans. On Fuzzy System, Vol. 10, PP 117-127, April, 2002.

Meng, Q. H., Yang, W. X., Wang, Y., & Zeng, M. (2011). Collective Odor Source Estimation and Search in Time-Variant Airflow Environments Using Mobile Robots. *Sensors (Basel)*, *11*(11), 10415–10443. doi:10.3390/s111110415 PMID:22346650

Nurmaini, S., Tutuko, B., & Rahman, A. (2014). A New Navigation of Behavior- based Olfacory Mobile Robot. *Applied Mechanics and Materials*, *446-447*, 1255–1260. doi:10.4028/www.scientific.net/ AMM.446-447.1255

Weigel, T., Gutmann, J. S., Dietl, M., Kleiner, A., & Nebel, B. (2006). "CS Freiburg: Coordinating Robots for Successful Soccer Playing", C.A.C. Parker and H. Zhang, Collective Robotic Site Preparation. *Adaptive Behavior*, *14*(1), 5–19.

KEY TERMS AND DEFINITIONS

Autonomous Robot: The autonomous robot can navigate and move without human intervention. It is equipped with sensors and has complex algorithm that enable it to work independently in unstructured environment.

Chamber Electronic Nose (MCE-Nose): A compound sensor that uses six identical sets of tin oxide sensors accommodated in separate chambers that alternate between sensing and recovery states, providing, as a whole, a device capable of sensing changes in chemical concentrations faster.

Collaborative Robot (Cobot): Cobot is a collaborative application where robots share the same workspace to do a common task. Robots should be trained to work together. Robots should be equipped with additional sensors to be able to achieve collaboration.

Drone: A drone, in a technological context, is an unmanned aircraft. Drones are more formally known as unmanned aerial vehicles (UAVs) or unmanned aircraft systems (UASes). Essentially, a drone is a flying robot. The aircrafts may be remotely controlled or can fly autonomously through software-controlled flight plans in their embedded systems working in conjunction with onboard sensors and GPS.

Evolutionary Robotics (ER): ER is a method for automatically generating artificial brains for autonomous robots. It uses evolutionary computation to develop such artificial brains for the robot. ER uses algorithms that frequently operate on populations of candidate controllers, initially selected from some distribution. This population is then repeatedly modified according to a fitness function.

Fuzzy Logic (FL): FL is a computing approach based on "degrees of truth" rather than the usual Boolean logic which is based on "true or false." LE was introduced by Dr. Lotfi Zadeh of the University of California at Berkeley in the 1960s. It is also called Type-1 fuzzy logic.

Internet of Things (IoT): IoT is new computation model in which we have machine-to-machine communication. It connects thousands of devices and is built on cloud computing and networks of data-gathering sensors; it's mobile, virtual, and instantaneous connection, and they say it's going to make everything in our lives from streetlights to seaports "smart."

Intervals Type-2 Fuzzy Logic (IT2FL): IT2FL incorporate uncertainty about the membership function into fuzzy set theory, which is an improvement on type-1 fuzzy logic. When there is no uncertainty, then a type-2 fuzzy set reduces to a type-1 fuzzy set.

Micro-Electro Mechanical Sensor (MEMS): MEMS is a technology that uses miniaturized mechanical and electro-mechanical elements made using microfabrication techniques. The physical dimensions of MEMS devices can vary from well below one micron to several millimeters.

Mobile Robot Olfaction (MRO): MRO is the discipline that studies the capabilities of mobile robots equipped with gas sensing. The olfactory robots sense and track the chemical odor plume. Many applications have emerged for olfactory robots in the area of monitoring hazardous area, gas leak detection, and detection of illegal materials.

Particle Swarm Optimization (PSO): PSO is an optimization technique that uses computational method that try iteratively to improve a candidate solution of a problem with regard to a given measure of quality. It solves a problem by having a population of candidate solutions. A candidate solution is moved around in the search-space according to mathematical formulae position until an optimal solution is found.

Pheromone: A pheromone is a chemical smell produced by an animal, mostly insects, that changes the behavior of another animal of the same species. Some describe pheromones as behavior-altering agents.

Robot Navigation: The robot's ability to determine its own position and then to plan a path towards some goal location. In order to navigate in its environment, the robot or any other mobility device needs to know the environment in which it is navigating. Robots need to be equipped with necessary sensors and use intelligent algorithms to achieve good navigation.

Security Robot: A robot equipped with a set of environmental monitoring sensors like temperature, humidity, smoke, flame, and MOS gas sensors for detection of explosive and toxic gases. With these sensors, the robot is able to perform a set of preventive tasks that have already made possible the detection of toxic gases, gas leaks, burning equipment, etc.

Short Living Navigational Markers (SLNM): SLNM is a kind of marker that can be temporary and is dissolved after a period of time. Such a marker can be generated using odor or heat. The lines marked in this way disappear after the robot completes its work, so it is not necessary to remove the markers after use.

Tele-Operated Robot: Tele-operated robot is controlled by a human operator. Tele-operated operator can be wired controlled or wireless controlled. Wireless controlled can be achieved using infrared, radio frequency, Bluetooth, WiFi, or cellular/GPRS signal.

Time Delay Estimation (TDE): TDE is used with an array of sensors to find the time differences of arrival between signals received from the array of sensors. Time delay estimation has been a research topic of significant practical importance in many fields in robotics, radar, sonar, geophysics, and ultrasonic.

Unmanned Aerial Vehicle (UAV): Is an aircraft with no pilot. UAVs can be remote controlled aircraft (i.e., flown by a pilot at a ground control station) or can fly autonomously based on pre-programmed flight plans or more complex dynamic automation systems. UAVs are currently used for a number of missions, including reconnaissance and attack roles. Recently, the UAV has been used for rescue operations. It can be flown much lower to the ground compared to helicopters, and can be outfitted with wide-angle lenses to get a better view of huge fields. Equipping UAVs with e-nose enables searching through precarious terrain, following bodily scents to recover casualties like a rescue dog. It could be used in situations where an animal's life might be at risk, or perhaps in areas that are constricted or volatile in the wake of a natural disaster.

Unmanned Ground Vehicle (UGV): Robot in the shape of vehicle that operates while in contact with the ground and without an onboard human presence. UGVs can be used for many applications where it may be inconvenient, dangerous, or impossible to have a human operator present. UGV can work autonomously or can be tele-operated, and generally, the vehicle will have a set of sensors to observe the environment and make decisions accordingly.

Unmanned Underwater Vehicle (UUV): Unmanned underwater vehicles (UUV), sometimes known as underwater drones, are any vehicles that can operate underwater without a human occupant. These vehicles may be divided into two categories, remotely operated underwater vehicles (ROVs), which are controlled by a remote human operator, and autonomous underwater vehicles (AUVs), which operate independently of direct human input. The latter category would constitute a kind of robot.

Chapter 12
Towards Odor-Sensitive Mobile Robots

Javier Monroy
University of Malaga, Spain

Javier Gonzalez-Jimenez
University of Malaga, Spain

ABSTRACT

Out of all the components of a mobile robot, its sensorial system is undoubtedly among the most critical ones when operating in real environments. Until now, these sensorial systems mostly relied on range sensors (laser scanner, sonar, active triangulation) and cameras. While electronic noses have barely been employed, they can provide a complementary sensory information, vital for some applications, as with humans. This chapter analyzes the motivation of providing a robot with gas-sensing capabilities and also reviews some of the hurdles that are preventing smell from achieving the importance of other sensing modalities in robotics. The achievements made so far are reviewed to illustrate the current status on the three main fields within robotics olfaction: the classification of volatile substances, the spatial estimation of the gas dispersion from sparse measurements, and the localization of the gas source within a known environment.

INTRODUCTION

In the last decade, the number of research works published in the area of artificial olfaction has increased notably, with important advances in chemical sensor technology, bio-inspired and engineering based e-noses, and a broad range of algorithms to counteract drift and environmental cross-sensitivity, as well as to improve efficiency in the recognition of chemical volatiles. Likewise, a promising transfer from laboratories to real world applications has started, which despite the long and challenging road ahead, vows for granting this forgotten sense the importance it has in the animal kingdom (Doty, 2015).

In this regard, mobile robot olfaction (MRO), the branch of robotics that addresses the integration of gas and chemical sensors on-board mobile platforms, has also gained substantial relevance in the scientific community due to the interesting advantages a mobile robot brings when compared with the

DOI: 10.4018/978-1-5225-3862-2.ch012

traditional approach based on networks of static e-noses (Tsujita, Yoshino, Ishida, & Moriizumi, 2005). First, a mobile robot usually carries only one e-nose, therefore a more sophisticated and powerful (and more expensive) model can be used, enabling the analysis of more complex compounds and the detection of faster changes in the gas concentration (Gonzalez-Jimenez, Monroy, Garcia, & Blanco, 2011; Ishida, Kobayashi, Nakamoto, & Moriizumi, 1999; Sanchez-Garrido, Monroy, & Gonzalez-Jimenez, 2014; Werle et al., 2002). The calibration phase of the sensing devices is greatly simplified because of the reduced number of e-noses, something that represents an important issue in large gas sensing networks (Esposito et al., 2016). Also, MRO systems permit sampling at higher (and adaptive) resolutions, while still providing the required accurate localization of each measurement. Finally, a mobile robot can leverage environmental information provided by other sensors on board (anemometers, cameras, laser scanners, etc.) to enhance the olfaction task, for example by detecting obstacles or changes in the environmental conditions, and to process such data in an online fashion, allowing decision making.

Three are the main fields where gas-sensitive mobile robots have been proposed: volatile chemical recognition, which deals with the problem of identifying which of a set of categories a new volatile sample belongs to, gas distribution mapping, where the objective is to obtain a truthful representation of how volatiles are dispersed in the inspected area and their respective concentrations, and gas source localization, where the robot is commanded to localize the emission sources. In this chapter, achievements made to each of these three fields are reviewed after a brief overview of the specific challenges of gas-sensitive mobile robots.

SPECIFIC CHALLENGES OF GAS-SENSITIVE MOBILE ROBOTS

The development of mobile robot olfaction systems is not a trivial problem, and despite recent achievements, the potential of gas-sensitive mobile robots has yet to be fully realized. Besides the inherent complexity of artificial olfaction, new difficulties emerge when performing olfaction with a mobile robot. In this section, a review of the main issues and technical solutions proposed so far is presented.

Chemical Sensors

While most animals, from simple bacteria to mammals, are empowered with a highly developed and sharp sense of smell, sensors for robots with capabilities close to those of animals are not yet available. One of their main drawbacks is related to the response speed. While the response time of an animal's chemoreceptor is in the order of 100ms (Beer & Ritzmann, 1993), typical gas sensors need several tens of seconds before their responses reach the steady state values (Pearce, Schiffman, Nagle, & Gardner, 2006). For illustration, Figure 1 shows the rise and recovery times of a conventional metal oxide gas sensor when exposed to a rapid excitation. As reported in (Monroy, Gonzalez-Jimenez, & Blanco, 2012) the adopted solution to palliate this negative effect has been, in most cases, to slow down the locomotion of the robot to a few cm/s, as in (Ishida, Suetsugu, Nakamoto, & Moriizumi, 1994). Yet, over the past years, different hardware and software approaches have been proposed to overcome to a certain extent this important limitation (Di Lello, Trincavelli, Bruyninckx, & De Laet, 2014; Fonollosa, Sheik, Huerta, & Marco, 2015; Gonzalez-Jimenez, Monroy, & Blanco, 2011), enabling a higher speed for the robot and consequently improving its effectiveness in real world applications.

Figure 1. Rise and recovery phases of a metal oxide gas sensor response to a step gas concentration. (A) 2D plot of the sensor response over time, where the shaded-blue region denotes the sensor exposure to the analyte. (B) 3D gas distribution map generated from the readings of the gas sensor carried by a mobile robot along a corridor. It can be appreciated how the recovery phase after the gas exposure last for tens of seconds.

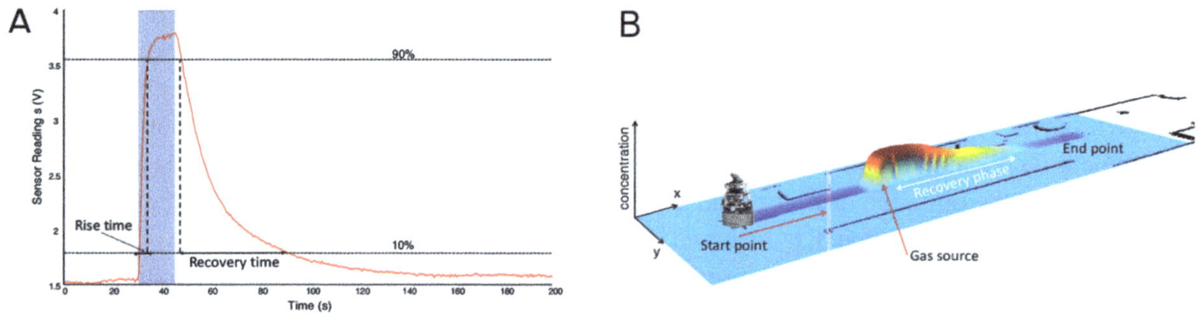

Slow sensor response also poses a significant challenge when performing volatile chemical discrimination. Gases disperse chaotically, led by turbulent advection, resulting in a concentration field that consists of fluctuating, intermittent patches of high gas concentration (Balkovsky & Shraiman, 2002). As a consequence, the sensor signals to be processed are noisy and dominated by the signal transient component. Patterns obtained are consequently distorted because sensors with different selectivities tend to have different response times (Schleif et al., 2016). Despite this, some recent works have addressed the issue, presenting different perspectives which will be covered.

Absence of Ground Truth

In real, uncontrolled environments, both indoor and outdoor, the dispersion of gases is dominated by turbulent flows. A turbulent flow is that in which fluid particles move in a random and chaotic way within the flow field (Sklavounos & Rigas, 2004). Furthermore, environmental variables such as temperature, pressure or humidity, as well as the airflow disruptions caused by the own robot movement, also have an important impact in the gas dispersion. All this entails the impossibility to know the exact behavior of a volatile release, and consequently, to have a *ground truth* of the nature and concentration of the chemical volatiles. This undoubtedly represents one of the main drawbacks for conducting real experimentation, and usually forces researchers to consider semi-controlled scenarios from which ground truth estimations can be reasonable assumed (Monroy & Gonzalez-Jimenez, 2017).

To cope with this problem, simulation tools with the capacity to properly handle the gas dispersion phenomenon (e.g. based on computation fluid dynamics) can be used to perform extensive evaluations before moving to experimental trials. While there exist a few implementations of robotics-oriented gas dispersion simulation frameworks, most of them consider simplified environments, are developed in outdated robotic software platforms or rely on expensive, external software packages (Cabrita, Sousa, & Marques, 2010; Monroy, Blanco, & Gonzalez-Jimenez, 2013; Pashami, Asadi, & Lilienthal, 2010). One exception is the newly developed GADEN simulator (Monroy, Hernandez-Bennetts, Fan, Lilienthal, & Gonzalez-Jimenez, 2017), an open source gas dispersion simulation framework aimed to mobile robotic olfaction applications. GADEN enables the simulation of gas dispersal in any 3D environment (including obstacles and realistic configurations), together with mobile robotic platforms and sensing devices (e.g. MOX sensors, anemometers, etc.). This framework is built upon the robot operating system (ROS), arguably, the most widespread robotics OS used in academia and industry.

Power Consumption, Weight, and Size

A common characteristic of most current mobile robots is the fact that they run on batteries. This entails the problem of balancing power consumption, payload and runtime, which becomes paramount for unmanned aerial vehicles (UAVs). In this context, and far from the bulky and power-hungry laboratory sensing devices (e.g. mass spectrometers), e-nose designs tailored for mobile robots have been proposed (Aleixandre et al., 2008; Gonzalez-Jimenez et al., 2011; Sanchez-Garrido et al., 2014). These robotic-oriented e-noses, some of which are illustrated in Figure 2, seek reducing energy usage and weight, while maintaining a decent efficiency in the detection and characterization of the chemical volatiles. However, the precision and reliability of these systems is, in most cases, far from their laboratory counterparts.

GAS CLASSIFICATION WITH MOBILE ROBOTS

The classification of volatile substances is, possibly, the most studied application of e-noses among the scientific community. It plays an important role for many industrial and medical applications including disease detection and diagnosis in medicine, quality control in food processing chains, finding drugs and explosives, or monitoring of pollution levels in air. Traditionally, this has been performed by analyzing the response of the array of gas sensors that compose the e-nose, when exposed to pulse-like gas excitations under well-controlled measurement conditions (i.e. temperature, humidity, exposure time, etc.) (Schaller, Bosset, & Escher, 1998; Vergara et al., 2012). Yet, from the variety of potential applications that may benefit from gas classification, some of them require to measure the environment continuously and at different locations. They include, among others, city odor mapping, pollution monitoring or leak detection. Mobile platforms can indeed contribute to this problem by trading off temporal coverage

Figure 2. Pictures of two portable e-nose designs developed for mobile robotics. (A) The multi-chamber electronic nose - MCE-nose, (B-C) prototype and detailed inspection of a modular e-nose containing eight smart gas detector modules.
(A) Gonzalez-Jimenez et al., 2011; (B-C) Sanchez-Garrido et al., 2014.

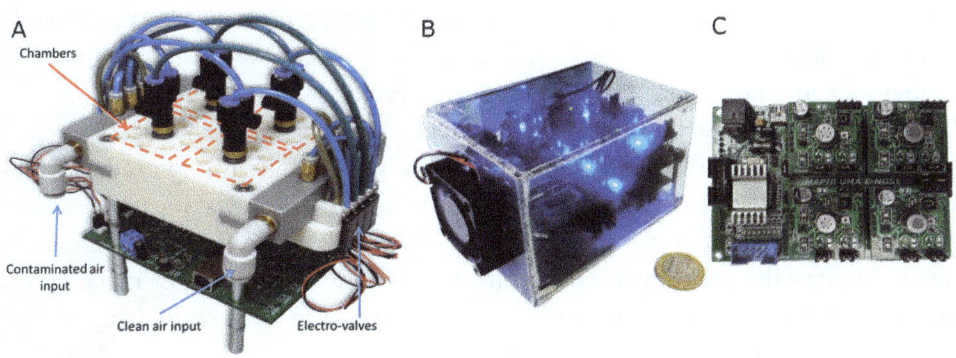

against spatial coverage, enabling a high spatial resolution across large areas without the need for a large number of sensing devices.

As mentioned, the discrimination of gases performed with a gas-sensitive robot presents a number of additional challenges when compared to standard analyte identification, mostly due to the differences in the measurement conditions and the fact that the sensor signals to be processed are noisy and dominated by the signal transient behavior. Most of the research in this direction has focused on three topics: impact analysis of the motion speed of the sensing device, study of which robot trajectory yields higher classification rates, and exploration of new feature sets to feed the pattern recognition algorithm. In this context, one of the first works was presented by Trincavelli and coauthors (Trincavelli, Lout, & Coradeschi, 2009), who used a robot carrying an e-nose and conducted a preliminary investigation on the most suitable path to optimize the classification accuracy, taking into account the possible effects of environmental variables on the signals collected along that path. The authors concluded that the movement strategy of the robot clearly affects the properties of the e-nose signals, therefore being possible to enhance the classification performance by optimizing the robot trajectory.

On this same topic, Monroy and Gonzalez-Jimenez (Monroy & Gonzalez-Jimenez, 2017) empirically analyzed the impact of the robot motion speed on the classification performance. The authors presented a large experimental dataset composed of 240 inspections of an indoor scenario, driving a mobile robot at four different speeds, under the presence of two gas sources. Two questions were addressed: (i) What is the relationship (if any) between the motion speed of an e-nose which is continuously sampling the environment and the classification accuracy?, and (ii) How must the classifier be trained to get the best possible performance?. The authors reaffirmed the conclusions from (Trincavelli et al., 2009), stating that an important deterioration in the classification performance (up to 30% in some of the configurations tested) was appreciated when the e-nose was sampling the environment in movement. Furthermore, they concluded that training the classifier with data collected at a few different motion speeds (from the range of possible velocities) should be enough to palliate this negative effect. Thus, and according to this study, training of the pattern recognition algorithm must be carried out with data collected in motion, in order to not deteriorate too much the classification performance.

An interesting work reaching similar conclusions is that presented by Vergara et al. (Vergara et al., 2013), which does not employ an e-nose in motion to gather the gas samples, but considers different wind velocities (which may be seen as the counterpart of moving the sensing device). The authors compiled an extensive dataset with nine static e-noses placed within a wind tunnel to evaluate the performance of several sensor arrays working in open sampling settings. Different locations of the e-noses, heater voltages, wind speeds, and chemical volatiles compose the list of variables that make this dataset one of the most complete currently available. Particularly, two important conclusions where highlighted by the authors with regard to the wind speed: the classification performance was affected by the wind speed used during training, and that, in order to increase the robustness of the system against air flow variations, one may want to train the system at all the expected system conditions.

Naturally, not only the motion of the gas sensing device must be taken into account. As reported in (Trincavelli & Loutfi, 2010), the feature selection process is also of great importance. In this work the authors analyzed multiple features to determine those which are more robust to changes in the sampling trajectory and less dependent of the experimental setup. Two feature selection algorithms were proposed and compared to improve the performance of an olfactory robotic system. Experiments were performed in three different scenarios (two indoor and one outdoor) with four moving strategies, attempting to vary the interaction of the robot with a possible plume.

The classification of multiple volatile substances with an e-nose in motion is still an open and challenging research topic. Recent works keep proposing new feature selection algorithms and feature sets to enhance the performance, as in (Hernandez-Bennetts, Schaffernicht, & Pomareda-Ses, 2014) where the authors proposed to employ the sensor amplitude as an additional feature, or the application of statistical tools to exploit the temporal correlation of the e-nose data (Monroy & Gonzalez-Jimenez, 2015; Monroy, Palomo, Lopez-Rubio, & Gonzalez-Jimenez, 2016). Furthermore, like in the case of a network of fixed e-nose, the calibration and drift compensation of the sensors composing the e-nose is also an important drawback to consider. In this regard, works like (De Vito, Massera, Piga, Martinotto, & Di Francia, 2008; Esposito et al., 2016) have presented different approaches to cross-calibrate different e-noses without the need to perform a tedious and costly laboratory calibration. These proposals are indeed fundamental when the MRO system is to be deployed for long times (drift and ageing), or when multiple mobile e-noses are set up simultaneously (Hasenfratz, 2015).

Related to environmental monitoring applications, there are works where a gas measuring device sensing the air quality is carried by a person (Zappi, Bales, Park, Griswold, & Šimuni, 2012), a bike (Elen et al., 2013; Monroy, Gonzalez-Jimenez, & Sanchez-Garrido, 2014), public transport vehicles (Hasenfratz et al., 2015) or even drones (Neumann, Bartholmai, Schiller, Wiggerich, & Manolov, 2010; Pobkrut, Eamsa-ard, & Kerdcharoen, 2016). Despite sampling the environment in motion, most of the works does not perform a classification phase to discriminate the type of gas, but rather employs an array of gas sensors with disjoint selectivity (i.e. one sensor for each analyte to monitor, and usually discarding the cross-selectivity among classes). An interesting remaining question is whether the concentration measurements of the different pollutants can also be improved by taking into account the motion speed of the sensing device.

GAS DISTRIBUTION MAPPING

Gas distribution mapping (GDM) is the process of creating a representation of how gases spread in an environment from a set of spatially and temporally distributed measurements of relevant variables. Foremost, these measurements include the gas concentration itself, but may also comprise wind flow, pressure or temperature, among other representative variables.

Many gas distribution models were developed back in the early 90s to tackle atmospheric dispersion (Holmes & Morawska, 2006). These models estimate the distribution of airborne materials at the different atmospheric-scales given a set of environmental and topographic conditions. Yet, such models are not suitable for local scales, not being designed to capture all the relevant aspects of gas propagation with a sufficient level of detail, as in the case of complex indoor and outdoor settings. Given that analytical solutions are intractable, it is then common practice to divide the space into a regular lattice of cells (gridmap), and then estimate a probability density function (pdf) of the gas concentration at each cell of the grid (see Figure 3). A crucial aspect when updating this gridmap is the fact that the majority of current e-noses are point sampling devices, that is, observations are only representative of the very near air around the e-nose. Moreover, given the ephemeral nature of gases (due to the mechanisms that rule gas dispersion (Shraiman & Siggia, 2000), the information conveyed by a given measurement quickly vanishes as time goes by. The latter is the reason why most GDM approaches aims at modeling the time-averaged gas distribution (Lilienthal, Reggente, Trincavelli, Blanco, & Gonzalez-Jimenez, 2009; Loutfi, Coradeschi, Lilienthal, & Gonzalez-Jimenez, 2008; Stachniss, Plagemann, & Lilienthal, 2009),

Figure 3. The 2D map is commonly represented by a lattice where each cell keeps the estimate of gas concentration by means of a probability density function, represented here as a Gaussian density in the vertical axis.
From (Blanco, Monroy, Gonzalez-Jimenez, & Lilienthal, 2013).

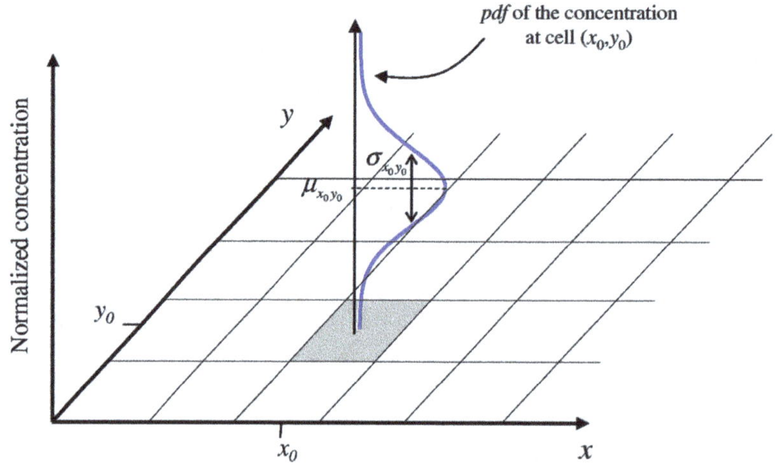

being the works presented in (Asadi, Pashami, Loutfi, Lilienthal, & Gouma, 2011; Marjovi & Marques, 2014; Monroy, Blanco, & Gonzalez-Jimenez, 2016) the only notable exceptions.

It is the task of a GDM algorithm to extrapolate sparse measurements, both spatially and temporally, to obtain an estimation of the gas dispersal for the entire environment. Several efforts addressing this problem have been proposed in the robotics literature. Farell et al., (Farrell, Pang, & Li, 2003) presented a hidden Markov model based approach to estimate the location of odor source based on mapping the plume. Marques et al. (Marques, Martins, & de Almeida, 2005) proposed a mapping methodology based on neural network regression and an advection-diffusion model by means of a reduced order Kalman filter. Lilienthal and Duckett (Lilienthal & Duckett, 2004) presented a grid-mapping technique based on Gaussian density functions, modeling the likelihood of each observation in the grid as a decreasing Normal distribution with respect to the distance from the point of measurement. This method was later extended for the case of multiple odor sources (Loutfi et al., 2008) and to the three-dimensional case (Reggente, Lilienthal, Pardo, & Sberveglieri, 2009). It was further shown how gas distribution mapping methods can be embedded into a Blackwellized particle filter to account for the uncertainty about the position of the robot (Lilienthal, Loutfi, Blanco, Galindo, & Gonzalez-Jimenez, 2007).

Stressing the importance that probability and uncertainty have on any mobile robotic system, recent approaches provide in addition to the most-likely value of the gas distribution, an estimate of the associated uncertainty (via a variance value as illustrated in Figure 4). Stachniss et al. (Stachniss et al., 2009) proposed an approach using Gaussian process mixture models (GPM), treating gas distribution modeling as a regression problem. The components of the mixture model and the gating function, which decides to what component a data point belongs, were learned using Expectation Maximization (EM). At the same time, in (Lilienthal, Reggente, Trincavelli, Blanco, & Gonzalez-Jimenez, 2009), Lilienthal et al. carried out two parallel estimation processes, one for the mean and another for the variance, understanding the latter as the variability of the gas readings, not the uncertainty in the estimation which is the standard in probabilistic estimators. Results demonstrated that although providing similar maps

Figure 4. Example of a gas distribution map generated by a mobile robot. (A) Occupancy map of the environment (white represents obstacles and the source location is depicted as a cross inscribed within a circle) and robot path during the inspection (white-solid line). (B) Map of the estimated gas concentration and (C) associated uncertainty.
Adapted from (Monroy, Blanco, et al., 2016).

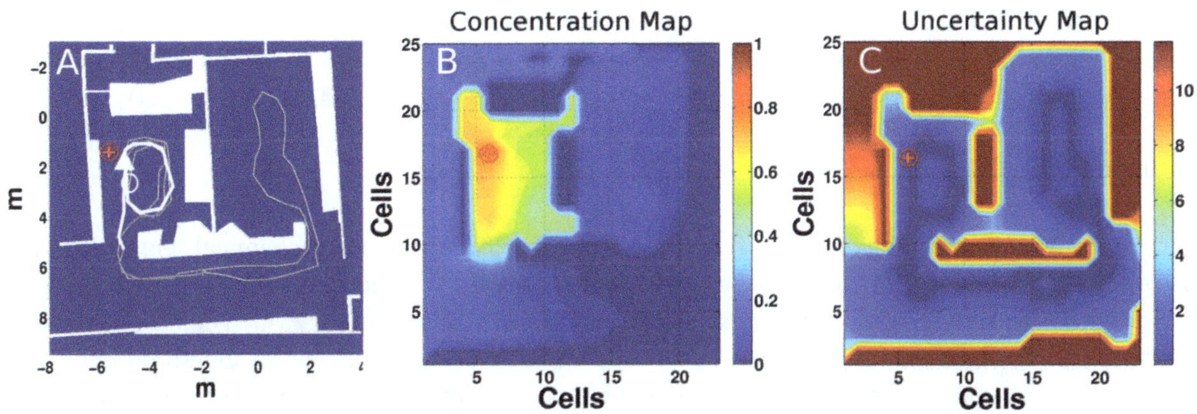

to previous approaches, this method had the advantage of scaling better to larger training datasets and to possess a simpler learning procedure. In (Blanco et al., 2013), an approach to obtain the variance of each map cell is proposed employing a sparsified Kalman filter.

Exploiting the availability of other sensors onboard the robot, some works have considered the presence of obstacles in the environment (Monroy, Blanco, et al., 2016), introduced wind flow measurements as secondary variables (Reggente & Lilienthal, 2010), and even considered the modeling of homologous wind maps to exploit the strong correlation between gas dispersion and wind flow conditions (Monroy, Jaimez, & Gonzalez-Jimenez, 2017). With respect the number of robots, it is common practice to face GDM with a swarm of robots (Marjovi & Marques, 2014; Nguyen, Kodagoda, Ranasinghe, & Dissanayake, 2014). This approach enables measuring multiple locations simultaneously, but introduces the problems of cross-calibration between the different e-noses, and the communication between the robots in the swarm. These problems are also shared by applications that mount multiple e-noses on other mobile platforms, such as buses (Hasenfratz, 2015), or bikes (Zappi et al., 2012) for creating pollution maps.

Up to date multiple approaches based on different mathematical and statistical models have been presented. Yet, a key hurdle to provide a formal and serious comparison among them is the lack of experimental datasets with ground truth. This is a one of the main limitations for the development of precise and trustable GDM algorithms in real robots. Waiting for a breakthrough in the chemical sensing technology that could provide such ground truth, the only intermediate solution is to employ computational fluid dynamic tools which, not modeling whole the dispersion phenomena, allow us to gather reasonable accurate simulated data with ground truth.

GAS SOURCE LOCALIZATION

Gas source localization is likely the most studied research area and direct application of an olfactory mobile robot. Traditionally, it has been divided into three stages: first, explore the environment looking for the presence of a chemical release, then search for the source guided by chemical and other sensing stimuli, and finally the so called source declaration, that is, verification of the identified source location. The first stage can be considered a trigger event, not being usually carried as an active stage, but run on the background while the robot is performing other non-related olfaction operations such as patrolling, exploration, delivering, etc. If while performing these tasks, the e-nose on-board the robot detects an abnormal concentration level, then the gas source localization task is triggered and the search process begin. Therefore, the rest of this section will focus on the other two stages, the search and declaration of a gas source. More detailed information can be found in the review (Kowadlo & Russell, 2008).

Source Search

This stage involves searching of the gas release point, primarily relying on the chemical sensed data, but also on the wind flow or the objects and structure of the scenario. The success of this stage heavily relies on how well the given algorithm adjust itself to the environmental conditions, which determine the way in which odor is dispersed. In this sense, gas source localization strategies can be classified into those designed to work under the presence of a chemical plume, also known as plume tracking strategies, and those which do not relay on the existence of a well formed, downwind plume.

Plume Tracking

At medium Reynolds values, odor dispersal occurs mainly through carriage by the background fluid currents (advection), causing an odor plume to form downflow of the source (Kowadlo & Russell, 2008). In most real scenarios, this plume is not straight and continuous, but given the time-varying nature of flow fields and the predominance of turbulence over diffusion, plumes tend to meander, become patchy and, to a far lesser extent, spread out. This results in peak concentration values much higher than the average, and fluctuations in the instantaneous gradients in magnitude and direction. Furthermore, the plume structure can even change if the direction of air or water flow shifts considerably. Therefore, occasional failures are almost inevitable in the tracking of plumes (Pearce et al., 2006), being a key for success not only the track of the plume but also the plume recovery mechanisms to relocate the lost plume in case of failure.

As in the case of gas distribution mapping, research on plume tracking robots started with pure chemotactic approaches (Kazadi, Goodman, Tsikata, Green, & Lin, 2000; Menzel & Goschnick, 2000). They involve taking measurements of the chemical concentration at more than one spatially separated position, and determining the chemical gradient, which is used to move towards the source (see Figure 5). In most occasions, algorithms for searching for an odor source are based on local sensing and the reactive behaviors of microbes, insects, or crustaceans. Such strategies involve tracking the plume along its entire length, which can be a limitation as it is time consuming and may not always be possible. Proposed strategies include Braitenberg approaches (Lilienthal & Duckett, 2003a; Russell, 2003), where a pair of bilateral chemical sensors, each directly controlling the speed of a wheel, either the opposite wheel in cross-coupling, or the wheel on the same side, are used to reach the source; Algorithms based on E. coli bacteria (Marques, Nunes, & de Almeida, 2002), rooted in gradient navigation through the plume; and other methods, like in (Ishida, Zhu, Johansson, & Moriizumi, 2004), exploited the airflow obstructions generated by the robot shape, and the consequent disparity between the responses of upwind and downwind sensors, to determine the moving direction. Experimental results have demonstrated that

Figure 5. Illustration of chemotactic plume tracking approaches. (A) Navigation based on concentration gradients detected by comparison of the left and right chemical sensor outputs. (B) Three different plume recovery algorithms in case of lost. Robot 1 is programmed to back up when neither sensor detects chemical. Robot 2 performs random walk. Robot 3 mimics the behavior of a male silkworm moth. When one of the sensors is stimulated, the robot surges in that direction to track a plume. When the chemical signal is lost, the robot performs zigzag walk and circling to relocate the lost plume.
Adapted from (Pearce et al., 2006).

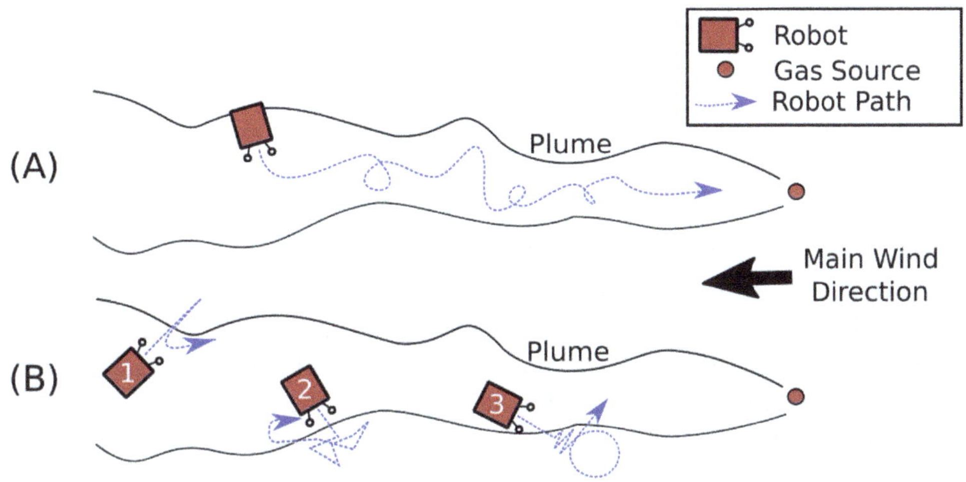

pure chemotactic approaches are relatively ineffective under real conditions where turbulence dominates, attributing their low success rate to the susceptibility of the algorithms to rapid fluctuations in the chemical concentration and the fact that the concentration gradient along the plume centerline is extremely small except in the vicinity of the source.

Performance was then improved by combining chemotaxis with anemotaxis, exploiting the strong directional cue that the flow direction brings when acting under turbulent flows. Several methods have been proposed to exploit information from chemical concentration and fluid flow measurements (anemometric data), among them is the dung beetle or zigzag method, which involves moving upwind within the odor plume in a zigzagging fashion (see Figure 6A). Each time the plume boundary is encountered, the robot turns back into the plume. Reported implementations include that in (Ishida et al., 1999), which employed a basic "odor probe" composed of four pairs of a semiconductor gas sensor and a hot wire anemometer, each pair spaced 90° with respect the other, (Farrell, Pang, & Li, 2005) for underwater robots, or (Russell, 2006) for the case of 3D. Other approaches exploiting anemotactic information are the plume-center upwind search (Ishida, Nakayama, Nakamoto, & Moriizumi, 2005; Marques, Almeida, & de Almeida, 2003) or the silkworm moth strategy, also known as the surge-cast algorithm (Lochmatter & Martinoli, 2009b). The former involves moving towards the center of the plume whilst tracking upwind, assuming that higher concentrations are likely to be found at the plume centerline. The latter is based on the studied behavior of how male silkworm moths locate female mates by tracking a pheromone release. In a nutshell, a robot in the plume moves straight upwind until it loses the plume (surge phase), then, it tries to reacquire the plume by moving cross-wind (cast phase), first on one side and then on the other (see Figure 6B). Different versions of this behavior have been presented both with procedural algorithms, and with neural networks (Pyk et al., 2006).

Figure 6. Illustration of (A) the zigzag algorithm and (B) the surge-cast algorithm. The robot continuously measures the gas concentration, and at some points (indicated by stars) also the wind direction.
(A) Ishida et al., 1994; (B) Lochmatter & Martinoli, 2009b, Adapted from (Neumann, Hernandez-Bennetts, Lilienthal, Bartholmai, & Schiller, 2013).

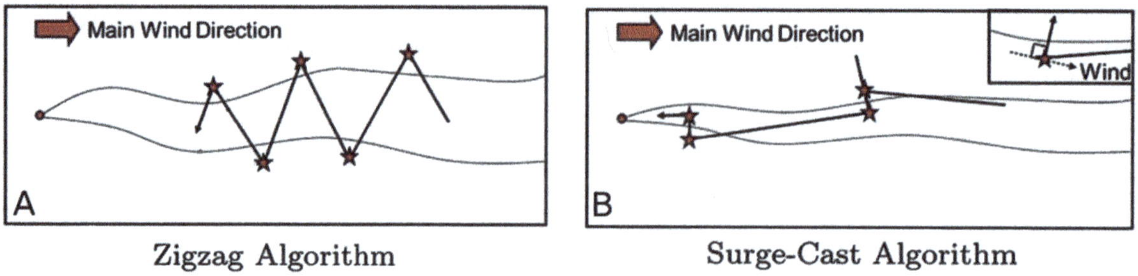

As reported in (Lochmatter & Martinoli, 2009a), the main drawback of these anemotactic strategies is the fact that under realistic, turbulent environments, a great difficulty exists in accurately determining the wind flow direction. The latter entails, in most occasions, a major impact on the success rate when compared with laboratory results. In addition, obstacles will not only negatively influence the performance, but require, in many cases, to substantially modify the algorithm (Lochmatter, Heiniger, Martinoli, Pardo, & Sberveglieri, 2009).

Turbulence Dominated Search

In most real environments, the strong predominance of turbulence as the dispersal mechanism, together with the presence of obstacles, lead to chaotic dispersion patterns where the chemical downwind plume breaks down into intermittent gas patches. Moreover, for indoor scenarios, the low strength of wind flows does not guarantee the formation of chemical plumes, being the movement of other entities (e.g. people, or the own robot) responsible of the chaotic dispersion of the gases. This entails that most plume tracking algorithms are unable to localize the source, as there is not such a thing as a chemical plume to follow.

Researchers have attempted to overcome this limitation by developing systems that go beyond purely reactive control, for example by exploiting other information sources available on the mobile robot. One of the first alternatives to appear was the combination of traditional strategies with vision based systems. This approach enables robots to identify candidates from a distance, thus dramatically diminishing the effective search space and greatly enhancing the ability to locate an odor source when a downwind plume is not well formed or even inexistent. Yet, only very basic algorithms have been proposed so far (Ishida, Tanaka, Taniguchi, & Moriizumi, 2004; Loutfi, Coradeschi, Karlsson, & Broxvall, 2004), most of them relaying on strong assumptions about knowledge of the source shape or color, for detection of the visual candidates.

Moreover, some engineered plume-tracing strategies have also been presented, such as fluxotaxis (Zarzhitsky, Spears, Thayer, & Spears, 2004), a multirobot based approach in which computational fluid dynamics techniques are applied, infotaxis (Vergassola, Villermaux, & Shraiman, 2007) a gradient-free method exploiting the expected entropy of future samples to guide the robot search, or the SPIRAL algorithm (Ferri et al., 2009), a source location strategy for indoor environments with no strong airflows (see Figure 7). More recently, and facing more challenging environmental conditions such as time-varying

Figure 7. Illustration of the SPIRAL algorithm for gas source localization. The robot performs spiral movements, stopping to sample the gas concentration. If current measurements suggest that the robot is closer to the source (referred as a hit), it starts a new spiral movement.
Adapted from (Ferri et al., 2009).

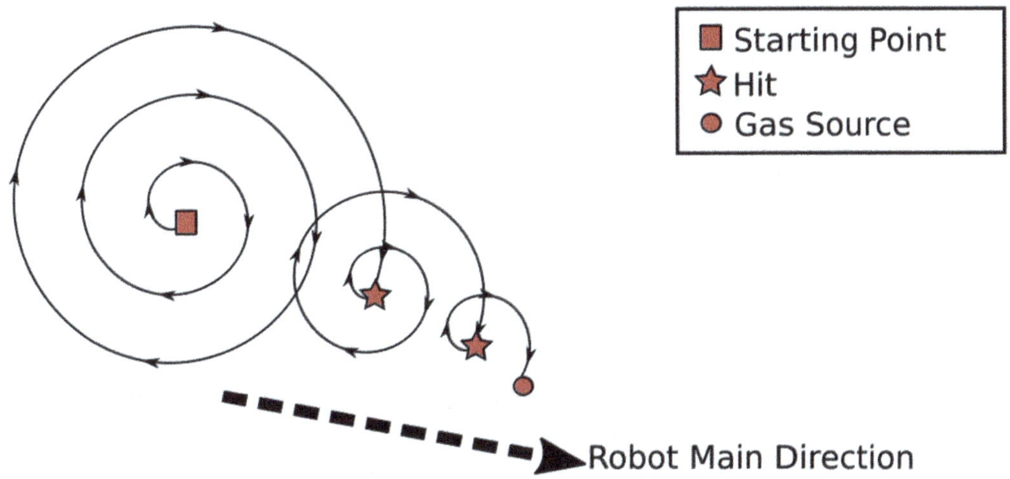

airflows, Li et al proposed a novel probabilistic approach based on a particle filter (J. G. Li, Meng, Wang, & Zeng, 2011). Their main contribution was the estimation of the chemical source location while the robot performs exploratory behaviors, not necessarily requiring it to navigate towards it. In this line, Ping et al proposed a behavior-action based method (Ping, Xiao-fang, & Ai-dong, 2014), combining multiple search strategies with visual recognition, and in (Hernandez-Bennetts, Schaffernicht, Stoyanov, Lilienthal, & Trincavelli, 2014) and (Bonow & Kroll, 2013), the authors proposed the use of a TDLAS range gas sensor to localize a gas source by applying tomography principles.

Finally, gas distribution mapping techniques have also been proposed to pinpoint the location of one or multiple chemical sources (Lilienthal et al., 2009; Monroy, Blanco, et al., 2016). These generic methods do not rely on the presence of a plume, neither on strong assumptions about the environmental conditions, however, their limitation resides in the time necessary to sweep the entire environment, and their bad scalability as the environment enlarges.

Source Declaration

The declaration of the chemical source corresponds to the last phase of the gas source localization task. Its purpose is to inspect the candidate locations provided by the search algorithm (second phase described previously), and to verify the presence of the gas source. In many cases, this task is relegated to a human operator, which analyzing the data offered by the robot (e.g. the robot path followed during the search stage, the gas distribution map created, or just the current gas and wind measurements) is in charge of declaring the gas source location, or a failure in the search stage. Yet, different approaches have been proposed to declare the gas source location autonomously, in most occasions derived on the basis of empirical studies.

Initial works based source declaration on the intuitive fact that gas concentration in the vicinity of the gas source should be higher. Approaches like (Grasso & Atema, 2002) proposed heuristic strategies based on gas sensor saturation over empirically set thresholds, while Cowen & Ward (Cowen & Ward, 2002) sharpen the declaration by comparing the chemical concentration at different heights. Hayes et al. (Hayes, Martinoli, & Goodman, 2002) mentioned a source declaration method which identified the source by detecting a series of contiguous odor hits, provided that such frequency of odor patches is likely to occur only near the source.

More sophisticated methods based on machine learning were proposed to improve the declaration of the gas source under more realistic environments. In (Weissburg et al., 2002) the authors based the declaration of the gas source attending to the studied chemical patterns a gas source generates, while in (Lilienthal et al., 2004) the declaration was approached as a classification problem, using neural networks and support vector machines. As reported in (Cabrita & Marques, 2013), the problem with these and similar strategies is the need of a training phase, and the assumption that the source characteristics during training would be similar to the odor sources found during normal operation. More general approaches, resorting to concentration gridmaps to achieve odor source declaration, were presented in (Lilienthal & Duckett, 2003b; Marques, Nunes, & De Almeida, 2003). Later, Li (W. Li, 2006) conducted experiments on underwater vehicles, constructing a source identification zone based on chemical detection points, and more recently (J.-G. Li, Meng, Wang, & Zeng, 2010), proposing a particle filter approach to locate and declare the gas source. Similarly, in (Neumann et al., 2013) the authors presented a novel pseudo-gradient-based plume tracking algorithm and a particle filter-based source declaration approach, testing the proposal with a gas-sensitive micro-drone.

Recent approaches like (Cabrita & Marques, 2013; Wang et al., 2016), have tackled the problem from a multi-robot approach and mass flux divergence theory, relaying in the simultaneous measurement of gas concentration and fluid flow at different locations of the environment to analyze the presence of sources.

CONCLUDING REMARKS

Gas sensitive mobile robots are not yet a reality. Most of the works mentioned through this chapter present only simulations or laboratory-based experimentation, where some control over the environmental conditions and the chemical sensors is performed. Development of chemical sensors designed for mobile robots is indeed a necessary subject for future work. More robust and efficient sensors are needed to overcome limitations such as the long response times, ageing or drift, drawbacks that hinder gas sensitive mobile robots to be deployed in real world applications.

Besides the limitations imposed by technology, that will hopefully banish with time, future work is needed to develop design strategies which enable a mobile robot with the capabilities to detect, recognize and locate chemical releases, to be useful in a variety of scenarios such as homes, manufactures or emergency areas.

REFERENCES

Aleixandre, M., Lozano, J., Gutiérrez, J., Sayago, I., Fernández, M. J., & Horrillo, M. C. (2008). Portable e-nose to classify different kinds of wine. *Sensors and Actuators. B, Chemical*, *131*(1), 71–76. doi:10.1016/j.snb.2007.12.027

Asadi, S., Pashami, S., Loutfi, A., Lilienthal, A., & Gouma, P. (2011). TD Kernel DM+ V: Time-Dependent Statistical Gas Distribution Modelling on Simulated Measurements. In AIP Conference Proceedings (Vol. 1362, pp. 281–282). AIP. doi: 10.1063/1.3651651

Balkovsky, E., & Shraiman, B. I. (2002). Olfactory search at high Reynolds number. *Proceedings of the National Academy of Sciences of the United States of America*, *99*(20), 12589–12593. doi:10.1073/pnas.192393499 PMID:12228727

Beer, R. D., & Ritzmann, R. E. (1993). *Biological neural networks in invertebrate neuroethology and robotics*. Academic Press.

Blanco, J.-L., Monroy, J., Gonzalez-Jimenez, J., & Lilienthal, A. (2013). A Kalman Filter Based Approach to Probabilistic Gas Distribution Mapping. *28th Symposium On Applied Computing (SAC)*. doi:10.1145/2480362.2480409

Bonow, G., & Kroll, A. (2013). Gas leak localization in industrial environments using a TDLAS-based remote gas sensor and autonomous mobile robot with the Tri-Max method. In *Robotics and Automation (ICRA), 2013 IEEE International Conference on* (pp. 987–992). IEEE. doi:10.1109/ICRA.2013.6630693

Cabrita, G., & Marques, L. (2013). Divergence-based odor source declaration. In *Control Conference (ASCC), 2013 9th Asian* (pp. 1–6). ASCC. doi: 10.1109/ASCC.2013.6606390

Cabrita, G., Sousa, P., & Marques, L. (2010). Player/stage simulation of olfactory experiments. In *Intelligent Robots and Systems (IROS), 2010 IEEE/RSJ International Conference on* (pp. 1120–1125). IEEE. doi: 10.1109/IROS.2010.5652518

Cowen, E. A., & Ward, K. B. (2002). Chemical plume tracing. *Environmental Fluid Mechanics*, *2*(1), 1–7. doi:10.1023/A:1016247516041

De Vito, S., Massera, E., Piga, M., Martinotto, L., & Di Francia, G. (2008). On field calibration of an electronic nose for benzene estimation in an urban pollution monitoring scenario. *Sensors and Actuators. B, Chemical*, *129*(2), 750–757. doi:10.1016/j.snb.2007.09.060

Di Lello, E., Trincavelli, M., Bruyninckx, H., & De Laet, T. (2014). Augmented Switching Linear Dynamical System Model for Gas Concentration Estimation with MOX Sensors in an Open Sampling System. *Sensors (Basel)*, *14*(7), 12533–12559. doi:10.3390/s140712533 PMID:25019637

Doty, R. L. (2015). *Handbook of olfaction and gustation*. John Wiley & Sons. doi:10.1002/9781118971758

Elen, B., Peters, J., Van Poppel, M., Bleux, N., Theunis, J., Reggente, M., & Standaert, A. (2013). The Aeroflex: A bicycle for mobile air quality measurements. *Sensors (Basel)*, *13*(1), 221–240. doi:10.3390/s130100221 PMID:23262484

Esposito, E., De Vito, S., Salvato, M., Bright, V., Jones, R. L., & Popoola, O. (2016). Dynamic neural network architectures for on field stochastic calibration of indicative low cost air quality sensing systems. *Sensors and Actuators. B, Chemical, 231*, 701–713. doi:10.1016/j.snb.2016.03.038

Farrell, J. A., Pang, S., & Li, W. (2003). Plume Mapping via Hidden Markov Methods. *IEEE Transactions on Systems, Man, and Cybernetics. Part B, Cybernetics, 33*(6), 850–863. doi:10.1109/TSMCB.2003.810873 PMID:18238238

Farrell, J. A., Pang, S., & Li, W. (2005). Chemical plume tracing via an autonomous underwater vehicle. *IEEE Journal of Oceanic Engineering, 30*(2), 428–442. doi:10.1109/JOE.2004.838066

Ferri, G., Caselli, E., Mattoli, V., Mondini, A., Mazzolai, B., & Dario, P. (2009). SPIRAL: A novel biologically-inspired algorithm for gas/odor source localization in an indoor environment with no strong airflow. *Robotics and Autonomous Systems, 57*(4), 393–402. doi:10.1016/j.robot.2008.07.004

Fonollosa, J., Sheik, S., Huerta, R., & Marco, S. (2015). Reservoir computing compensates slow response of chemosensor arrays exposed to fast varying gas concentrations in continuous monitoring. *Sensors and Actuators. B, Chemical, 215*, 618–629. doi:10.1016/j.snb.2015.03.028

Gonzalez-Jimenez, J., Monroy, J., & Blanco, J.-L. (2011). The Multi-Chamber Electronic Nose—An Improved Olfaction Sensor for Mobile Robotics. *Sensors (Basel), 11*(12), 6145–6164. doi:10.3390/s110606145 PMID:22163947

Gonzalez-Jimenez, J., Monroy, J., Garcia, F., & Blanco, J.-L. (2011). The Multi-Chamber Electronic Nose (MCE-nose). *IEEE International Conference on Mechatronics*. doi:10.1109/ICMECH.2011.5971193

Grasso, F. W., & Atema, J. (2002). Integration of flow and chemical sensing for guidance of autonomous marine robots in turbulent flows. *Environmental Fluid Mechanics, 2*(1), 95–114. doi:10.1023/A:1016275516949

Hasenfratz, D. (2015). *Enabling Large-Scale Urban Air Quality Monitoring with Mobile Sensor Nodes*. Academic Press. https://doi.org/10.3929/ethz-a-010361120

Hasenfratz, D., Saukh, O., Walser, C., Hueglin, C., Fierz, M., Arn, T., … Thiele, L. (2015). Deriving high-resolution urban air pollution maps using mobile sensor nodes. *Pervasive and Mobile Computing, 16*(PB), 268–285. 10.1016/j.pmcj.2014.11.008

Hayes, A. T., Martinoli, A., & Goodman, R. M. (2002). Distributed odor source localization. *IEEE Sensors Journal, 2*(3), 260–271. doi:10.1109/JSEN.2002.800682

Hernandez-Bennetts, V., Schaffernicht, E., & Pomareda-Ses, V. (2014). A Novel Approach for Gas Discrimination in Natural Environments with Open Sampling Systems. In IEEE Sensors (pp. 3–6). IEEE. doi: 10.1109/ICSENS.2014.6985437

Hernandez-Bennetts, V., Schaffernicht, E., Stoyanov, T., Lilienthal, A., & Trincavelli, M. (2014). Robot assisted gas tomography—Localizing methane leaks in outdoor environments. In *Robotics and Automation (ICRA), 2014 IEEE International Conference on* (pp. 6362–6367). IEEE. doi: 10.1109/ICRA.2014.6907798

Holmes, N. S., & Morawska, L. (2006). A review of dispersion modelling and its application to the dispersion of particles: An overview of different dispersion models available. *Atmospheric Environment*, *40*(30), 5902–5928. doi:10.1016/j.atmosenv.2006.06.003

Ishida, H., Kobayashi, A., Nakamoto, T., & Moriizumi, T. (1999). Three-dimensional odor compass. *IEEE Transactions on Robotics and Automation*, *15*(2), 251–257. doi:10.1109/70.760346

Ishida, H., Nakayama, G., Nakamoto, T., & Moriizumi, T. (2005). Controlling a gas/odor plume-tracking robot based on transient responses of gas sensors. *IEEE Sensors Journal*, *5*(3), 537–545. doi:10.1109/JSEN.2004.839597

Ishida, H., Suetsugu, K., Nakamoto, T., & Moriizumi, T. (1994). Study of autonomous mobile sensing system for localization of odor source using gas sensors and anemometric sensors. *Sensors and Actuators. A, Physical*, *45*(2), 153–157. doi:10.1016/0924-4247(94)00829-9

Ishida, H., Tanaka, H., Taniguchi, H., & Moriizumi, T. (2004). Mobile robot navigation using vision and olfaction to search for a gas/odor source. In *Intelligent Robots and Systems, 2004.(IROS 2004). Proceedings. 2004 IEEE/RSJ International Conference on* (Vol. 1, pp. 313–318). IEEE. doi: 10.1109/IROS.2004.1389370

Ishida, H., Zhu, M., Johansson, K., & Moriizumi, T. (2004). Three-dimensional gas/odor plume tracking with blimp. In *Proc. ICEE* (pp. 117–120). ICEE.

Kazadi, S., Goodman, R., Tsikata, D., Green, D., & Lin, H. (2000). An autonomous water vapor plume tracking robot using passive resistive polymer sensors. *Autonomous Robots*, *9*(2), 175–188. doi:10.1023/A:1008970418316

Kowadlo, G., & Russell, R. A. (2008). Robot Odor Localization: A Taxonomy and Survey. *The International Journal of Robotics Research*, *27*(8), 869–894. doi:10.1177/0278364908095118

Li, J.-G., Meng, Q.-H., Wang, Y., & Zeng, M. (2010). Single odor source declaration in outdoor time-variant airflow environments. In *Robotics and Biomimetics (ROBIO), 2010 IEEE International Conference on* (pp. 143–148). IEEE. doi:10.1109/ROBIO.2010.5723317

Li, J. G., Meng, Q.-H., Wang, Y., & Zeng, M. (2011). Odor source localization using a mobile robot in outdoor airflow environments with a particle filter algorithm. *Autonomous Robots*, *30*(3), 281–292. doi:10.1007/s10514-011-9219-2

Li, W. (2006). Abstraction of odor source declaration algorithm from moth-inspired plume tracing strategies. In *Robotics and Biomimetics, 2006. ROBIO'06. IEEE International Conference on* (pp. 1024–1029). IEEE. doi:10.1109/ROBIO.2006.340369

Lilienthal, A., & Duckett, T. (2003a). Experimental analysis of smelling Braitenberg vehicles. *Environment*, *5*(10).

Lilienthal, A., & Duckett, T. (2003b). Gas source localisation by constructing concentration gridmaps with a mobile robot. *Proceedings of the European Conference on Mobile Robots (ECMR 2003).*

Lilienthal, A., & Duckett, T. (2004). Building gas concentration gridmaps with a mobile robot. *Robotics and Autonomous Systems*, *48*(1), 3–16. doi:10.1016/j.robot.2004.05.002

Lilienthal, A., Loutfi, A., Blanco, J., Galindo, C., & Gonzalez-Jimenez, J. (2007). A Rao-Blackwellisation Approach to GDM-SLAM – Integrating SLAM and Gas Distribution Mapping (GDM). *European Conference on Mobile Robots*, 126–131.

Lilienthal, A., Reggente, M., Trincavelli, M., Blanco, J.-L., & Gonzalez-Jimenez, J. (2009). A statistical approach to gas distribution modelling with mobile robots - The Kernel DM+V algorithm. In *Intelligent Robots and Systems (IROS). IEEE/RSJ International Conference on* (pp. 570–576). IEEE. doi:10.1109/IROS.2009.5354304

Lilienthal, A., Reggente, M., Trincavelli, M., Blanco, J.-L., & Gonzalez-Jimenez, J. (2009). A statistical approach to gas distribution modelling with mobile robots - The Kernel DM+V algorithm. *2009 IEEE/RSJ International Conference on Intelligent Robots and Systems*. doi:10.1109/IROS.2009.5354304

Lilienthal, A., Ulmer, H., Frohlich, H., Stutzle, A., Werner, F., & Zell, A. (2004). Gas source declaration with a mobile robot. In *Robotics and Automation, 2004. Proceedings. ICRA'04. 2004 IEEE International Conference on* (Vol. 2, pp. 1430–1435). IEEE. doi:10.1109/ROBOT.2004.1308025

Lochmatter, T., Heiniger, N., Martinoli, A., Pardo, M., & Sberveglieri, G. (2009). Localizing an odor source and avoiding obstacles: Experiments in a wind tunnel using real robots. In AIP Conference Proceedings (Vol. 1137, pp. 69–72). AIP. https://doi.org/10.1063/1.3156629

Lochmatter, T., & Martinoli, A. (2009a). Theoretical Analysis of Three Bio-Inspired Plume Tracking Algorithms. *Proceedings of the IEEE International Conference on Robotics and Automation (2009)*, 2661–2668. doi:10.1109/ROBOT.2009.5152686

Lochmatter, T., & Martinoli, A. (2009b). Tracking odor plumes in a laminar wind field with bio-inspired algorithms. *Experimental Robotics*, 473–482. 10.1007/978-3-642-00196-3_54

Loutfi, A., Coradeschi, S., Karlsson, L., & Broxvall, M. (2004). Putting olfaction into action: Using an electronic nose on a multi-sensing mobile robot. In *Intelligent Robots and Systems, 2004 (IROS 2004). Proceedings. 2004 IEEE/RSJ International Conference on* (Vol. 1, pp. 337–342). IEEE. doi: 10.1109/IROS.2004.1389374

Loutfi, A., Coradeschi, S., Lilienthal, A., & Gonzalez-Jimenez, J. (2008). Gas distribution mapping of multiple odour sources using a mobile robot. *Robotica*, *27*(2), 311. doi:10.1017/S0263574708004694

Marjovi, A., & Marques, L. (2014). Multi-Robot Odor Distribution Mapping in Realistic Time-Variant Conditions. In *IEEE International Conference on Robotics and Automation (ICRA 2014)* (pp. 3720–3727). IEEE. doi:10.1109/ICRA.2014.6907398

Marques, L., Almeida, N., & de Almeida, A. T. (2003). *Olfactory sensory system for odour-plume tracking and localization* (Vol. 1). IEEE. doi:10.1109/ICSENS.2003.1278971

Marques, L., Martins, A., & de Almeida, A. T. (2005). Environmental monitoring with mobile robots. In *Intelligent Robots and Systems, 2005 (IROS 2005). 2005 IEEE/RSJ International Conference on* (pp. 3624–3629). IEEE. doi:10.1109/IROS.2005.1545133

Marques, L., Nunes, U., & de Almeida, A. T. (2002). Olfaction-based mobile robot navigation. *Thin Solid Films*, *418*(1), 51–58. doi:10.1016/S0040-6090(02)00593-X

Marques, L., Nunes, U., & De Almeida, A. T. (2003). Odour searching with autonomous mobile robots: An evolutionary-based approach. In *Proceedings of the IEEE Int. Conf. on Advanced Robotics* (pp. 494–500). IEEE.

Menzel, R., & Goschnick, J. (2000). Gradient gas sensor microarrays for on-line process control — a new dynamic classification model for fast and reliable air quality assessment. *Sensors and Actuators. B, Chemical*, *68*(1–3), 115–122. doi:10.1016/S0925-4005(00)00470-6

Monroy, J., Blanco, J.-L., & Gonzalez-Jimenez, J. (2013). An Open Source Framework for Simulating Mobile Robotics Olfaction (Abstract only). *15th International Symposium on Olfaction and Electronic Nose (ISOEN).*

Monroy, J., Blanco, J.-L., & Gonzalez-Jimenez, J. (2016). Time-Variant Gas Distribution Mapping with Obstacle Information. *Autonomous Robots*, *40*(1), 1–16. doi:10.1007/s10514-015-9437-0

Monroy, J., & Gonzalez-Jimenez, J. (2015). Real-Time Odor Classification Through Sequential Bayesian Filtering. *16th International Symposium on Olfaction and Electronic Nose (ISOEN).* Retrieved from http://mapir.isa.uma.es/mapirwebsite/index.php/mapir-downloads/papers/210

Monroy, J., & Gonzalez-Jimenez, J. (2017). Gas Classification in Motion: An Experimental Analysis. *Sensors and Actuators. B, Chemical*, *240*, 1205–1215. doi:10.1016/j.snb.2016.09.013

Monroy, J., Gonzalez-Jimenez, J., & Blanco, J.-L. (2012). Overcoming the slow recovery of MOX gas sensors through a system modeling approach. *Sensors (Basel)*, *12*(10), 13664–13680. doi:10.3390/s121013664 PMID:23202015

Monroy, J., Gonzalez-Jimenez, J., & Sanchez-Garrido, C. (2014). Monitoring Household Garbage Odors in Urban Areas Through Distribution Maps. In IEEE Sensors (pp. 1364–1367). IEEE. doi:10.1109/ICSENS.2014.6985265

Monroy, J., Hernandez-Bennetts, V., Fan, H., Lilienthal, A., & Gonzalez-Jimenez, J. (2017). GADEN: A 3D gas dispersion simulator for mobile robot olfaction in realistic environments. *MDPI Sensors*, *17*(7), 1–16. doi:10.3390/s17071479

Monroy, J., Jaimez, M., & Gonzalez-Jimenez, J. (2017). Online Estimation of 2D Wind Maps for Olfactory Robots. *International Symposium on Olfaction and Electronic Nose (ISOEN).* doi:10.1109/ISOEN.2017.7968883

Monroy, J., Palomo, E., Lopez-Rubio, E., & Gonzalez-Jimenez, J. (2016). Continuous Chemical Classification in Uncontrolled Environments with Sliding Windows. *Chemometrics and Intelligent Laboratory Systems*, *158*, 117–129. doi:10.1016/j.chemolab.2016.08.011

Neumann, P. P., Bartholmai, M., Schiller, J. H., Wiggerich, B., & Manolov, M. (2010). Micro-drone for the characterization and self-optimizing search of hazardous gaseous substance sources: A new approach to determine wind speed and direction. In *Robotic and Sensors Environments (ROSE), 2010 IEEE International Workshop on* (pp. 1–6). IEEE. doi:10.1109/ROSE.2010.5675265

Neumann, P. P., Hernandez-Bennetts, V., Lilienthal, A., Bartholmai, M., & Schiller, J. H. (2013). Gas source localization with a micro-drone using bio-inspired and particle filter-based algorithms. *Advanced Robotics*, *27*(9), 725–738. doi:10.1080/01691864.2013.779052

Nguyen, L. V, Kodagoda, S., Ranasinghe, R., & Dissanayake, G. (2014). *Mobile Robotic Wireless Sensor Networks for Efficient Spatial Prediction*. Academic Press.

Pashami, S., Asadi, S., & Lilienthal, A. (2010). Integration of OpenFOAM flow simulation and filament-based gas propagation models for gas dispersion simulation. *Open Source CFD International Conference*. Retrieved from http://scholar.google.com/scholar?hl=en&btnG=Search&q=intitle:Integration+of+OpenFOAM+Flow+Simulation+and+Filament-Based+Gas+Propagation+Models+for+Gas+Dispersion+Simulation#0

Pearce, T. C., Schiffman, S. S., Nagle, H. T., & Gardner, J. W. (2006). *Handbook of machine olfaction: electronic nose technology*. John Wiley & Sons. doi:10.1002/3527601597

Ping, J., Xiao-fang, H., & Ai-dong, G. E. (2014). Mobile Robot Gas Source Localization Based on Behavior Strategies. *Proceedings of The 33rd Chinese Control Conference*, 8304–8308.

Pobkrut, T., Eamsa-ard, T., & Kerdcharoen, T. (2016). Sensor drone for aerial odor mapping for agriculture and security services. In *Electrical Engineering/Electronics, Computer, Telecommunications and Information Technology (ECTI-CON), 2016 13th International Conference on* (pp. 1–5). Academic Press. doi:10.1109/ECTICon.2016.7561340

Pyk, P., Badia, S. B., Bernardet, U., Knüsel, P., Carlsson, M., Gu, J., … Verschure, P. F. M. J. (2006). An artificial moth: Chemical source localization using a robot based neuronal model of moth optomotor anemotactic search. *Autonomous Robots*, *20*(3), 197–213. doi:10.1007/s10514-006-7101-4

Reggente, M., & Lilienthal, A. (2010). The 3D-Kernel DM+V/W algorithm: Using wind information in three dimensional gas distribution modelling with a mobile robot. *2010 IEEE Sensors*, 999–1004. 10.1109/ICSENS.2010.5690924

Reggente, M., Lilienthal, A., Pardo, M., & Sberveglieri, G. (2009). Three-Dimensional Statistical Gas Distribution Mapping in an Uncontrolled Indoor Environment. *AIP Conference Proceedings*, *109–112*, 109–112. doi:10.1063/1.3156484

Russell, R. A. (2003). Chemical source location and the robomole project. *Proceedings Australian Conference on Robotics and Automation*.

Russell, R. A. (2006). Tracking chemical plumes in 3-dimensions. In *Robotics and Biomimetics, 2006. ROBIO'06. IEEE International Conference on* (pp. 31–36). IEEE.

Sanchez-Garrido, C., Monroy, J., & Gonzalez-Jimenez, J. (2014). A Configurable Smart E-Nose for Spatio-Temporal Olfactory Analysis. In IEEE Sensors (pp. 1968–1971). IEEE. doi:10.1109/ICSENS.2014.6985418

Schaller, E., Bosset, J. O., & Escher, F. (1998). "Electronic noses" and their application to food. *Lebensmittel-Wissenschaft + Technologie*, *31*(4), 305–316. doi:10.1006/fstl.1998.0376

Schleif, F.-M., Hammer, B., Monroy, J., Gonzalez-Jimenez, J., Blanco, J.-L., Biehl, M., & Petkov, N. (2016). Odor recognition in robotics applications by discriminative time series modeling. *Pattern Analysis & Applications*, *19*(1), 207–220. doi:10.1007/s10044-014-0442-2

Shraiman, B. I., & Siggia, E. D. (2000). Scalar Turbulence. *Nature*, *405*(6787), 639–646. doi:10.1038/35015000 PMID:10864314

Sklavounos, S., & Rigas, F. (2004). Validation of turbulence models in heavy gas dispersion over obstacles. *Journal of Hazardous Materials*, *108*(1–2), 9–20. doi:10.1016/j.jhazmat.2004.01.005 PMID:15081160

Stachniss, C., Plagemann, C., & Lilienthal, A. (2009). Learning gas distribution models using sparse Gaussian process mixtures. *Autonomous Robots*, *26*(2–3), 187–202. doi:10.1007/s10514-009-9111-5

Trincavelli, M., Lout, A., & Coradeschi, S. (2009). Classification of Odours for Mobile Robots Using an Ensemble of Linear Classifiers. In AIP Conference Proceedings (pp. 475–478). AIP. doi:10.1063/1.3156587

Trincavelli, M., & Loutfi, A. (2010). Feature selection for gas identification with a mobile robot. In *Robotics and Automation (ICRA), 2010 IEEE International Conference on* (pp. 2852–2857). IEEE. doi:10.1109/ROBOT.2010.5509617

Tsujita, W., Yoshino, A., Ishida, H., & Moriizumi, T. (2005). Gas sensor network for air-pollution monitoring. *Sensors and Actuators. B, Chemical*, *110*(2), 304–311. doi:10.1016/j.snb.2005.02.008

Vergara, A., Fonollosa, J., Mahiques, J., Trincavelli, M., Rulkov, N., & Huerta, R. (2013). On the performance of gas sensor arrays in open sampling systems using Inhibitory Support Vector Machines. *Sensors and Actuators. B, Chemical*, *185*(0), 462–477. doi:10.1016/j.snb.2013.05.027

Vergara, A., Vembu, S., Ayhan, T., Ryan, M. A., Homer, M. L., & Huerta, R. (2012). Chemical gas sensor drift compensation using classifier ensembles. *Sensors and Actuators B: Chemical, 166–167*(0), 320–329. https://doi.org/10.1016/j.snb.2012.01.074

Vergassola, M., Villermaux, E., & Shraiman, B. I. (2007). "Infotaxis" as a strategy for searching without gradients. *Nature*, *445*(7126), 406–409. doi:10.1038/nature05464 PMID:17251974

Wang, Y., Meng, Q. H., Luo, B., Zeng, M., Xue, L., & Ma, S. G. (2016). Multi-robot based odor source declaration in 3D airflow fields. In *2016 12th World Congress on Intelligent Control and Automation (WCICA)* (pp. 384–388). Academic Press. doi:10.1109/WCICA.2016.7578671

Weissburg, M. J., Dusenbery, D. B., Ishida, H., Janata, J., Keller, T., Roberts, P. J. W., & Webster, D. R. (2002). A multidisciplinary study of spatial and temporal scales containing information in turbulent chemical plume tracking. *Environmental Fluid Mechanics*, *2*(1–2), 65–94. doi:10.1023/A:1016223500111

Werle, P., Slemr, F., Maurer, K., Kormann, R., Mücke, R., & Jänker, B. (2002). Near-and mid-infrared laser-optical sensors for gas analysis. *Optics and Lasers in Engineering*, *37*(2), 101–114. doi:10.1016/S0143-8166(01)00092-6

Zappi, P., Bales, E., Park, J. H., Griswold, W., & Šimuni, T. (2012). The CitiSense Air Quality Monitoring Mobile Sensor Node. *Proceedings of the 11th ACM/IEEE Conference on Information Processing in Sensor Networks*.

Zarzhitsky, D., Spears, D., Thayer, D., & Spears, W. (2004). Agent-based chemical plume tracing using fluid dynamics. In *International Workshop on Formal Approaches to Agent-Based Systems* (pp. 146–160). Academic Press. doi:10.1007/978-3-540-30960-4_10

Chapter 13
Integrating Smell Sense in Robots Using Artificial Neural Networks

Yousif Albastaki
University of Bahrain, Bahrain

Fatema Albalooshi
University of Bahrain, Bahrain

ABSTRACT

One of the approaches to mimic the remarkable abilities of the human olfactory system is by the design of computer-controlled sensor arrays that are capable of detecting and distinguishing a different range of smells and odors with consistent monitoring, referred to as electronic noses. This chapter introduces the opportunity of integrating smell sense in robots by the use of artificial neural networks. The study proposes a structure for integrating electronic noses in robots to add the capabilities of smell-related assignments, typically to recognize hazardous substances such as sampling the air and decide its actions based on this information. Utilizing the proposed algorithm allows experts in this field to be aware of gas leakage areas and thus reduce unexpected incidences. The effectiveness of the algorithm is demonstrated using real-world samples, and the performance is examined via quantitative metrics and analysis. The results show that the proposed algorithmic framework outperforms state-of-the-art methods with an error rate of only 0.0999%.

BACKGROUND

A tremendous amount of researches have been conducted in the area of robotics, resulting in innumerable advancements in this technology. The developments in robotics can range from solving Rubik's cube in less than a second time to deep-diving robots zap, to dancing droids that dance simultaneously.

Along these lines, robots can turn into the eyes, ears, and even noses of humans permitting an appropriate intervention when required.

DOI: 10.4018/978-1-5225-3862-2.ch013

On the other hand, during the last few decades, the electronic nose has been applied to a number of disciplines. Many systems have been developed to explore, for example, the odor highlight of tests, and human evaluation has been the fundamental instrument of diagnosis. From an evolutionary perspective, the chemical senses, particularly olfaction, are coined as the oldest sensory systems. They remain in many ways the least understood systems of the sensory modalities. The olfactory system is the most thoroughly studied component of the chemo-sensory triad and processes information about the identity, concentration, and quality of a wide range of chemical stimuli. There are many models available that simulate human sensors. These stimuli, called odorants, interact with olfactory receptor neurons within an epithelial sheet, the olfactory epithelium that lines the interior of the nose. Similar to the way road atlases map out different real-world locations, the brain configures itself a map of our physical world. One such map is the olfactory system, which represents one of the oldest sense models in the history of mammals. This system has been long believed to map in such a way that those groups of chemically related odorants index with clusters of cells that are situated next to one another.

Our focus in this chapter is odor-sensing robots that have the ability to distinguish between different significant odors, as this technology has a great advantage for the humanity. For example robots with the capability of sensing odors can replace trained sniffer dogs to locate different substances such as illegal drugs, victims of natural disasters, escaped prisoners, explosives, earthquakes, wildlife scat, chemical leaks, currency, or blood. The shocking fact that the behaviour of those trained dogs might be affected by their trainers believes (Li and Hopfield, 1989) leads us to emphasise and rethink in other alternatives such as odor-sensing robots. The magnificent number of mistaken alarms distinguished crosswise over conditions affirms that handler convictions influence performance. Moreover, the coordinated example of cautions in conditions containing a marker contrasted with the pattern of alarms in the condition with unmarked fake fragrance proposes that human impact on handler convictions influences alerts to a more prominent degree than dog impact on handler convictions. That is, the otal number of alarms distinguished by handlers did not vary crosswise over conditions.

Researches in the area of robotics have begun to examine the challenges and capabilities of electronic noses that utilize chemical sensors in the late 1980s and early 1990s. Various applications utilize chemical sensors to provide a guide for odor-sensing robotics.

At the point when confronting a bomb risk, officers can survey the circumstance through the camera introduced on the robot. On the off chance that the hazardous gadget can be examined by the robot, the administrator can utilize the robot's hand to grab the gadget and take it to a sheltered place for explosion.

Robots nowadays are widely used as they are robust, easy to set up, have low maintenance, low risk of technology failure, and easy to troubleshoot.

Moreover, increasing the sensitivity of the sensor opens up numerous trends. The sensor will likewise be utilized as a part of an unconstrained situation and therefore should resist being damaged by the chemicals it will encounter in its environment.

The reminder of the chapter is organized in the following manner. The chapter starts by introducing the concept of electronic noses. In the second part, this chapter illustrates artificial neural networks and a review of related literature is introduced. Finally, in this chapter data sampling and the methodology used with the findings of the research work are described.

MODELS OF THE ELFACTORY SYSTEM

The objective of a great part of the examination with respect to the olfactory framework is to see how singular smells are recognized. Numerous analysts have created numerical models of the olfactory framework. These models regularly incorporate recreations of the neurobiological data preparing frameworks (biological neural networks). The olfactory data is handled in both the olfactory bulb and in the olfactory cortex. Figure 1 demonstrates the fundamental data preparing structures inside the brain. The olfactory cortex performs design grouping and acknowledgment of the detected smells. Once recognized, scent data is transmitted to the hippocampus, limbic framework and the cerebral cortex. The connection to the hippocampus explains why odor can sub-consciously evoke memories. Conscious perception of the odor and how to act on the odor takes place in the cerebral cortex (Lit et al. 2016). The mammalian olfactory system uses a variety of chemical sensors, known as olfactory receptors, combined with signal processing in the olfactory bulb and automated pattern recognition in the olfactory cortex of the brain.

ELECTRONIC NOSES

Electronic noses (Wilson, 2012, Bikov, 2015 and Kea-Tiong, 2013), are being created as frameworks for the automated recognition and characterization of smells, vapours, and gasses. The two principal parts of an electronic nose are the odor detection framework and the computerized pattern recognition framework. The odor detection framework can be an array of several different sensing elements (e.g., chemical sensors), where each element measures a different property of the sensed odor, or it can be a single sensing device (e.g., spectrometer) that produces an array of measurements for each odor, or it can be a combination of both chemical sensors and a spectrometer. By showing a wide range of smells to the sensor array, a database of signatures is developed. This database of marked odor signatures is utilized

Figure 1. The major processes of the olfactory system

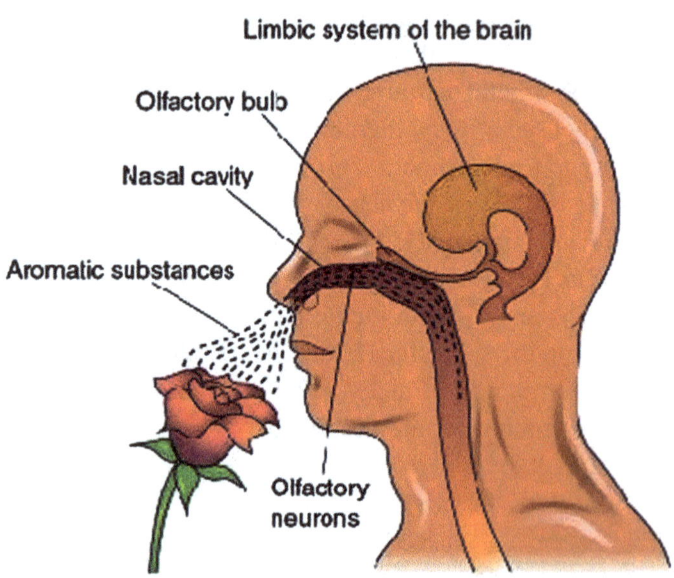

to train the pattern recognition system. The objective of this preparation procedure is to arrange and configure the recognition system to deliver extraordinary mappings of every odor so that an automated distinguishing proof can be executed (Keller and Kangas, 1994). Albeit every sensor is intended for a particular chemical, each reacts to a wide assortment of chemical vapours. On the whole, these sensors react with one of a kind signature to various chemicals. Amid the preparation procedure, different chemicals with known blends are exhibited to the system.

ARITIFICIAL NEURAL NETWORKS AND ELECTRONIC NOSES

ANNs, which have been utilized to examine complex data for pattern recognition, are indicating promising outcomes in chemical vapor recognition. Figure 3 illustrates an electronic nose prototype with the help ANN to identify odors from several common chemicals.

RELATED WORKS

Odor-sensing robotics had gained great interest from researchers and numerous frameworks accessible for identifying and examining chemicals. Chemical sensors continue to develop and become more appropriate for robots. The following paragraph describes a briefer review of previous works in the field of robotics and electronic noses.

(Rozas et al., 1991) designed inorganic semiconductor sensors for robotic navigation. Their design is based on fixing the sensors on a self-directed platform.

(Yuan et al., 2015) introduced a Petri net model for independent tracking of chemical plumes. Their study clarified the effort of utilizing the change relationships of chemical concentrations for tracking chemical plumes under both diffusive and turbulent airflow environments where they showed satisfying results, however, their study lacked optimizing partial parameters, and the chemical plume tracing needed improved modelling.

(Blanco et al., 2012) proposed an odor sensing mobile robot for detecting an ethanol odor field by using Metal Oxide Semiconductors sensors to measure the odor. They implemented an explore exploit-based algorithm in the mobile robot. This makes use of this environment information by combining the spiral and the plume-centered upwind search algorithms.

Figure 2. Electronic nose prototype using ANN

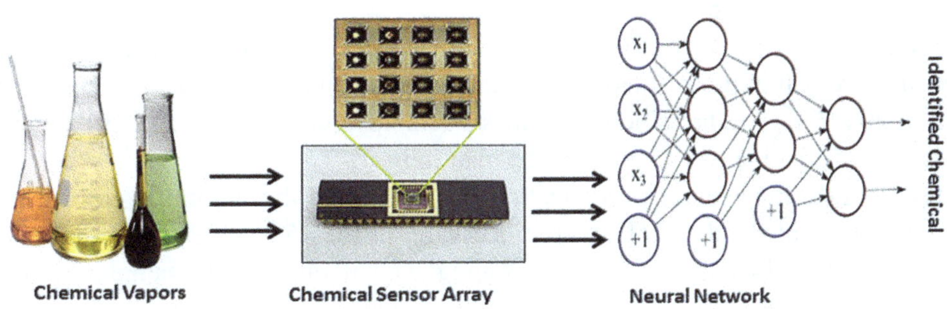

(Russell, 2014) introduced a sensor system that can monitor volatile chemicals and airflow. The system is modelled on characteristics of the human body that are thought to have a significant influence on the human odor and airflow senses. A heated sensor system was constructed to investigate these effects and the resulting sensor was mounted on a mobile robot.

(Kok et al., 2013) proposed a quadrotor based flying sniffer robot under two selection criteria; degree of locomotion maneuverability and airflow disturbances during flight. Kok claims that both criteria are crucial, as the high degree of locomotion maneuverability can ensure that the odour plume tracking follows closely in all three dimensions, while the low airflow disturbance is designed to minimize the disturbance during the sniffing process. Secondly, the study of airflow simulation of quadrotor by using computational fluid dynamics (CFD) software was carried out to identify the odour sensing region and determine sensor mounting position of quadrotor based flying sniffer robot in order to avoid the disturbance of airflow during sniffing process.

(Loutfi and Coradeschi, 2006) investigated the integration of an electronic nose and its odour discrimination functionalities into a multi-sensing robotic system which works over an extended period of time. This research work considered experimental platforms which operate over an extended period of time, a number of issues related to odour discrimination arise such as the drift in the sensor data, online learning of new odours, and the correct association of odour properties related to objects. In addition to an electronic nose, this proposed robotic system consists of other sensing modalities (vision and sonar), behavior-based control and a high-level symbolic planner.

As a result of reviewing relative research in this field, it obvious that robots are usually programmed to have convenient analogy and digital methods of recording images and sounds so that it can see and hear them and behave accordingly, but not much of research is conducted to enable robots to capture the scent of a dish at a restaurant, a favourite perfume, or any other smell that moves us in some way. Therefore, this research work proposes a methodology for integrating smell sense in robots using the concept of electronic noses and artificial numeral network.

DATA SAMPLING

A handheld OMX-GR odometer sensor (Figure 4) is used in this research study to collect data samples for three different materials: White Board Marker, Muscle Cream and Herbal Inhalant (these materials are used because of their strong smell). The sensor includes two semiconductor gas sensors with the capability to detect a variety of odors from combustible gases. This sensor measures odor with different strength and classifications. The odor strength ranges from 0 to 999 and a classification feature range from 0 to 89. Additionally, the sensor supports two types of sampling measurement modes: real-time sampling and memory samplings.

During this phase of the research work, we measured approximately 1800 data representing the odor classification and strength of three different materials. For each material 15 data samples are collected in an interval of 20 seconds each. Table 1 describes a sample of data collected for each material representing the odor classification and strength of the material.

Table 2 describes the range of the classification and strength of the entire data collected.

In the second phase of the data collection process, all of the 15 samples of the data for the three materials were graphed to identify the similarities and the overlapping of the samples before feeding it to the ANN. These similarities and overlapping are illustrated in Figure 4.

Figure 3. OMX-GR odometer sensor and the materials used for data sampling

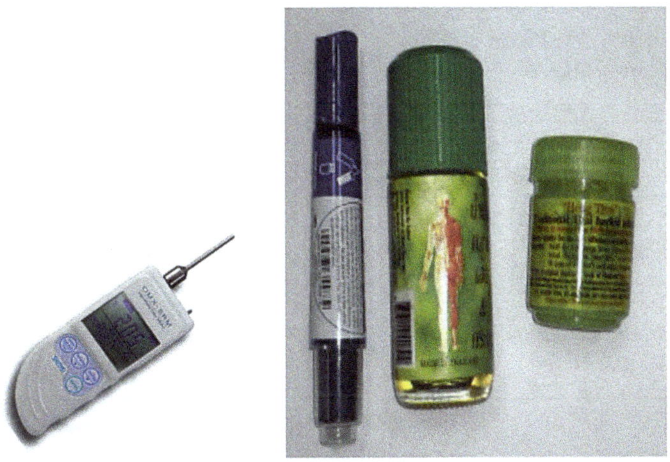

Table 1. Sample data representing the classification and strength of odors measured

White Board Marker	88,564	88,574	88,585	89,585	89,595	89,595	89,595	89,595	89,585	89,585
Muscle Cream	65,143	65,142	65,142	65,142	65,142	65,142	65,142	65,142	65,142	65,142
Herbal Inhalant	69,528	69,528	69,528	69,528	69,528	69,528	69,527	69,527	69,527	69,527

Table 2. The range of the measured odor classification and strength

White Board Marker			Muscle Cream		Herbal Inhalant	
Sample #	**Class**	**Strength**	**Class**	**Strength**	**Class**	**Strength**
1	89-89	889-906	65-66	138-152	71-72	481-502
2	89-89	857-906	64-65	138-152	71-71	502-522
3	89-89	826-873	64-66	143-157	70-71	522-524
4	88-88	640-769	65-66	147-157	69-70	525-526
5	88-88	650-694	65-66	138-147	69-69	526-527
6	87-88	592-650	65-65	138-142	69-69	526-527
7	88-89	564-595	64-65	148-162	68-69	510-523
8	88-89	535-574	65-65	162-163	68-68	510-511
9	88-88	508-544	65-65	157-162	68-68	510-511
10	88-88	446-476	65-66	162-167	68-68	511-511
11	87-88	412-446	65-66	162-167	68-68	511-511
12	86-87	387-412	64-65	152-162	68-68	511-511
13	88-89	703-729	64-65	148-162	68-68	511-511
14	88-89	611-691	65-66	162-167	67-68	493-512
15	88-88	553-611	65-66	158-167	67-68	493-494
All Data	86-89	387-906	64-66	138-167	67-72	481-528

Figure 4. Similarities and overlapping of the odor data samples for each material

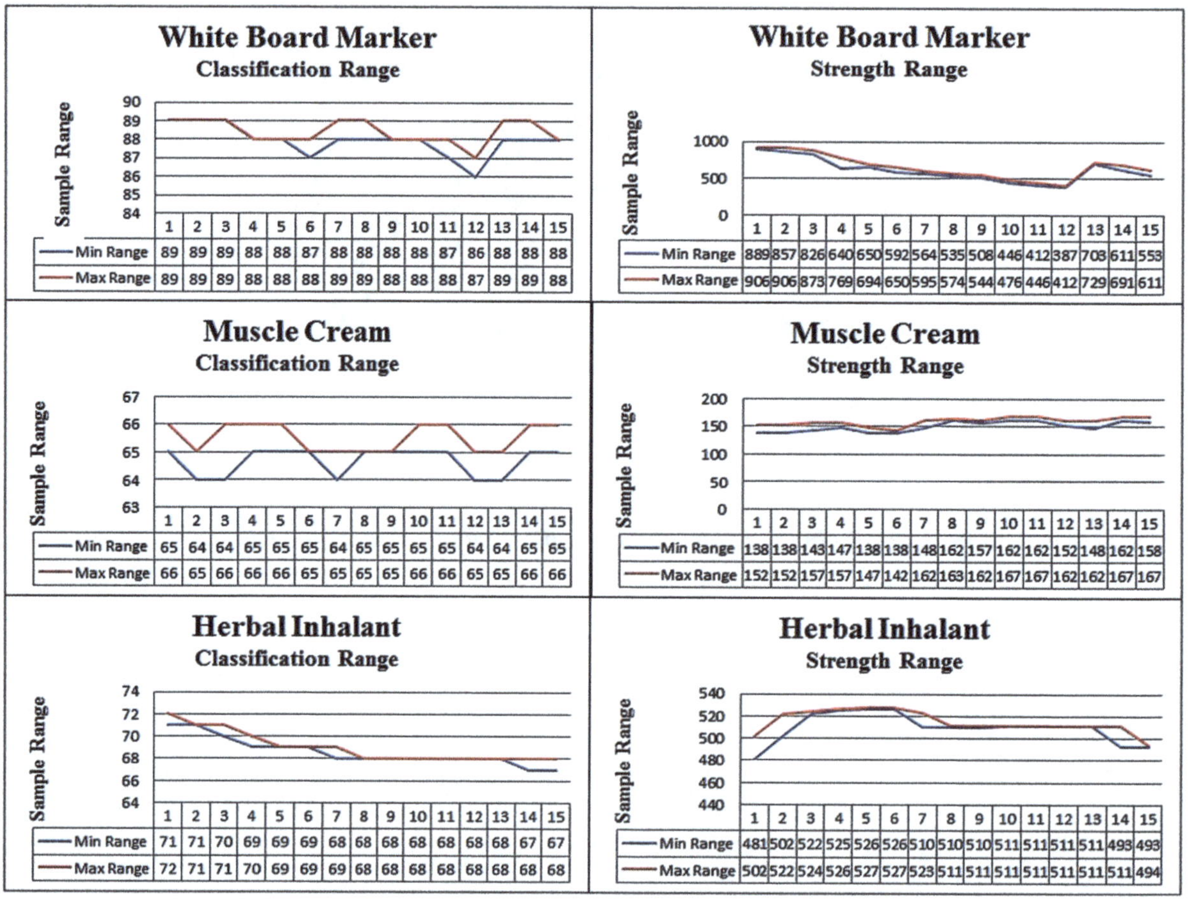

METHODOLOGY AND RESULTS

The two main components of an electronic nose are the sensing system and the automated pattern recognition system. In this research study, the sensing system is a pair of semiconductor gas sensor to classify odors. This is a popular simplified handheld OMX-GR odor meter used as a tool for odor analysis to enable the indication of relative strength and odor classification numerically by comparing odor gases and purified air. The sensor produced an array of measurements for three different materials (White Board Marker, Muscle Cream, and Herbal Inhalant). Each material vapor presented to the sensor array produced a signature or pattern characteristic of the vapor as shown in Table 1.

To generalize the ANN, the data is divided into three sets: a training set, validation set, and testing set. Figure 6 illustrates data divisions.

A backpropagation supervised learning technique is used and resulted in 0.0999 training error, 0.05312 validations, and 0% testing. A total of 3178 epoch was used. Figure 7 shows the setup to use sigmoid function as a nonlinear with S shape curved function. The figure then illustrates the training graph and main results generated after feeding the data to the ANN program.

Figure 5. Data distribution

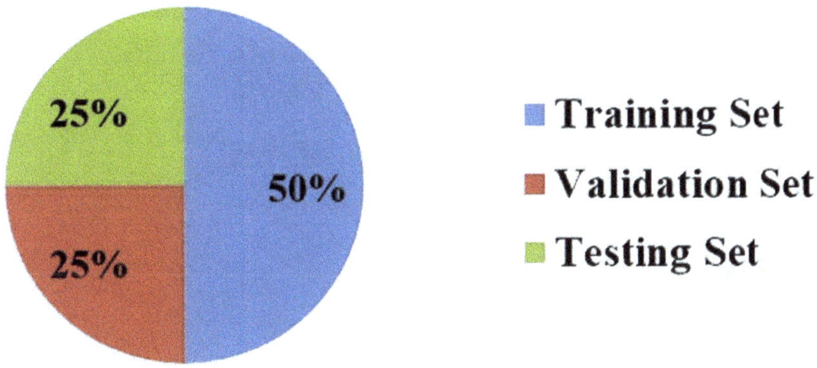

Figure 6. Setups and results

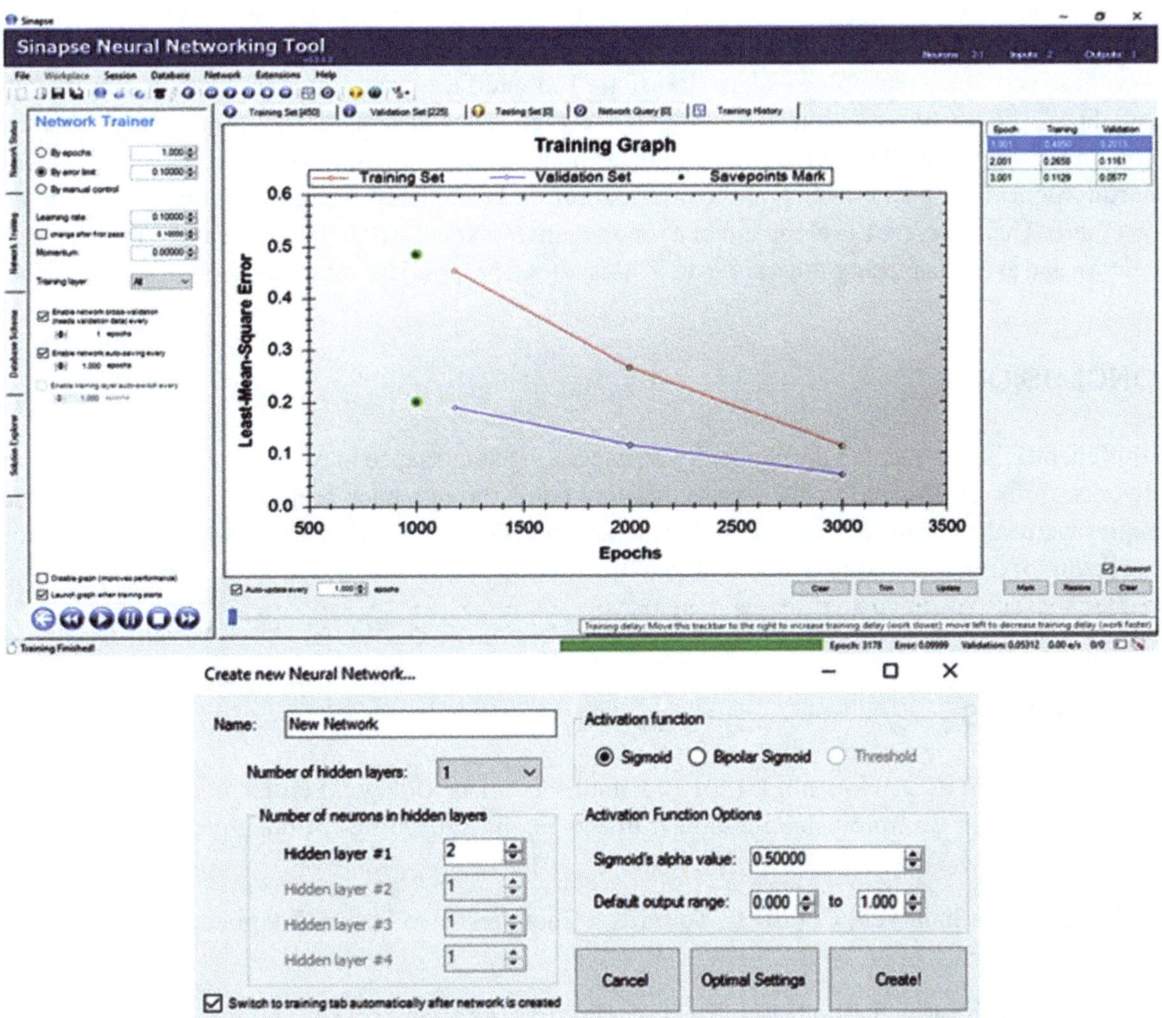

Figure 7. Integrating smell sense in robots

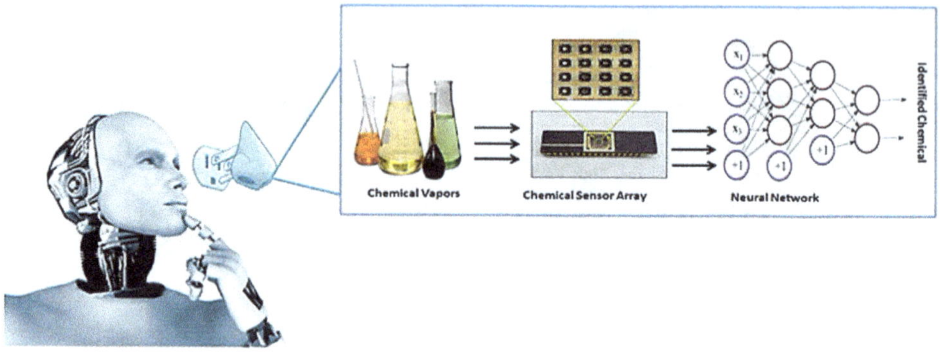

INTEGRATING SMELL SENSE IN ROBOTS

Modern Robotics research relies heavily on computer science and AI techniques. Therefore, many of the known issues in these technologies have transferred to interface programming of robots. On the other hand, the e-nose can be programmed to recognize several chemical signatures, like gas, blood, chemical, and even human sweat. Once it has collected data about a certain smell, the nose radios it to a computer that runs algorithms such as ANN on the information, and calculates where it is coming from and how to get there. Therefore, the development of a robotic sense of smell will help with search-and-rescue, gas detection and even diagnosing illness. Figure 8 illustrates a proposal for integrating smell sense in robots.

CONCLUSION

In this chapter, we investigate the capability of integrating smell sense in robotics with the aid of neural networks. ANN was utilized to distinguish different odors through back propagation network. Our data samples were taken from whiteboard markers of different companies. Experimental results show promising future of this technology with only 0.0999% of error. A possible future work would be to expand the odor sensor to distinguish diseases in humans.

REFERENCES

Bikov, A., La'za'r, Z., & Horvath, I. (2015). Established methodological issues in electronic nose research: how far are we from using these instruments in clinical settings of breath analysis? Journal of Breath Research, 9(3).

Kea-Tiong, T., & Shih-Wen, C. (2013). Towards a Chemiresistive Sensor-Integrated Electronic Nose: A Review. Sensors, 13(10), 14214–14247.

Keller, P. K., & Kangas, L. J. (1994). Three neural network based sensor systems for environmental monitoring. *Proceeding of the IEEE Electro 94 Conference on Combined Volumes*, 377-382. doi:10.1109/ELECTR.1994.472686

Kok, S. E., Kian, M. Y., Tiam, H. T., Soo Yong, L., & Beek, Y. (2013). Paper. *IEEE International Conference on Robotics, Biomimetics, & Intelligent Computational Systems (ROBIONETICS)*, 188-93.

Li, Z., & Hopfield, J. J. (1989). Modeling the olfactory bulb and its neural oscillatory processing. *Biological Cybernetics*, *61*(5), 379–392. doi:10.1007/BF00200803 PMID:2551392

Lit, L., Schweitzer, J. B., & Oberbauer, A. M. (2016). Handler Beliefs Affect Scent Detection Dog Outcomes. *Animal Cognition*, *14*(3), 387–394.

Rozas, R., Morales, J., & Vega, D. (1991). Artificial smell detection for robotic navigation. *Advanced Robotics. 'Robots in Unstructured Environments', 91 ICAR., Fifth International Conference on*, 1730-1733. doi:10.1109/ICAR.1991.240354

Russell, R. A. (2014). Article. Journal of Bionic Engineering, 11(1), 82-89.

Wilson, A. D. (2012). Review of Electronic-nose Technologies and Algorithms to Detect Hazardous Chemicals in the Environment. *First World Conference on Innovation and Computer Sciences (INSODE 2011), Procedia Technology*, 453-463. doi:10.1016/j.protcy.2012.02.101

Yuan, J., Oswald, D., & Li, W. (2015). Autonomous tracking of chemical plumes developed in both diffusive and turbulent airflow environments using Petri nets. *Expert Systems with Applications*, *42*(1), 527–538. doi:10.1016/j.eswa.2014.08.005

Compilation of References

Adnan, K. N. A. K., Yusuf, N., Maamor, H. N., Rashid, F. N. A., Ismail, S. W. M., Thriumani, R., . . . Ahmad, M. N. (2014, August). *Water quality classification and monitoring using e-nose and e-tongue in aquaculture farming*. Paper presented at the 2nd International Conference on Electronic Design, Penang, Malaysia. doi:10.1109/ICED.2014.7015827

Ai, N., Liu, H., Wang, J., Zhang, X., Zhang, H., Chen, H., ... Sun, B. (2015). Triple-channel comparative analysis of volatile flavour composition in raw whole and skim milk via electronic nose, GC-MS and GC-O. *Analytical Methods*, *7*(10), 4278–4284. doi:10.1039/C4AY02751E

Aishima, T. (1991). Aroma discrimination by pattern recognition analysis of responses from semiconductor gas sensor array. *Journal of Agricultural and Food Chemistry*, *39*(4), 752–756. doi:10.1021/jf00004a027

Al Maskari, S., Kumar, D., & Chiffings, T. (2012). Data mining for environment monitoring. In *International Conference on Software Technology and Engineering (ICSTE 2012)*. ASME Press. doi:10.1115/1.860151_ch87

Al-Bastaki, Y. (2009). An Artificial Neural Networks-Based on-Line Monitoring Odor Sensing System. *Journal of Computational Science*, *11*, 878–882.

Aleixandre, M., Lozano, J., Gutiérrez, J., Sayago, I., Fernández, M. J., & Horrillo, M. C. (2008). Portable e-nose to classify different kinds of wine. *Sensors and Actuators. B, Chemical*, *131*(1), 71–76. doi:10.1016/j.snb.2007.12.027

Alexandrakis, D., Brunton, N. P., Downey, G., & Scannell, A. G. M. (2012). Identification of Spoilage Marker Metabolites in Irish Chicken Breast Muscle Using HPLC, GC–MS Coupled with SPME and Traditional Chemical Techniques. *Food and Bioprocess Technology*, *5*(5), 1917–1923. doi:10.1007/s11947-010-0500-8

Ali, M. A., Geng, Y., Cavaye, H., Burn, P. L., Gentle, I. R., Meredith, P., & Shaw, P. E. (2015). Molecular versus exciton diffusion in fluorescence-based explosive vapour sensors. *Chemical Communications*, *51*(98), 17406–17409. doi:10.1039/C5CC06367A PMID:26465590

Ali, M. A., Shoaee, S., Fan, S., Burn, P. L., Gentle, I. R., Meredith, P., & Shaw, P. E. (2016). Detection of explosive vapors: The roles of exciton and molecular diffusion in real-time sensing. *ChemPhysChem*, *17*(21), 3350–3353. doi:10.1002/cphc.201600767 PMID:27583839

Al-Maskari, S., B'elisle, E., Li, X., Le Digabel, S., Nawahda, A., & Zhong, J. (2016). Classification with quantification for air quality monitoring. Academic Press. doi:10.1007/978-3-319-31753-3_46

Al-Maskari, S., Guo, W., & Zhao, X. (2016). Biologically inspired pattern recognition for e-nose sensors. In *Advanced data mining and applications: 12th international conference* (pp. 142–155). Academic Press. doi:10.1007/978-3-319-49586-6_10

Al-Maskari, S., Li, X., & Liu, Q. (2014). An effective approach to handling noise and drift in electronic noses. In *Databases theory and applications - 25th australasian database conference* (pp. 223–230). Academic Press. doi:10.1007/978-3-319-08608-8_21

Al-Maskari, S., Saini, D., & Omar, W. (2010). Cyber infrastructure and data quality for environmental pollution control - in Oman. *Proceedings of the 2010 DAMD International Conference on Data Analysis, Data Quality and Metada.* doi:10.5176/978-981-08-6308-1_D-038

Amoore, J. E. (1952). The stereochemical specificities of human olfactory receptors, perfumes. *Essent Oil Rec.,* (43), 321-323, 330.

Amoore, J. E., & Venstrom, D. (1967). Correlations between stereochemical assessment and organoleptic analysis of odorous compounds. Olfaction and Taste, (2), 3 – 17.

Ampuero, S., & Bee, G. (2006). The potential to detect boar tainted carcasses by using an electronic nose based on mass spectrometry. *Acta Veterinaria Scandinavica, 48*(Suppl 1), 1. doi:10.1186/1751-0147-48-S1-P1

Ampuero, S., & Bosset, J. O. (2003). The electronic nose applied to dairy products: A review. *Sensors and Actuators. B, Chemical, 94*(1), 1–12. doi:10.1016/S0925-4005(03)00321-6

Andresen, O. (2006). Boar taint related compounds: Androstenone/skatole/other substances. *Acta Veterinaria Scandinavica, 48*(Suppl 1), S5. doi:10.1186/1751-0147-48-S1-S5

Andretta, M., Coppola, F., & Seccia, L. (2016). Investigation on the interaction between the outdoor environment and the indoor microclimate of a historical library. *Journal of Cultural Heritage, 17*, 75–86. doi:10.1016/j.culher.2015.07.002

Andrew Russel, R. (1999). Odour Detection by Mobile Robots. *World Scientific Series in Robotics and Intelligent System, 22*, 197-217.

Andrew Russell, R. (2001). Survey of Robotic Applications for Odor-Sensing Technology. *The International Journal of Robotics Research, 20*(2), 144–162. doi:10.1177/02783640122067318

Andrew Russell, R. (2006). Taste and Search in a Robotics Context. *Proc. Australasian Conf. on Robotics and Automation*, 1–6.

Andrew Russell, R., Thiel, D., & Mackay-Sim, A. (1997). Recruiting swarm robots using coded odour trails. *Proceedings of the International Conference on Field and Service Robotics*, 472–476.

Andrew Russel, R. (1995). Laying and sensing odor markings as a strategy for assisting mobile robot navigation task. *IEEE Robotics & Automation Magazine, 2*(3), 3–9. doi:10.1109/100.414920

Andrew, T. L., & Swager, T. M. (2007). A fluorescence turn-on mechanism to detect high explosives RDX and PETN. *Journal of the American Chemical Society, 129*(23), 7254–7255. doi:10.1021/ja071911c PMID:17508754

An, K.-A., Arshad, M. S., Jo, Y., Chung, N., & Kwon, J.-H. (2017). E-Beam Irradiation for Improving the Microbiological Quality of Smoked Duck Meat with Minimum Effects on Physicochemical Properties During Storage. *Journal of Food Science, 82*(4), 865–872. doi:10.1111/1750-3841.13671 PMID:28267865

Arora, K., Chand, S., & Malhotra, B. D. (2006). Recent developments in bio-molecular electronics techniques for food pathogens. *Analytica Chimica Acta, 568*(1-2), 259–274. doi:10.1016/j.aca.2006.03.078 PMID:17761267

Artursson, T., Ekl¨ov, T., & Lundstr¨om, I., Martensson, P., Sj¨ostr¨om, M., & Holmberg, M. (2000). Drift correction for gas sensors using multivariate methods. *Journal of Chemometrics, 14*(5-6), 711–723. doi:10.1002/1099-128X(200009/12)14:5/6<711::AID-CEM607>3.0.CO;2-4

Asadi, S., Pashami, S., Loutfi, A., Lilienthal, A., & Gouma, P. (2011). TD Kernel DM+ V: Time-Dependent Statistical Gas Distribution Modelling on Simulated Measurements. In AIP Conference Proceedings (Vol. 1362, pp. 281–282). AIP. doi: 10.1063/1.3651651

Atema, J. (1996). Eddy chemotaxis and odor landscapes: Exploration of nature with animal sensors. *The Biological Bulletin*, *191*(1), 129–138. doi:10.2307/1543074

Azeriee, A. A., Hashim, H., Jarmin, R., & Ahmad, A. (2009, March). *A study on freshness of fish by using fish freshness meter*. Paper presented at the 5th International Colloquim o Signal Processing and Its Applications, Kuala Lumpur, Malaysia.

Bai, H., & Shi, G. (2007). Gas Sensors Based on Conducting Polymers. *Sensors (Basel)*, *7*(3), 267–307. doi:10.3390/s7030267

Balasubramanian, S., Amamcharla, J., & Shin, J.-E. (2016). Possible application of electronic nose systems for meat safety: An Overview. In M. L. Rodriguez-Mendez (Ed.), *Electronic Noses and Tongues in Food Science*. London, UK: Academic Press. doi:10.1016/B978-0-12-800243-8.00007-X

Balasubramanian, S., Panigrahi, S., Logue, C. M., Doetkott, C., Marchello, M., & Sherwood, J. S. (2008). Independent component analysis-processed electronic nose data for predicting salmonella typhimurium populations in contaminated beef. *Food Control*, *19*(3), 236–246. doi:10.1016/j.foodcont.2007.03.007

Balestra, F., Laghi, L., Taneyo Saa, D., Gianotti, A., Rocculi, P., & Pinnavaia, G. (2015). Physico-chemical and metabolomic characterization of KAMUT® Khorasan and durum wheat fermented dough. *Food Chemistry*, *187*, 451–459. doi:10.1016/j.foodchem.2015.04.041 PMID:25977050

Balkin, S. D., & Ord, J. K. (2000). Automatic neural network modeling for univariate time series. *International Journal of Forecasting*, *16*(4), 509–515. doi:10.1016/S0169-2070(00)00072-8

Balkovsky, E., & Shraiman, B. I. (2002). Olfactory search at high Reynolds number. *Proceedings of the National Academy of Sciences of the United States of America*, *99*(20), 12589–12593. doi:10.1073/pnas.192393499 PMID:12228727

Banerjee, R., Tudu, B., Bandyopadhyay, R., & Bhattacharyya, N. (2016). A review on combined odor and taste sensor systems. *Journal of Food Engineering*, *190*, 10–21. doi:10.1016/j.jfoodeng.2016.06.001

Barczak, R., Kulig, A., & Szyłak-Szydłowski, M. (2012). Olfactometric methods application for odour nuisance assessment of wastewater treatment facilities in Poland. *Chemical Engineering Transactions*, *30*, 187–192.

Barker M. (2010). *Partial least squares for discrimination.* Saarbrucken, Deutschland: LAP LAMBERT Academic Publishing.

Bartheld, C., Bahney, J., & Herculano-Houzel, S. (2016). The search for true numbers of neurons and glial cells in the human brain: A review of 150 years of cell counting. *The Journal of Comparative Neurology*, *524*(18), 3865–3895. doi:10.1002/cne.24040 PMID:27187682

Bartlett, M. S. (1937). Properties of sufficiency and statistical tests. *Proceedings of the Royal Society of London. Series A, Mathematical and Physical Sciences*, *160*(901), 268–282. doi:10.1098/rspa.1937.0109

Batoga, P., & Wołczowski, A. (2012). Odor markers detection system for mobile robot navigation. *Proc. Eurosensors XXVI*, 1442-1445. doi:10.1016/j.proeng.2012.09.429

Baxter, M. J., & Jackson, M. (2001). Variable selection artefact compositional studies. *Archaeometry*, *43*(2), 253–268. doi:10.1111/1475-4754.00017

Beer, R. D., & Ritzmann, R. E. (1993). *Biological neural networks in invertebrate neuroethology and robotics*. Academic Press.

Bell, A. (1914). *Discovery and Invention*. Washington, DC: National Geographic Society.

Bender, F., Barié, N., Romoudis, G., Voigt, A., & Rapp, M. (2003). Development of a preconcentration unit for a SAW sensor micro array and its use for indoor air quality monitoring. *Sensors and Actuators. B, Chemical*, *93*(1-3), 135–141. doi:10.1016/S0925-4005(03)00239-9

Benedetti, S., Buratti, S., Spinardi, A., Mannino, S., & Mignani, I. (2008). Electronic nose as a non-destructive tool to characterise peach cultivars and to monitor their ripening stage during shelf-life. *Postharvest Biology and Technology*, *47*(2), 181–188. doi:10.1016/j.postharvbio.2007.06.012

Berna, A.Z., Lammertyn, J., Saevels, S., Natale, C.D., Nicolaï, B.M. (2004). Electronic nose systems to study shelf life and cultivar effect on tomato aroma profile. *Sens. Actuat. B: Chem.*, (97), 324–333.

Berna, A. (2010). Metal Oxide Sensors for Electronic Noses and Their Application to Food Analysis. *Sensors (Basel)*, *10*(4), 3882–3910. doi:10.3390/s100403882 PMID:22319332

Berna, A. Z., Trowell, S., Clifford, D., Cynkar, W., Cozzolino, D., Berna, A. Z., ... Cozzolino, D. (2009). Geographical origin of Sauvignon Blanc wines predicted by mass spectrometry and metal oxide based electronic nose. *Analytica Chimica Acta*, *648*(2), 146–152. doi:10.1016/j.aca.2009.06.056 PMID:19646576

Bertoncelj, J., Golob, T., Kropf, U., & Korosec, M. (2011). Characterisation of Slovenian honeys on the basis of sensory and physicochemical analysis with a chemometric approach. *International Journal of Food Science & Technology*, *46*(8), 1661–1671. doi:10.1111/j.1365-2621.2011.02664.x

Bhattacharya, N., Tudu, B., Jana, A., Ghosh, D., Bandhopadhyaya, R., & Bhuyan, M. (2008b). Preemptive identification of optimum fermentation time for black tea using electronic nose. *Sensors and Actuators. B, Chemical*, *131*(1), 110–116. doi:10.1016/j.snb.2007.12.032

Bhattacharya, N., Tudu, B., Jana, A., Ghosh, D., Bandhopadhyaya, R., & Saha, A. (2008a). Illumination heating and physical raking for increasing sensitivity of electronic nose measurements with black tea. *Sensors and Actuators. B, Chemical*, *131*(1), 37–42. doi:10.1016/j.snb.2007.12.031

Bhuyan, M. (2006). *Measurement and Control in Food Processing*. Taylor & Francis. doi:10.1201/9781420006797

Bicego, M., Tessari, G., Tecchiolli, G., & Bettinelli, M. (2002). A comparative analysis of basic pattern recognition techniques for the development of small size electronic nose. *Sensors and Actuators. B, Chemical*, *85*(1), 137–144. doi:10.1016/S0925-4005(02)00065-5

Bielecki, Z., Janucki, J., Kawalec, A., Mikolajczyk, J., Palka, N., Pasternak, M., ... Wojtas, J. (2012). Sensors and systems for the detection of explosive devices – an overview. *Metrology and Measurement Systems*, *19*(1), 3–28. doi:10.2478/v10178-012-0001-3

Bikov, A., La'za'r, Z., & Horvath, I. (2015). Established methodological issues in electronic nose research: how far are we from using these instruments in clinical settings of breath analysis? Journal of Breath Research, 9(3).

Bishop, C. (2006). *Pattern recognition and Machine Learning*. Springer.

Bishop, C. M. (1995). *Neural networks for pattern recognition*. New York: Oxford University Press, Inc.

Bishop, J. R., & White, C. H. (1986). Assessment of dairy product quality and potential shelf lifesa review. *Journal of Food Protection*, *49*(9), 739–753. doi:10.4315/0362-028X-49.9.739

Blanco, J.-L., Monroy, J., Gonzalez-Jimenez, J., & Lilienthal, A. (2013). A Kalman Filter Based Approach to Probabilistic Gas Distribution Mapping. *28th Symposium On Applied Computing (SAC)*. doi:10.1145/2480362.2480409

Blom, G. (1958). *Statistical estimates and transformed beta variables*. New York, NY: John Wiley & Sons.

Bochenkov, V., & Sergeev, G. (2010). Sensitivity, selectivity, and stability of gas-sensitive metal-oxide nanostructures. *Metal Oxide Nanostructures and Their Applications, 3*, 31-52.

Bocksrocker, T., Preinfalk, J. B., Asche-Tauscher, J., Pargner, A., Eschenbaum, C., Maier-Flaig, F., & Lemmer, U. (2012). White organic light emitting diodes with enhanced internal and external outcoupling for ultra-efficient light extraction and Lambertian emission. *Optics Express*, *20*(56), A932–A940. doi:10.1364/OE.20.00A932

Boilot, P., Hines, E. L., John, S., Mitchell, J., Lopez, F., Gardner, J. W., ... Gonogora, M. A. (2000). Detection of bacteria causing eye infections using a neural network based electronic nose system. In J. W. Gardner & K. C. Persaud (Eds.), *Electronic nose and olfaction 2000* (pp. 189–196). Bristol, UK: IOP Publishing.

Bolse, N., Schend, M., Bell, L., Eschenbaum, C., & Lemmer, U. (2015). *An embedded system for the evaluation of ultra-sensitive optical sensor arrays based on photoluminescence quenching*. Paper presented at the 16th International Symposium on Olfaction and Electronic Noses (ISOEN 2015), Dijon, France.

Bolse, N., Chouikh, G., Sellner, M., Eschenbaum, C., & Lemmer, U. (2015a). Temperature-control for optical vapor sensors based on photoluminescence quenching. In *Proceedings of OSA Advanced Photonics 2015, SeT2D.6*. Boston, MA: OSA. doi:10.1364/SENSORS.2015.SeT2D.6

Bolse, N., Chouikh, G., Sellner, M., Eschenbaum, C., & Lemmer, U. (2015b). Temperaturstabilisierung für Phosphoreszenz-Quenching-Sensoren zur Sauerstoffdetektion. In *Proceedings of MikroSystemTechnik Kongress 2015*. Karlsruhe, Germany: VDE Verlag.

Bolse, N., Eckstein, R., Schend, M., Habermehl, A., Eschenbaum, C., Hernandez-Sosa, G., & Lemmer, U. (2017a). A digitally printed optoelectronic nose for the selective trace detection of nitroaromatic explosive vapours using fluorescence quenching. *Flexible and Printed Electronics*, *2*(2), 024001. doi:10.1088/2058-8585/aa6601

Bolse, N., Eckstein, R., Schend, M., Habermehl, A., Hernandez-Sosa, G., Eschenbaum, C., & Lemmer, U. (2017b). Discrimination of trace nitroaromatics using linear discriminant analysis on aerosol jet printed fluorescent sensor arrays. *Proceedings of the Society for Photo-Instrumentation Engineers*, *10231*, 102310R. doi:10.1117/12.2265522

Bolse, N., Schend, M., Habermehl, A., Eschenbaum, C., & Lemmer, U. (2017c). Transient response of a microfibrous web array for the discrimination of nitroaromatic explosive vapors. In *2017 ISOCS/IEEE International Symposium on Olfaction and Electronic Nose (ISOEN)*. Montreal, Canada: IEEE. doi:10.1109/ISOEN.2017.7968850

Bonow, G., & Kroll, A. (2013). Gas leak localization in industrial environments using a TDLAS-based remote gas sensor and autonomous mobile robot with the Tri-Max method. In *Robotics and Automation (ICRA), 2013 IEEE International Conference on* (pp. 987–992). IEEE. doi:10.1109/ICRA.2013.6630693

Boothe, D. D. H., & Arnold, J. W. (2002). Electronic nose analysis of volatile compounds from poultry meat samples, fresh and after refrigerated storage. *Journal of the Science of Food and Agriculture*, *82*(3), 315–322. doi:10.1002/jsfa.1036

Borras, E., Ferré, J., Boqué, R., Mestres, M., Acena, L., & Busto, O. (2015). Data fusion methodologies for food and beverage authentication and quality assessment - A review. *Analytica Chimica Acta*, *891*, 1–14. doi:10.1016/j.aca.2015.04.042 PMID:26388360

Bougrini, M., Tahri, K., Haddi, Z., El Bari, N., Llobet, E., Jaffrezic-Renault, N., & Bouchikhi, B. (2014). Aging time and brand determination of pasteurized milk using a multisensor e-nose combined with a voltammetric e-tongue. *Materials Science and Engineering C*, *45*, 348–358. doi:10.1016/j.msec.2014.09.030 PMID:25491839

Bourgeois, W., & Stuetz, R. M. (2002). Use of a chemical sensor array for detecting pollutants in domestic wastewater. *Water Research*, *36*(18), 4505–4512. doi:10.1016/S0043-1354(02)00183-5 PMID:12418653

Box, G. E. P. (1949). A general distribution theory for a class of likelihood criteria. *Biometrika*, *36*(3/4), 317–346. doi:10.2307/2332671 PMID:15402070

Branciari, R., Valiani, A., Trabalza-Marinucci, M., Miraglia, D., Ranucci, D., Acuti, G., ... Mughetti, L. (2012). Consumer acceptability of ovine cheese from ewes fed extruded linseed-enriched diets. *Small Ruminant Research*, *106*, 43–48. doi:10.1016/j.smallrumres.2012.04.032

Breen, H. (1992). Odor recognition and second messenger signaling in olfactory receptor neurons. *Seminars in Cell Biology*, (5): 25–32. PMID:8186393

Brezmes, J., Llobet, E., Vilanova, X., Saiz, G., Correig, X., (2000). Fruit ripeness monitoring using an electronic nose. *Sens. Actuat. B: Chem.*, (69), 223–229.

Brezmes, J., Llobet, E., Vilanova, X., Orts, J., Saiz, G., & Correig, X. (2001). Correlation between electronic nose signals and fruit quality indicators on shelf-life measurements with pinklady apples. *Sensors and Actuators. B, Chemical*, *80*(1), 41–50. doi:10.1016/S0925-4005(01)00867-X

Bright, C. J., Nallon, E. C., Polcha, M. P., & Schnee, M. P. (2015). Quantum dot and polymer composite cross-reactive array for chemical vapor detection. *Analytical Chemistry*, *87*(24), 12270–12275. doi:10.1021/acs.analchem.5b03559 PMID:26548712

Brink, S. A. (1996). Bofors schnauzer - a biosensor for detection of explosives. *The EUREL International Conference on The Detection of Abandoned Land Mines Edinburgh*, *431*, 33-36. doi:10.1049/cp:19961074

Broomhead, D. S., & Lowe, D. (1988). *Radial basis functions, multi-variable functional interpola- tion and adaptive networks (Tech. Rep.)*. DTIC Document.

Brown, M. L., Holbrook, D. M., Hoerning, E. F., Legendre, M. G., & St. Angelo, A. J. (1986). Volatile indicators of deterioration in liquid egg products. *Poultry Science*, *65*(10), 1925–1933. doi:10.3382/ps.0651925

Buck, L., & Axel, R. (1991). A novel multigene family may encode odorant receptors, a molecular basis for odor recognition. *Cell*, *65*(1), 175–187. doi:10.1016/0092-8674(91)90418-X PMID:1840504

Byliński, H., Lewkowska, P., Gębicki, J., Dymerski, T., & Namieśnik, J. (2016). Identification of aromatic compounds in odours mixture by gas chromatography and field olfactometry techniques. *Proceedings of the 12th Modern Analytical Chemistry*, 108–112.

Cabrita, G., & Marques, L. (2013). Divergence-based odor source declaration. In *Control Conference (ASCC), 2013 9th Asian* (pp. 1–6). ASCC. doi: 10.1109/ASCC.2013.6606390

Cabrita, G., Sousa, P., & Marques, L. (2010). Player/stage simulation of olfactory experiments. In *Intelligent Robots and Systems (IROS), 2010 IEEE/RSJ International Conference on* (pp. 1120–1125). IEEE. doi: 10.1109/IROS.2010.5652518

Canhoto, O., & Magan, N. (2005). Electronic nose technology for the detection of microbial and chemical contamination of potable water. *Sensors and Actuators, B: Chemical*, *106*(1), 3-6.

Capelli, L., Sironi, S., Del Rosso, R., & Centola, P. (2009). Predicting odour emissions from wastewater treatment plants by means of odour emission factors. *Water Research*, *43*(7), 1977–1985. doi:10.1016/j.watres.2009.01.022 PMID:19232669

Capelli, L., Sironi, S., Del Rosso, R., Centola, P., & Il Grande, M. (2008). A comparative and critical evaluation of odour assessment methods on a landfill site. *Atmospheric Environment*, *42*(30), 7050–7058. doi:10.1016/j.atmosenv.2008.06.009

Capelli, S., Sironi, P., Centola, R., del Rosso, R., & Il Grande, M. (2008). Electronic noses for the continuous monitoring of odours from a wastewater treatment plant at specific receptors: Focus on training methods. *Sensors and Actuators. B, Chemical*, *131*(1), 53–62. doi:10.1016/j.snb.2007.12.004

Capone, S., Epifani, M., Quaranta, F., Siciliano, P., Taurino, A., & Vasanelli, L. (2001). Monitoring of rancidity of milk by means of an electronic nose and a dynamic PCA analysis. *Sensors and Actuators. B, Chemical*, *78*(1-3), 174–179. doi:10.1016/S0925-4005(01)00809-7

Capua, E., Cao, R., Sukenik, C. N., & Naaman, R. (2009). Detection of triacetone triperoxide (TATP) with an array of sensors based on non-specific interactions. *Sensors and Actuators. B, Chemical*, *140*(1), 122–127. doi:10.1016/j.snb.2009.04.045

Carpenter, G. A., Grossberg, S., & Reynolds, J. H. (1991). Artmap: Supervised real-time learning and classification of nonstationary data by a self-organizing neural network. *Neural Networks, 4*(5), 565 - 588. Retrieved from http://www.sciencedirect.com/ science/article/pii/089360809190012T

Casale, M., Casolino, C., Oliveri, P., & Forina, M. (2010). The potential of coupling information using three analytical techniques for identifying the geographical origin of Liguria extra virgin olive oil. *Food Chemistry*, *118*(1), 163–170. doi:10.1016/j.foodchem.2009.04.091

Caygill, J. S., Davis, F., & Higson, S. P. J. (2012). Current trends in explosive detection techniques. *Talanta*, *88*(15), 14–29. doi:10.1016/j.talanta.2011.11.043 PMID:22265465

Cevoli, C., Cerretani, L., Gori, A., Gallina Toschi, T., & Fabbri, A. (2011). Classification of Pecorino cheeses using electronic nose combined with artificial neural network and comparison with GC–MS analysis of volatile compounds. *Food Chemistry*, *129*(3), 1315–1319. doi:10.1016/j.foodchem.2011.05.126 PMID:25212373

Chang, C.-C., Yang, M.-H., Wen, H.-M., & Chern, J.-C. (2002). Estimation of total flavonoid content in propolis by two complementary colorimetric methods. *Journal of Food and Drug Analysis*, *10*(3), 178–182.

Chao, C. Y., & Chan, G. Y. (2001). Quantification of indoor VOCs in twenty mechanically ventilated buildings in Hong Kong. *Atmospheric Environment*, *35*(34), 5895–5913. doi:10.1016/S1352-2310(01)00410-1

Chastrette, M. (1997). Trends in structure – odor relationships. *SAR and QSAR in Environmental Research*, *6*(3-4), 215–254. doi:10.1080/10629369708033253 PMID:9487700

Chatterjee, D., Bhattacharjee, P., & Bhattacharyya, N. (2014). Development of methodology for assessment of shelf-life of fried potato wedges using electronic noses: Sensor screening by fuzzy logic analysis. *Journal of Food Engineering*, *133*, 23–29. doi:10.1016/j.jfoodeng.2014.02.009

Chawla, N. V. (2005). Data Mining for imbalanced datasets: An overview. In Data Mining and Knowledge Discovery Handbook. Boston: Springer. doi:10.1007/0-387-25465-X_40

Cheng, H., Qin, Z. H., Guo, X. F., Hu, X. S., & Wu, J. H. (2013). Geographical origin identification of propolis using GC-MS and electronic nose combined with principal component analysis. *Food Research International*, *51*(2), 813–822. doi:10.1016/j.foodres.2013.01.053

Cheng, J., Sun, D.-W., Zeng, X.-A., & Liu, D. (2015). Recent advances in methods and techniques for freshness quality determination and evaluation of fish and fish fillets: A review. *Critical Reviews in Food Science and Nutrition*, *55*(7), 1012–1225. doi:10.1080/10408398.2013.769934 PMID:24915394

Chen, Q., Liu, A., Zhao, J., & Ouyang, Q. (2013). Classification of tea category using a portable electronic nose based on an odor imaging sensor array. *Journal of Pharmaceutical and Biomedical Analysis*, *84*, 77–83. doi:10.1016/j.jpba.2013.05.046 PMID:23810847

Comon, P. (1994). Independent component analysis, a new concept? *Signal Processing*, *36*(3), 287–314. doi:10.1016/0165-1684(94)90029-9

Concina, I., Falasconi, M., Gobbi, E., Musci, M., Mattarozzi, M., Pardo, M., ... Sberveglieri, G. (2009). Early detection of microbial contamination in processed tomatoes by electronic nose. *Food Control*, *20*(10), 873–880. doi:10.1016/j.foodcont.2008.11.006

Costa, G., Noferini, M., Fiori, G., Orlandi, A., & Miserocchi, O. (2003). Non-destructive technique to assess internal fruit quality. *International Conference on Quality in Chains. An Integrated View on Fruit and Vegetable Quality*, (604), 571–576. doi:10.17660/ActaHortic.2003.604.68

Cover, T., & Hart, P. (1967). Nearest neighbor pattern classification. *Information Theory. IEEE Transactions on*, *13*(1), 21–27.

Cowen, E. A., & Ward, K. B. (2002). Chemical plume tracing. *Environmental Fluid Mechanics*, *2*(1), 1–7. doi:10.1023/A:1016247516041

Cozzolino, D., Corbella, E., & Smyth, H. E. (2011). Quality control of honey using infrared spectroscopy - A review. *Applied Spectroscopy Reviews*, *46*(7), 523–538. doi:10.1080/05704928.2011.587857

Cubero, S., Aleixos, N., Moltó, E., Gómez-Sanchis, J., & Blasco, J. (2011). Advances in machine vision applications for automatic inspection and quality evaluation of fruits and vegetables. *Food and Bioprocess Technology*, *4*(4), 487–504. doi:10.1007/s11947-010-0411-8

Cynkar, W., Dambergs, R., Smith, P., & Cozzolino, D. (2010). Classification of Tempranillo wines according to geographic origin: Combination of mass spectrometry based electronic nose and chemometrics. *Analytica Chimica Acta*, *660*(1-2), 227–231. doi:10.1016/j.aca.2009.09.030 PMID:20103167

Czerny, M., Christlbauer, M., Christlbauer, M., Fischer, A., Granvogl, M., Hammer, M., ... Schieberle, P. (2008). Re-investigation on odour thresholds of key food aroma compounds and development of an aroma language based on odour qualities of defined aqueous odorant solutions. *European Food Research and Technology*, *228*(2), 265–273. doi:10.1007/s00217-008-0931-x

d'Acampora Zellner, B., Dugo, P., Dugo, G., & Mondello, L. (2011). Milk and dairy foods. In L. M. L. Nollet & F. Toldrà (Eds.), *Sensory analysis of foods of animal origin*. Boca Raton, FL: CRC Press.

D'Amico, A., Pennazza, G., Santonico, M., Martinelli, E., Roscioni, C., Galluccio, G., & Di Natale, C. (2010). An investigation on electronic nose diagnosis of lung cancer. *Lung Cancer (Amsterdam, Netherlands)*, *68*(2), 170–176. doi:10.1016/j.lungcan.2009.11.003 PMID:19959252

Dai, J. J., Lieu, L., & Rocke, D. (2006). Dimension reduction for classification with gene expression microarray data. *Statistical Applications in Genetics and Molecular Biology*, *5*(1), 1–19. doi:10.2202/1544-6115.1147 PMID:16646870

Dascalaki, E. G., Lagoudi, A., Balaras, C. A., & Gaglia, A. G. (2008). Air quality in hospital operating rooms. *Building and Environment*, *43*(11), 1945–1952. doi:10.1016/j.buildenv.2007.11.015

Davide, F. A., Di Natale, C., & D'Amico, A. (1994). Self-organizing multisensor systems for odour classification: Internal categorization, adaptation and drift rejection. *Sensors and Actuators. B, Chemical*, *18*(1), 244–258. doi:10.1016/0925-4005(94)87090-X

de Sousa, J. M. B., de Souza, E. L., Marques, G., Benassi, M. D. T., Gullon, B., Pintado, M. M., & Magnani, M. (2016). Sugar profile, physicochemical and sensory aspects of monofloral honeys produced by different stingless bee species in Brazilian semi-arid region. *Lebensmittel-Wissenschaft + Technologie*, *65*, 645–651. doi:10.1016/j.lwt.2015.08.058

De Vito, S., Massera, E., Piga, M., Martinotto, L., & Di Francia, G. (2008). On field calibration of an electronic nose for benzene estimation in an urban pollution monitoring scenario. *Sensors and Actuators. B, Chemical*, *129*(2), 750–757. doi:10.1016/j.snb.2007.09.060

Delgado-Rodriguez, M., Ruiz-Montoya, M., Giraldez, I., Lopez, R., Madejon, E., & Diaz, M. J. (2012). Use of electronic nose and GC-MS in detection and monitoring some VOC. *Atmospheric Environment*, *51*, 278–285. doi:10.1016/j.atmosenv.2012.01.006

Demuth, H. B., Beale, M. H., De Jess, O., & Hagan, M. T. (2014). *Neural network design*. Martin Hagan.

Deshmukh, S., Jana, A., Bhattacharyya, N., Bandyopadhyay, R., & Pandey, R. A. (2014). Quantitative determination of pulp and paper industry emissions and associated odor intensity in methyl mercaptan equivalent using electronic nose. *Atmospheric Environment*, *82*, 401–409. doi:10.1016/j.atmosenv.2013.10.041

Deveza, R., Russell, R. A., & Thiel, D. (1993). Robot navigation by smell: Problems and some solutions. *Proceedings of the Robots for Competitive Industries Conference*, 458–466.

Dewettinck, T., van Hege, K., & Verstraete, W. (2001). The electronic nose as a rapid sensor for volatile compounds in treated domestic wastewater. *Water Research*, *35*(10), 2475–2483. doi:10.1016/S0043-1354(00)00530-3 PMID:11394782

Di Lello, E., Trincavelli, M., Bruyninckx, H., & De Laet, T. (2014). Augmented Switching Linear Dynamical System Model for Gas Concentration Estimation with MOX Sensors in an Open Sampling System. *Sensors (Basel)*, *14*(7), 12533–12559. doi:10.3390/s140712533 PMID:25019637

Di Natale, C., Macagnano, A., Martinelli, E., Paolesse, R., Proietti, E., D'Amico, A., (2001). The evaluation of quality of post-harvest oranges and apples by means of an electronic nose. *Sens. Actuat. B: Chem.*, (78), 26–31.

Di Natale, C., Macagnano, A., D'Amico, A., & Davide, F. (1997). Electronic-nose modelling and data analysis using a self-organizing map. *Measurement Science & Technology*, *8*(11), 1236–1243. doi:10.1088/0957-0233/8/11/004

Di Natale, C., Macagnano, A., Martinelli, E., Paolesse, R., D'Arcangelo, G., Roscioni, C., & D'Amico, A. (2003). Lung cancer identification by the analysis of breath by means of an array of non-selective gas sensors. *Biosensors & Bioelectronics*, *18*(10), 1209–1218. doi:10.1016/S0956-5663(03)00086-1 PMID:12835038

Di Natale, C., Olafsdottir, G., Einarsson, S., Martinelli, E., Paolesse, R., & D'Amico, A. (2001). Comparison and integration of different electronic noses for freshness evaluation of cod-fish fillets. *Sensors and Actuators. B, Chemical*, *77*(1-2), 572–578. doi:10.1016/S0925-4005(01)00692-X

Di Rosa, A. R., Leone, F., Cheli, F., & Chiofalo, V. (2017). Fusion of electronic nose, electronic tongue and computer vision for animal source food authentication and quality assessment – A review. *Journal of Food Engineering*, *210*, 62–75. doi:10.1016/j.jfoodeng.2017.04.024

Di Rosa, A. R., Leone, F., Scattareggia, C., & Chiofalo, V. (2018). Botanical origin identification of Sicilian honeys based on artificial senses and multi-sensor data fusion. *European Food Research and Technology*, *244*(1), 117–125. doi:10.1007/s00217-017-2945-8

Dias, L. G., Veloso, A. C. A., Sousa, M. E. B. C., Estevinho, L., Machado, A. A. S. C., & Peres, A. M. (2015). A novel approach for honey pollen profile assessment using an electronic tongue and chemometric tools. *Analytica Chimica Acta*, *900*, 36–45. doi:10.1016/j.aca.2015.10.014 PMID:26572837

Dickinson, J., White, J., Kauer, J., & Walt, D. (1996). A chemical-detecting system based on a cross reactive optical sensor array. *Nature*, *382*(6593), 697–700. doi:10.1038/382697a0 PMID:8751439

Dillon, W. R., & Goldstein, M. (1984). *Multivariate analysis, methods and applications*. New York: John Wiley & Sons.

Dincer, F., Odabasi, M., & Muezzinoglu, A. (2006). Chemical characterization of odorous gases at a landfill site by gas chromatography-mass spectrometry. *Journal of Chromatography. A*, *1122*(1-2), 222–229. doi:10.1016/j.chroma.2006.04.075 PMID:16697389

Distante, C., Indiveri, G., & Reina, G. (2009). An application of mobile robotics for olfactory monitoring of hazardous industrial sites. *Industrial Robot: An International Journal*, *36*(1), 51–59. doi:10.1108/01439910910924675

Doeswijk, T. G., Smilde, A. K., Hagemann, J. A., Westerhuis, J. A., & van Eeuwijk, F. A. (2011). On the increase of predictive performance with high-level data fusion. *Analytica Chimica Acta*, *705*(1-2), 41–47. doi:10.1016/j.aca.2011.03.025 PMID:21962346

Doty, R. L. (2015). *Handbook of olfaction and gustation*. John Wiley & Sons. doi:10.1002/9781118971758

Dragonieri, S., Schot, R., Mertens, B. J. A., Le Cessie, S., Gauw, S. A., Spanevello, A., ... Sterk, P. J. (2007). An electronic nose in the discrimination of patients with asthma and controls. *The Journal of Allergy and Clinical Immunology*, *120*(4), 856–862. doi:10.1016/j.jaci.2007.05.043 PMID:17658592

Dutta, R., Kashwan, K., Bhuyan, M., Hines, E. L., & Gardner, J. (2003). Electronic nose based tea quality standardization. *Neural Networks*, *16*(5), 847–853. doi:10.1016/S0893-6080(03)00092-3 PMID:12850043

Elen, B., Peters, J., Van Poppel, M., Bleux, N., Theunis, J., Reggente, M., & Standaert, A. (2013). The Aeroflex: A bicycle for mobile air quality measurements. *Sensors (Basel)*, *13*(1), 221–240. doi:10.3390/s130100221 PMID:23262484

Elmenreich, W. (2002). *Sensor fusion in time-triggered systems*. Retrieved April 26, 2017, from https://mobile.aau.at/~welmenre/papers/elmenreich_Dissertation_sensorFusionInTimeTriggeredSystems.pdf

Eom, K. H., Hyun, K. H., & Sen, L. (2014). The Meat Freshness Monitoring System Using the Smart RFID Tag. *International Journal of Distributed Sensor Networks*, *3*, 1–9.

Esposito, E., De Vito, S., Salvato, M., Bright, V., Jones, R. L., & Popoola, O. (2016). Dynamic neural network architectures for on field stochastic calibration of indicative low cost air quality sensing systems. *Sensors and Actuators. B, Chemical*, *231*, 701–713. doi:10.1016/j.snb.2016.03.038

Falasconi, M., Concina, I., Gobbi, E., Sberveglieri, V., Pulvirenti, A., & Sberveglieri, G. (2012). Electronic nose for microbiological quality control of food products. *International Journal of Electrochemistry*, *2012*, 1–12. doi:10.1155/2012/715763

Fang, J. J., Yang, N., Cen, D. Y., Shao, L. M., & He, P. J. (2012). Odor compounds from different sources of landfill: Characterization and source identification. *Waste Management (New York, N.Y.)*, *32*(7), 1401–1410. doi:10.1016/j.wasman.2012.02.013 PMID:22480726

Farrell, J. A., Pang, S., & Li, W. (2003). Plume Mapping via Hidden Markov Methods. *IEEE Transactions on Systems, Man, and Cybernetics. Part B, Cybernetics*, *33*(6), 850–863. doi:10.1109/TSMCB.2003.810873 PMID:18238238

Farrell, J. A., Pang, S., & Li, W. (2005). Chemical plume tracing via an autonomous underwater vehicle. *IEEE Journal of Oceanic Engineering*, *30*(2), 428–442. doi:10.1109/JOE.2004.838066

Fayyad, U., & Uthurusamy, R. (1996). Data Mining and Knowledge Discovery in Databases (Introduction to the Special Section). *Communications of the ACM*, *45*(8), 24–26. doi:10.1145/240455.240463

Ferri, G., Caselli, E., Mattoli, V., Mondini, A., Mazzolai, B., & Dario, P. (2009). SPIRAL: A novel biologically-inspired algorithm for gas/odor source localization in an indoor environment with no strong airflow. *Robotics and Autonomous Systems*, *57*(4), 393–402. doi:10.1016/j.robot.2008.07.004

Figaro U. S. A. Inc. (n.d.). Retrieved from http://www.figarosensor.com

Fisher, R. A. (1936). The use of multiple measurements in taxonomic problems. *Annals of Eugenics*, *7*(2), 179–188. doi:10.1111/j.1469-1809.1936.tb02137.x

Fleischer, M. (2011). Solid State and Spectroscopic Chemical Sensors and Applications. *Sensor Letters*, *9*(2), 706–709. doi:10.1166/sl.2011.1597

Fonollosa, J., Sheik, S., Huerta, R., & Marco, S. (2015). Reservoir computing compensates slow response of chemosensor arrays exposed to fast varying gas concentrations in continuous monitoring. *Sensors and Actuators. B, Chemical*, *215*, 618–629. doi:10.1016/j.snb.2015.03.028

Foody, G., & Arora, M. (1997). An evaluation of some factors affecting the accuracy of classification by an artificial neural network. *International Journal of Remote Sensing*, *18*(4), 799–810. doi:10.1080/014311697218764

Frank, J. M., & Wawilow, S. J. (1931). Über die Wirkungssphäre der Auslöschungsvorgänge in den fluoreszierenden Flüssigkeiten. *Zeitschrift fur Physik*, *69*(1), 100–110. doi:10.1007/BF01391516

Friedman, J. (2001). Greedy function approximation: A gradient boosting machine. *Annals of Statistics*, *29*(5), 1189–1232. doi:10.1214/aos/1013203451

Fryder, M., Holmberg, M., Winquist, F., & Lundstrom, I. (1995, Jun). A calibration technique for an electronic nose. In *Solid-state sensors and actuators, 1995 and eurosensors ix. transducers '95. the 8th international conference on* (Vol. 1, p. 683-686). Academic Press. .717323 doi:10.1109/SENSOR.1995.717323

Gallego, E., Perales, J. F., Roca, F. J., & Guardino, X. (2014). Surface emission determination of volatile organic compounds (VOC) from a closed industrial waste landfill using a self-designed static flux chamber. *The Science of the Total Environment*, *470-471*, 587–599. doi:10.1016/j.scitotenv.2013.09.105 PMID:24176707

Gan, Z., Yang, Y., Li, J., Wen, X., Zhu, M., Jiang, Y., & Ni, Y. (2016). Using sensor and spectral analysis to classify botanical origin and determine adulteration of raw honey. *Journal of Food Engineering*, *178*, 151–158. doi:10.1016/j.jfoodeng.2016.01.016

Gao, L., Liu, T., An, X., Zhang, J., Ma, X., & Cui, J. (2017). Analysis of volatile flavor compounds influencing Chinese-type soy sauces using GC–MS combined with HS-SPME and discrimination with electronic nose. *Journal of Food Science and Technology*, *54*(1), 130–143. doi:10.1007/s13197-016-2444-0 PMID:28242911

Gardner, J., & Bartlett, P. (1992). Pattern recognition in odour sensing. *Sensors and sensory systems for an electronic nose*, 161-179.

Gardner, J. (1987). Pattern recognition in the Warwick Electronic Nose. *8th International Congress of European Chemoreception Research Organization.*

Gardner, J. W. (1991). Detection of vapours and odours from a multisensor array using pattern recognition part 1. principal component and cluster analysis. *Sensors and Actuators. B, Chemical*, *4*(1), 109–115. doi:10.1016/0925-4005(91)80185-M

Gardner, J. W., & Bartlet, P. N. (1999). *Electronic Noses, Principles and Applications*. Oxford, UK: Oxford University Press.

Gardner, J. W., Shin, H. W., & Hines, E. L. (2000). An electronic nose system to diagnose illness. *Sensors and Actuators. B, Chemical*, *70*(1), 19–24. doi:10.1016/S0925-4005(00)00548-7

Gardner, J., & Barlett, P. (1994). A Brief History of Electronic Noses. *Sensors and Actuators. B, Chemical*, *18*(1-3), 211–220. doi:10.1016/0925-4005(94)87085-3

Gares, K. L., Hufziger, K. T., Bykov, S. V., & Asher, S. A. (2016). Review of explosive detection methodologies and the emergence of standoff deep UV resonance Raman. *Journal of Raman Spectroscopy: JRS*, *47*(1), 124–141. doi:10.1002/jrs.4868

Gasbarrini, A., Lauritano, E. C., Gabrielli, M., Scarpellini, E., Lupascu, A., Ojetti, V., & Gasbarrini, G. (2007). Small Intestinal Bacterial Overgrowth: Diagnosis and Treatment. *Digestive Diseases (Basel, Switzerland)*, *25*(3), 237–240. doi:10.1159/000103892 PMID:17827947

Gaspardo, B., Procida, G., Toso, B., & Stefanon, B. (2008). Determination of volatile compounds in San Daniele ham using headspace GC-MS. *Meat Science*, *80*(2), 204–209. doi:10.1016/j.meatsci.2007.11.021 PMID:22063323

Gębicki, J., Dymerski, T., & Namieśnik, J. (2017). Investigation of Air Quality beside a Municipal Landfill: The Fate of Malodour Compounds as a Model VOC. *Environments*, *4*(1).

Gebicki, J. (2016). Application of electrochemical sensors and sensor matrixes for measurement of odorous chemical compounds. *Trends in Analytical Chemistry*, *77*, 1–13. doi:10.1016/j.trac.2015.10.005

Gębicki, J., Dymerski, T., & Namieśnik, J. (2016). Evaluation of Ambient Air Odour Quality in Vicinity of Municipal Landfill Using Electronic Nose Technique. *Chemical Engineering Transactions*, *54*, 253–258.

Germain, M. E., & Knapp, M. J. (2008). Turn-on fluorescence detection of H_2O_2 and TATP. *Inorganic Chemistry Communications*, *47*(21), 9748–9750. doi:10.1021/ic801317x PMID:18828585

Ghasemi-Varnamkhasti, M., & Aghbashlo, M. (2014). Electronic nose and electronic mucosa as innovative instruments for real-time monitoring of food dryers. *Trends in Food Science & Technology*, *38*(2), 158–166. doi:10.1016/j.tifs.2014.05.004

Ghasemi-Varnamkhasti, M., Mohtasebi, S. S., Rodriguez-Mendez, M. L., Lozano, J., Razavi, S. H., & Ahmadi, H. (2011). Potential application of electronic nose technology in brewery. *Trends in Food Science & Technology*, *22*(4), 165–174. doi:10.1016/j.tifs.2010.12.005

Giungato, P., de Gennaro, G., Barbieri, P., Briguglio, S., Amodio, M., de Gennaro, L., & Lasigna, S. (2016). Improving recognition of odors in a waste management plant by using electronic noses with different technologies, gas chromatography-mass spectrometry/olfactometry and dynamic olfactometry. *Journal of Cleaner Production*, *133*, 1395–1402. doi:10.1016/j.jclepro.2016.05.148

Gliszczyńska-Świglo, A., & Chmielewski, J. (2016). Electronic nose as a tool for monitoring the authenticity of food - A review. *Food Analytical Methods*, *10*(6), 1800–1816. doi:10.1007/s12161-016-0739-4

Goldsmith, B. R., Mitala, J. J. Jr, Josue, J., Castro, A., Lerner, M. B., Bayburt, T. H., ... Johnson, C. (2011). Biomimetic chemical sensors using nanoelectronic readout of olfactory receptor proteins. *ACS Nano*, *5*(7), 5408–5416. doi:10.1021/nn200489j PMID:21696137

Gomes, M. T. S. R. (2016). Electronic nose in dairy products. In M. L. Rodriguez-Mendez (Ed.), *Electronic noses and tongues in food science*. London, UK: Academic Press. doi:10.1016/B978-0-12-800243-8.00003-2

Gómez, A.H., Hu, G., Wang, J., & Pereira, A.G. (2006). Evaluation of tomato maturity by electronic nose. *Comput. Electron. Agricult.,* (54), 44–52.

Gómez, A.H., Wang, J., Hu, G., & Pereira, A.G. (2006). Electronic nose technique potential monitoring mandarin maturity.*Sens. Actuat. B: Chem.,* (113), 347–353.

Gomez, A. H., Wang, J., Hu, G., & Pereira, A. G. (2008). Monitoring storage shelf life of tomato using electronic nose technique. *Journal of Food Engineering, 85*(4), 625–631. doi:10.1016/j.jfoodeng.2007.06.039

Gonzalez-Jimenez, Monroy & Blanco. (2011). The Multi-Chamber Electronic Nose— An Improved Olfaction Sensor for Mobile Robotics. *Sensors (Basel), 11*(6), 6145–6164. PMID:22163947

González-Jimenez, J., Monroy, J. G., & Blanco, J. L. (2013). Robots that can smell: motivation and problems. *The International Symposium on Olfaction and Electronic Nose (ISOEN).*

Gonzalez-Jimenez, J., Monroy, J., & Blanco, J.-L. (2011). The Multi-Chamber Electronic Nose—An Improved Olfaction Sensor for Mobile Robotics. *Sensors (Basel), 11*(12), 6145–6164. doi:10.3390/s110606145 PMID:22163947

Gonzalez-Jimenez, J., Monroy, J., Garcia, F., & Blanco, J.-L. (2011). The Multi-Chamber Electronic Nose (MCE-nose). *IEEE International Conference on Mechatronics.* doi:10.1109/ICMECH.2011.5971193

Górska-Horczyczak, E., Horczyczak, M., Guzek, D., Wojtasik-Kalinowska, I., & Wierzbicka, A. (2017). Chromatographic fingerprints supported by artificial neural network for differentiation of fresh and frozen pork. *Food Control, 73*, 237–244. doi:10.1016/j.foodcont.2016.08.010

Gouma, P., Prasad, A., & Stanacevic, S. A. (2011). Selective nanosensor device for exhaled breath analysis. *Journal of Breath Research, 5*(3), 037110. doi:10.1088/1752-7155/5/3/037110 PMID:21896971

Gram, L., Ravn, L., Rasch, M., Bruhn, J. B., Christensen, A. B., & Givskov, M. (2002). Food spoilage - Interactions between food spoilage bacteria. *International Journal of Food Microbiology, 78*(1-2), 79–97. doi:10.1016/S0168-1605(02)00233-7 PMID:12222639

Grasso, F. W., & Atema, J. (2002). Integration of flow and chemical sensing for guidance of autonomous marine robots in turbulent flows. *Environmental Fluid Mechanics, 2*(1), 95–114. doi:10.1023/A:1016275516949

Guney, S., & Atasoy, A. (2013, July). *Fish freshness assessment by using electronic nose.* Paper presented at the 36th International Conference on Telecommunications and Signal Processing, Rome, Italy. doi:10.1109/TSP.2013.6614036

Guney, S., & Atasoy, A. (2015, July). *Discrimination of different fish species by E-nose.* Paper presented at the 38th International Conference on Telecommunications and Signal Processing, Prague, Czech Republic. doi:10.1109/TSP.2015.7296458

Gunn, M., Allen, P., Bonneau, M., Byrne, D. V., Cinotti, S., & Fredriksen, B. (2004). Welfare aspects of the castration of piglets. Scientific report on the scientific panel for animal health and welfare on a request from the Commission related to welfare aspects of the castration of piglets (Question No. EFSA-Q-2003-091). *EFSA Journal*, 91.

Guohua, H., Lvye, W., Yanhong, M., & Lingxia, Z. (2012). Study of grass carp (*Ctenopharyngodon idellus*) quality predictive model based on electronic nose. *Sensors and Actuators. B, Chemical, 166-167*, 301–308. doi:10.1016/j.snb.2012.02.066

Gurney, K. (1997). *An introduction to neural networks.* CRC Press. doi:10.4324/9780203451519

Gutierrez-Osuna, R. (2000). Drift reduction for metal-oxide sensor arrays using canonical correlation regression and partial least squares. In *Proceedings of the 7th international symposium on olfaction & electronic nose* (pp. 1–7). Academic Press.

Haddad, R., Medhanie, A., Roth, Y., Harel, D., & Sobel, N. (2010). Predicting odor pleasantness with an electronic nose. *PLoS Computational Biology*, *6*(4), e1000740. doi:10.1371/journal.pcbi.1000740 PMID:20418961

Hall, D. L., & Llinas, J. (1997). An introduction to multisensor data fusion. *Proceedings of the IEEE*, *85*(1), 6–23. doi:10.1109/5.554205

Hand, D. J. (1997). *Construction and assessment of classification rules*. Chichester, UK: John Wiley & Sons.

Han, F., Huang, X., Teye, E., Gu, F., & Gu, H. (2014). Nondestructive detection of fish freshness during its preservation by combining electronic nose and electronic tongue techniques in conjunction with chemometric analysis. *Analytical Methods*, *6*(2), 529–536. doi:10.1039/C3AY41579A

Han, H. J., Lee, S. H., Moon, J. Y., Park, S., Dong, H., & Noh, B. S. (2016). Discrimination of the cultivar, growing region, and geographical origin of rice (Oryza sativa) using a mass spectrometer-based electronic nose. *Food Science and Biotechnology*, *25*(3), 695–700. doi:10.1007/s10068-016-0121-8

Hansen, M. J., Jonassen, K. E. N., & Feilberg, A. (2014). Evaluation of Abatement Technologies for Pig Houses by Dynamic Olfactometry and On-site Mass Spectrometry. *Chemical Engineering Journal*, *40*, 253–258.

Hanson, C. W., & Thaler, E. R. (2005). Electronic nose prediction of a clinical pneumonia score: Biosensors and microbes. *The Journal of the American Society of Anesthesiologists*, *102*(1), 63–68. PMID:15618788

Harynuk, J., & Górecki, T. (2004). Comprehensive two-dimensional gas chromatography in stop-flow mode. *Journal of Separation Science*, *27*(5-6), 431–441. doi:10.1002/jssc.200301649 PMID:15335078

Hasan, N. U., Naveed, E., & Waleed, E. (2012). Meat and fish freshness inspection system based on odor sensing. *Sensors (Basel)*, *12*(11), 15542–15557. doi:10.3390/s121115542 PMID:23202222

Hasenfratz, D. (2015). *Enabling Large-Scale Urban Air Quality Monitoring with Mobile Sensor Nodes*. Academic Press. https://doi.org/10.3929/ethz-a-010361120

Hasenfratz, D., Saukh, O., Walser, C., Hueglin, C., Fierz, M., Arn, T., … Thiele, L. (2015). Deriving high-resolution urban air pollution maps using mobile sensor nodes. *Pervasive and Mobile Computing*, *16*(PB), 268–285. 10.1016/j.pmcj.2014.11.008

Hatfield, J. V., Neaves, P., Hicks, P. J., Persaud, K., & Travers, P. (1994). Towards an integrated Electronic Nose using conducting polymer sensors. *Sensors and Actuators. B, Chemical*, *18*(1-3), 221–228. doi:10.1016/0925-4005(94)87086-1

Haugen, J. E. (2001). Electronic Noses in Food Analysis. Headspace Analysis of Foods and Flavors. *Headspace Analysis of Foods and Flavors*, *488*, 43–57. doi:10.1007/978-1-4615-1247-9_4

Haugen, J.-E. (2006). The use of chemical sensor array technology, the electronic nose, for detection of boar taint. *Acta Veterinaria Scandinavica*, *48*(Suppl 1), S15. doi:10.1186/1751-0147-48-S1-S15

Haugen, J.-E., Tomic, O., & Kvaal, K. (2000). A calibration method for handling the temporal drift of solid state gas-sensors. *Analytica Chimica Acta*, *407*(12), 23–39. doi:10.1016/S0003-2670(99)00784-9

Hayes, A. T., Martinoli, A., & Goodman, R. M. (2002). Distributed odor source localization. *IEEE Sensors Journal*, *2*(3), 260–271. doi:10.1109/JSEN.2002.800682

Haykin, S. (2009). *Neural networks and learning machines*. Upper Saddle River, NJ: Pearson.

Haykin, S. (2011). *Neural Networks and Learning Machines* (3rd ed.). Pearson.

Heidarbeigi, K., Mohtasebi, S. S., Foroughirad, A., Ghasemi-Varnamkhasti, M., Rafiee, S., & Rezaei, K. (2015). Detection of adulteration in saffron samples using electronic nose. *International Journal of Food Properties*, *18*(7), 1391–1401. doi:10.1080/10942912.2014.915850

Herculano-Houzel, S. (2009). The human brain in numbers: A linearly scaled-up primate brain. *Frontiers in Human Neuroscience*, *3*(31). PMID:19915731

Hernandez-Bennetts, V., Schaffernicht, E., & Pomareda-Ses, V. (2014). A Novel Approach for Gas Discrimination in Natural Environments with Open Sampling Systems. In IEEE Sensors (pp. 3–6). IEEE. doi: 10.1109/ICSENS.2014.6985437

Hernandez-Bennetts, V., Schaffernicht, E., Stoyanov, T., Lilienthal, A., & Trincavelli, M. (2014). Robot assisted gas tomography—Localizing methane leaks in outdoor environments. In *Robotics and Automation (ICRA), 2014 IEEE International Conference on* (pp. 6362–6367). IEEE. doi: 10.1109/ICRA.2014.6907798

Herrick, R. F., Stewart, J. H., & Allen, J. G. (2016). Review of PCBs in US schools: A brief history, an estimate of the number of impacted schools, and an approach for evaluating indoor air samples. *Environmental Science and Pollution Research International*, *23*(3), 1975–1985. doi:10.1007/s11356-015-4574-8 PMID:25940477

Hidayat, W., Shakaff, A. Y. M., Ahmad, M. N., & Adom, A. H. (2010). Classification of agarwood oil using an electronic nose. *Sensors (Basel)*, *10*(5), 4675–4685. doi:10.3390/s100504675 PMID:22399899

Hinrichsen, L. L., & Pedersen, S. B. (1995). Relationship among flavour, volatile compounds, chemical changes, and microflora in Italian-type dry-cured ham during processing. *Journal of Agricultural and Food Chemistry*, *43*(11), 2932–2940. doi:10.1021/jf00059a030

Hoffmann, K., Schulze, M. B., Schienkiewitz, A., Nothlings, U., & Boeing, H. (2004). Application of a new statistical method to derive dietary patterns in nutritional epidemiology. *American Journal of Epidemiology*, *159*(10), 935–944. doi:10.1093/aje/kwh134 PMID:15128605

Holmberg, M., Winquist, F., Lundstr¨om, I., Gardner, J., & Hines, E. (1995). Identification of paper quality using a hybrid electronic nose. *Sensors and Actuators. B, Chemical*, *27*(1), 246–249. doi:10.1016/0925-4005(94)01595-9

Holmes, N. S., & Morawska, L. (2006). A review of dispersion modelling and its application to the dispersion of particles: An overview of different dispersion models available. *Atmospheric Environment*, *40*(30), 5902–5928. doi:10.1016/j.atmosenv.2006.06.003

Honeybee' robots. (2014). Retrieved from https://phys.org/news/2014-09-honeybee-robots-replicate-swarm-behaviour.html#jCp

Hong, X., & Wang, J. (2011). Discrimination and Prediction of Pork Freshness by Electronic nose. *Computer and Computing Technologies in Agriculture V*, *370*, 1–14.

Howe, P. E. (1950). Foods of animal origin. *Journal of the American Medical Association*, *143*(15), 1337–1342. doi:10.1001/jama.1950.82910500017008 PMID:15428269

Huang & Wang. (Eds.). (2016). *Advances in knowledge discovery and data mining: 20th pacific-asia conference*. Cham: Springer International Publishing. doi: 4610.1007/978-3-319-31753-3

Huang, M.-C. (2004). Wave parameters and functions in wavelet analysis with filtering. *Ocean Engineering, 31*(7), 813 - 831. Retrieved from http://www.sciencedirect.com/ science/article/pii/S0029801803002129

Huang, L., Liu, H., Zhang, B., & Wu, D. (2015). Application of electronic nose with multivariate analysis and sensor selection for botanical identification and quality determination of honey. *Food and Bioprocess Technology*, *8*(2), 359–370. doi:10.1007/s11947-014-1407-6

Huang, L., Zhao, J., Chen, Q., & Zhang, Y. (2014). Nondestructive measurement of total volatile basic nitrogen (TVB-N) in pork meat by integrating near infrared spectroscopy, computer vision and electronic nose techniques. *Food Chemistry*, *145*, 228–236. doi:10.1016/j.foodchem.2013.06.073 PMID:24128472

Huang, X., He, Y., Chen, Z., & Hu, C. (2009). Colorimetric sensor for anion recognition based on the proton transfer signaling mechanism. *Chinese Journal of Chemistry*, *27*(8), 1526–1530. doi:10.1002/cjoc.200990257

Huang, X., Lv, R., Yao, L., Guan, C., Han, F., & Teye, E. (2015). Non-destructive evaluation of total volatile basic nitrogen (TVB-N) and K-values in fish using colorimetric sensor array. *Analytical Methods*, *7*(4), 1615–1621. doi:10.1039/C4AY02623C

Huang, X., Xin, J., & Zhao, J. (2011). A novel technique for rapid evaluation of fish freshness using colorimetric sensor array. *Journal of Food Engineering*, *105*(4), 632–637. doi:10.1016/j.jfoodeng.2011.03.034

Hu, G., Zheng, Y., Liu, Z., & Deng, Y. (2016). Effects of UV-C and single- and multiple-cycle high hydrostatic pressure treatments on flavour evolution of cow milk: Gas chromatography-mass spectrometry, electronic nose, and electronic tongue analysis. *International Journal of Food Properties*, *10*, 1–12.

Hui, D., Jun-hua, L., & Zhong-ru, S. (2003). Drift reduction of gas sensor by wavelet and principal component analysis. *Sensors and Actuators B: Chemical*, *96*(12), 354 - 363. Retrieved from http://www.sciencedirect.com/science/article/pii/

Hung, Y. T., Wang, L. K., & Shammas, N. K. (Eds.). (2012). Handbook of Environment and Waste Management: Air and Water Pollution Control. World Scientific Publishing Co. Pte. Ltd.

Hunter, G. W., Xu, J. C., Biaggilabiosa, A. M., Laskowski, D., Dutta, P. K., Mondal, S. P., ... Dweik, R. A. (2011). Smart sensor systems for human health breath monitoring applications. *Journal of Breath Research*, *5*(3), 037111. doi:10.1088/1752-7155/5/3/037111 PMID:21896970

Huss, H. H. (1995). *Quality and quality changes in fresh fish*. Retrieved April 26, 2017, from http://www.fao.org/docrep/v7180e/v7180e00.htm

Ijong, F. G., & Ohta, Y. (1995). Microflora and chemical assessment of an Indonesian traditional fermented fish sauce "Bakasang". *Journal of the Faculty of Applied Biological Science*, *34*, 95–100.

Institute of Economics and Peace. (2016). *Global Terrorism Index*. Retrieved April 10, 2017, from http://economicsandpeace.org/research/iep-indices-data/global-terrorism-index

Ishida, H., Tanaka, H., Taniguchi, H., & Moriizumi, T. (2004). Mobile robot navigation using vision and olfaction to search for a gas/odor source. In *Intelligent Robots and Systems, 2004.(IROS 2004). Proceedings. 2004 IEEE/RSJ International Conference on* (Vol. 1, pp. 313–318). IEEE. doi: 10.1109/IROS.2004.1389370

Ishida, H., Ushiku, T., & Toyama, S. (2004). Mobile Robot Path Planning Using Vision and Olfaction to Search for a Gas Source. *International Conference on Intelligent Robots and Systems*, 313-318.

Ishida, H., Kobayashi, A., Nakamoto, T., & Moriizumi, T. (1999). Three-Dimensional Odor Compass. *IEEE Transactions on Robotics and Automation*, *15*(2), 328–339. doi:10.1109/70.760346

Ishida, H., Nakayama, G., Nakamoto, T., & Moriizumi, T. (2005). Controlling a gas/odor plume-tracking robot based on transient responses of gas sensors. *IEEE Sensors Journal*, *5*(3), 537–545. doi:10.1109/JSEN.2004.839597

Ishida, H., Suetsugu, K., Nakamoto, T., & Moriizumi, T. (1994). Study of autonomous mobile sensing system for localization of odor source using gas sensors and anemometric sensors. *Sensors and Actuators. A, Physical*, *45*(2), 153–157. doi:10.1016/0924-4247(94)00829-9

Ishida, H., Zhu, M., Johansson, K., & Moriizumi, T. (2004). Three-dimensional gas/odor plume tracking with blimp. In *Proc. ICEE* (pp. 117–120). ICEE.

Jackson, D. A. (1993). Stopping rules in components analysis: A comparison of heuristical and statistical approaches. *Ecology*, *74*(8), 2204–2214. doi:10.2307/1939574

Jatmiko, W. (2005). Robotic applications for odor-sensing technology: Progress and challenge. *WSEAS Transactions on Systems*, *4*(7), 1134–1141.

Jatmiko, W. (2011). Robots implementation for odor source localization using PSO algorithm. *WSEAS Transactions on Circuits and Systems*, *10*(4), 115–125.

Jiang, J. J., Zeng, Q. X., Zhu, Z. W., & Zhang, L. Y. (2007). Chemical and sensory changes associated Yu-lu fermentation process - A traditional Chinese fish sauce. *Food Chemistry*, *104*(4), 1629–1634. doi:10.1016/j.foodchem.2007.03.024

Jiang, J., Li, J., Zheng, F., Lin, H., & Guohua, H. (2016). Rapid freshness analysis of mantis shrimps (*Oratosquilla oratoria*) by using electronic nose. *Journal of Food Measurement and Characterization*, *10*(1), 48–55. doi:10.1007/s11694-015-9275-y

Jin, H. J., Lee, S. H., Kim, T. H., Song, H. S., Park, T. H., & Hong, S. (2012). Nanovesicle-based bioelectronic nose platform mimicking human olfactory signal transduction. *Biosensors & Bioelectronics*, *35*(1), 335–341. doi:10.1016/j.bios.2012.03.012 PMID:22475887

Jinks, A., & Laing, D. G. (2001). The analysis of odor mixtures: Evidence for a configurational process. *Physiology & Behavior*, *72*(1-2), 51–63. doi:10.1016/S0031-9384(00)00407-8 PMID:11239981

Jollife, I. T. (2002). *Principal component analysis* (2nd ed.). New York: Springer.

Jun, F., Canqin, H., Jianguo, X., & Junbao, Z. (2012). Pattern classification using an olfactory model with PCA feature selection in electronic nose: Study and application. *Sensors (Basel)*, *12*(12), 2818–2830. doi:10.3390/s120302818 PMID:22736979

Kachele, R., Zhang, M., Gao, Z., & Adhikari, B. (2017). Effect of vacuum packaging on the shelf-life of silver carp (Hypophthalmichthys molitrix) fillets stored at 4 °C. *Lebensmittel-Wissenschaft + Technologie*, *80*, 163–168. doi:10.1016/j.lwt.2017.02.012

Kalita, P., Saikia, M., & Singh, N. (2015). Electronic-Nose Technology and its Application -A Systematic Survey. *International Journal of Innovative Research in Electrical Electronics Instrumentation and Control Engineering*, *3*(1), 123–128. doi:10.17148/IJIREEICE.2015.3126

Karoui, R., Kemps, B., Bamelis, F., De Ketelaere, B., Decuypere, E., & De Baerdemaeker, J. (2006). Methods to evaluate egg freshness in research and industry: A review. *European Food Research and Technology*, *222*(5-6), 727–732. doi:10.1007/s00217-005-0145-4

Kazadi, S., Goodman, R., Tsikata, D., Green, D., & Lin, H. (2000). An autonomous water vapor plume tracking robot using passive resistive polymer sensors. *Autonomous Robots*, *9*(2), 175–188. doi:10.1023/A:1008970418316

Kea-Tiong, T., & Shih-Wen, C. (2013). Towards a Chemiresistive Sensor-Integrated Electronic Nose: A Review. Sensors, 13(10), 14214–14247.

Keller, P. K., & Kangas, L. J. (1994). Three neural network based sensor systems for environmental monitoring. *Proceeding of the IEEE Electro 94 Conference on Combined Volumes*, 377-382. doi:10.1109/ELECTR.1994.472686

Khan, M. I., Jo, C., & Tariq, M. R. (2015). Meat flavor precursors and factors influencing flavour precursors - A systematic review. *Meat Science*, *110*, 278–284. doi:10.1016/j.meatsci.2015.08.002 PMID:26319308

Khot, L. R., Panigrahi, S., Doetkott, C., Chang, Y., Glower, J., Amamcharla, J., ... Sherwood, J. (2012). Evaluation of technique to overcome small dataset problems during neural-network based contamination classification of packaged beef using integrated olfactory sensor system. *Lebensmittel-Wissenschaft + Technologie*, *45*(2), 233–240. doi:10.1016/j.lwt.2011.06.011

Kiani, S., Minaei, S., & Ghasemi-Varnamkhasti, M. (2016). Fusion of artificial senses as a robust approach to food quality assessment. *Journal of Food Engineering*, *171*, 230–239. doi:10.1016/j.jfoodeng.2015.10.007

Kim, D. (2001). Performance comparison of neural network models: Backpropagation vs. fuzzy artmap. *International Journal of Computer Mathematics*, *78*(3), 365–382. doi:10.1080/00207160108805117

Kim, E., Lee, S., Kim, J. H., Kim, C., Byun, Y. T., Kim, H. S., & Lee, T. (2012). Pattern recognition for selective odor detection with gas sensor arrays. *Sensors (Basel)*, *12*(12), 16262–16273. doi:10.3390/s121216262 PMID:23443378

Kim, T. H., Lee, S. H., Lee, J., Song, H. S., Oh, E. H., Park, T. H., & Hong, S. (2008). Single-carbon-atomic-resolution detection of odorant molecules using a human olfactory receptor-based bioelectronic nose. *Advanced Materials*, *21*(1), 91–94. doi:10.1002/adma.200801435

Kodogiannis, V. S. (2017). Application of an Electronic Nose Coupled with Fuzzy-Wavelet Network for the Detection of Meat Spoilage. *Food and Bioprocess Technology*, *10*(4), 730–749. doi:10.1007/s11947-016-1851-6

Kok, S. E., Kian, M. Y., Tiam, H. T., Soo Yong, L., & Beek, Y. (2013). Paper. *IEEE International Conference on Robotics, Biomimetics, & Intelligent Computational Systems (ROBIONETICS)*, 188-93.

Korotkova, O. V., Salinas, Y. E., Vasilieva, E. A., Kozlov, A. V., Yashina, N. V., Loginov, I. A., & Dalin, M. V. (2013). Hydrogen respiratory test: Pilot examinations for evaluation of the small intestinal colonization by normal microflora. *Bulletin of Experimental Biology and Medicine*, *154*(6), 818–820. doi:10.1007/s10517-013-2064-x PMID:23658932

Kośmider, J., & Krajewska, B. (2007). Determining temporary odour concentration under field conditions - Comparison of methods. *Polish Journal of Environmental Studies*, *16*, 215–225.

Kovács, Z., Dalmadi, I., Lukács, L., Sipos, L., Sza'ntai-Ko˝hegyi, K., Kókai, Z., & Fekete, A. (2010). Geographical origin identification of pure Sri Lanka tea infusions with electronic nose, electronic tongue and sensory profile analysis. *Journal of Chemometrics*, *24*(3-4), 121–130. doi:10.1002/cem.1280

Kowadlo, G., & Russell, R. A. (2008). Robot Odor Localization: A Taxonomy and Survey. *The International Journal of Robotics Research*, *27*(8), 869–893. doi:10.1177/0278364908095118

Krupińska, B., Van Grieken, R., & De Wael, K. (2013). Air quality monitoring in a museum for preventive conservation: Results of a three-year study in the Plantin-Moretus Museum in Antwerp, Belgium. *Microchemical Journal*, *110*, 350–360. doi:10.1016/j.microc.2013.05.006

Kumar, R. V. K., & Murali, G. (2016). A survey on the present state-of-the-art of explosives, detection methods and automatic explosive detection using wireless sensor network. *International Journal of Applied Engineering Research*, *11*(1), 504–510.

Kuske, M., Romain, A.-C., & Nicolas, J. (2005). Microbial volatile organic compounds as indicators of fungi. Can an electronic nose detect fungi in indoor environments? *Building and Environment*, *40*(6), 824–831. doi:10.1016/j.buildenv.2004.08.012

Labreche, S., Bazzo, S., Cade, S., & Chanie, E. (2005). Shelf life determination by electronic nose: Application to milk. *Sensors and Actuators, B: Chemical 106*(1), 199-206.

Labreche, S., Bazzo, S., Cade, S., & Chanie, E. (2005). Shelf life determination by electronic nose: Application to milk. *Sensors and Actuators. B, Chemical*, *106*(1), 199–206. doi:10.1016/j.snb.2004.06.027

Lakowicz, J. R. (2006). Principles of fluorescence spectroscopy (3rd ed.). Baltimore, MD: Springer Science+Business Media. doi:10.1007/978-0-387-46312-4

Lang, H. P., Braun, T., Schmid, D., Hunziker, P., Jung, C., Hegner, M., & Gerber, C. (2007). An Artificial Nose Based on Microcantilever Array Sensors. *Journal of Physics: Conference Series*, *61*, 663–667. doi:10.1088/1742-6596/61/1/133

Larsen, C. S. (2003). Animal source foods and human health during evolution. *The Journal of Nutrition*, *133*(11), 3893–3897. PMID:14672287

Laura, T., & Thomas, L. (2011). Classification with correlated features: Unreliability of feature ranking and solutions. *Bioinformatics (Oxford, England)*, *27*(14), 1986–1994. doi:10.1093/bioinformatics/btr300 PMID:21576180

Laureati, M., Buratti, S., Giovanelli, G., Corazzin, M., Lo Fiego, D. P., & Pagliarini, E. (2014). Characterization and differentiation of Italian Parma, San Daniele and Toscano dry-cured hams: A multi-disciplinary approach. *Meat Science*, *96*(1), 288–294. doi:10.1016/j.meatsci.2013.07.014 PMID:23927917

Lee, S. H., Jin, H. J., Song, H. S., Hong, S., & Park, T. H. (2012b). Bioelectronic nose with high sensitivity and selectivity using chemically functionalized carbon nanotube combined with human olfactory receptor. *Journal of Biotechnology*, *157*(4), 467–472. doi:10.1016/j.jbiotec.2011.09.011 PMID:21945089

Lee, S. H., Kwon, O. S., Song, H. S., Park, S. J., Sung, J. H., Jang, J., & Park, T. H. (2012). Mimicking the human smell sensing mechanism with an artificial nose platform. *Biomaterials*, *33*(6), 1722–1729. doi:10.1016/j.biomaterials.2011.11.044 PMID:22153868

Lehrer, S. S. (1971). Solute perturbation of protein fluorescence. Quenching of the tryptophyl fluorescence of model compounds and of Iysozyme by iodide ion. *Biochemistry*, *10*(17), 3254–3263. doi:10.1021/bi00793a015 PMID:5119250

Lehtinen, J., & Veijanen, A. (2011). Odour monitoring by combined TD-GC-MS-sniff technique and dynamic olfactometry at the wastewater treatment plant of low H2S concentration. *Water, Air, and Soil Pollution*, *218*(1–4), 185–196. doi:10.1007/s11270-010-0634-3

Lewkowska, P., Dymerski, T., Gębicki, J., & Namieśnik, J. (2016). The Use of Sensory Analysis Techniques to Assess the Quality of Indoor Air. *Critical Reviews in Analytical Chemistry*, *47*(1), 37–50. doi:10.1080/10408347.2016.1176888 PMID:27105173

Li, C., Heinemann, P., Irudayaraj, J., & Peterson, D. (2005). Detection of apple defects using an electronic nose and znose. *ASAE Annual International Meeting.*

Li, J.-G., Meng, Q.-H., Wang, Y., & Zeng, M. (2010). Single odor source declaration in outdoor time-variant airflow environments. In *Robotics and Biomimetics (ROBIO), 2010 IEEE International Conference on* (pp. 143–148). IEEE. doi:10.1109/ROBIO.2010.5723317

Li, S. (2009). Overview of Odor Detection Instrumentation and the Potential for Human Odor Detection in Air Matrices. *MITRE Innovation Program and US Government Nano-enabled Technology Initiative*, 5-43.

Li, W. (2006). Abstraction of odor source declaration algorithm from moth-inspired plume tracing strategies. In *Robotics and Biomimetics, 2006. ROBIO'06. IEEE International Conference on* (pp. 1024–1029). IEEE. doi:10.1109/ROBIO.2006.340369

Li, J. G., Meng, Q.-H., Wang, Y., & Zeng, M. (2011). Odor source localization using a mobile robot in outdoor airflow environments with a particle filter algorithm. *Autonomous Robots*, *30*(3), 281–292. doi:10.1007/s10514-011-9219-2

Lilienthal, A., Reggente, M., Trincavelli, M., Blanco, J.-L., & Gonzalez-Jimenez, J. (2009). A statistical approach to gas distribution modelling with mobile robots - The Kernel DM+V algorithm. In *Intelligent Robots and Systems (IROS). IEEE/RSJ International Conference on* (pp. 570–576). IEEE. doi:10.1109/IROS.2009.5354304

Lilienthal, A., Ulmer, H., Frohlich, H., Stutzle, A., Werner, F., & Zell, A. (2004). Gas source declaration with a mobile robot. In *Robotics and Automation, 2004. Proceedings. ICRA'04. 2004 IEEE International Conference on* (Vol. 2, pp. 1430–1435). IEEE. doi:10.1109/ROBOT.2004.1308025

Lilienthal, A., & Duckett, T. (2003a). Experimental analysis of smelling Braitenberg vehicles. *Environment*, *5*(10).

Lilienthal, A., & Duckett, T. (2003b). Gas source localisation by constructing concentration gridmaps with a mobile robot. *Proceedings of the European Conference on Mobile Robots (ECMR 2003).*

Lilienthal, A., & Duckett, T. (2004). Building gas concentration gridmaps with a mobile robot. *Robotics and Autonomous Systems*, *48*(1), 3–16. doi:10.1016/j.robot.2004.05.002

Lilienthal, A., Loutfi, A., Blanco, J., Galindo, C., & Gonzalez-Jimenez, J. (2007). A Rao-Blackwellisation Approach to GDM-SLAM – Integrating SLAM and Gas Distribution Mapping (GDM). *European Conference on Mobile Robots*, 126–131.

Limbo, S., Sinelli, N., Torri, L., & Riva, M. (2009). Freshness decay and shelf life predictive modelling of European sea bass (Dicentrarchus labrax) applying chemical methods and electronic nose. *Lebensmittel-Wissenschaft + Technologie*, *42*(5), 977–984. doi:10.1016/j.lwt.2008.12.011

Lindh, E. M., Sandström, A., Andersson, M. R., & Edman, L. (2016). Luminescent line art by direct-write patterning. *Light, Science & Applications*, *5*, e16050. doi:10.1038/lsa.2016.50

Lin, E. C., & Massey, B. T. (2016). Scintigraphy Demonstrates High Rate of False-positive Results From Glucose Breath Tests for Small Bowel Bacterial Overgrowth. *Clinical Gastroenterology and Hepatology*, *14*(2), 203–208. doi:10.1016/j.cgh.2015.07.032 PMID:26241509

Lippolis, V., Ferrara, M., Cervellieri, S., Damascelli, A., Epifani, F., Pascale, M., & Perrone, G. (2016). Rapid prediction of ochratoxin A-producing strains of *Penicillium* on dry-cured meat by MOS-based electronic nose. *International Journal of Food Microbiology*, *218*, 71–77. doi:10.1016/j.ijfoodmicro.2015.11.011 PMID:26619315

Lit, L., Schweitzer, J. B., & Oberbauer, A. M. (2016). Handler Beliefs Affect Scent Detection Dog Outcomes. *Animal Cognition*, *14*(3), 387–394.

Littarru, P. (2007). Environmental odours assessment from waste treatment plants: Dynamic olfactometry in combination with sensorial analysers "electronic noses,". *Waste Management (New York, N.Y.)*, *27*(2), 302–309. doi:10.1016/j.wasman.2006.03.011 PMID:16766175

Liu, P., & Tu, K. (2012). Prediction of TVB-N content in eggs based on electronic nose. *Food Control*, *23*(1), 177–183. doi:10.1016/j.foodcont.2011.07.006

Liu, X., Klinkhammer, S., Sudau, K., Mechau, N., Vannahme, C., Kaschke, J., & Lemmer, U. (2012). Ink-jet-printed organic semiconductor distributed feedback laser. *Applied Physics Express*, *5*(7), 072101. doi:10.1143/APEX.5.072101

Li, Z., & Hopfield, J. J. (1989). Modeling the olfactory bulb and its neural oscillatory processing. *Biological Cybernetics*, *61*(5), 379–392. doi:10.1007/BF00200803 PMID:2551392

Lochmatter, T., & Martinoli, A. (2009b). Tracking odor plumes in a laminar wind field with bio-inspired algorithms. *Experimental Robotics*, 473–482. 10.1007/978-3-642-00196-3_54

Lochmatter, T., Heiniger, N., Martinoli, A., Pardo, M., & Sberveglieri, G. (2009). Localizing an odor source and avoiding obstacles: Experiments in a wind tunnel using real robots. In AIP Conference Proceedings (Vol. 1137, pp. 69–72). AIP. https://doi.org/10.1063/1.3156629

Lochmatter, T., & Martinoli, A. (2009a). Theoretical Analysis of Three Bio-Inspired Plume Tracking Algorithms. *Proceedings of the IEEE International Conference on Robotics and Automation (2009)*, 2661–2668. doi:10.1109/ROBOT.2009.5152686

Lopetcharat, K., Choi, Y. J., Park, J. W., & Daeschel, M. A. (2001). Fish sauce products and manufacturing: A Review. *Food Reviews International*, *17*(1), 65–88. doi:10.1081/FRI-100000515

Lorenzen, P. C., Walte, H.-G., & Bosse, B. (2013). Development of a method for butter type differentiation by electronic nose technology. *Sensors and Actuators. B, Chemical*, *181*, 690–693. doi:10.1016/j.snb.2013.01.034

Loutfi, A., Coradeschi, S., Karlsson, L., & Broxvall, M. (2004). Putting olfaction into action: Using an electronic nose on a multi-sensing mobile robot. In *Intelligent Robots and Systems, 2004 (IROS 2004). Proceedings. 2004 IEEE/RSJ International Conference on* (Vol. 1, pp. 337–342). IEEE. doi: 10.1109/IROS.2004.1389374

Loutfi, A., Coradeschi, S., Lilienthal, A., & Gonzalez-Jimenez, J. (2008). Gas distribution mapping of multiple odour sources using a mobile robot. *Robotica*, *27*(2), 311. doi:10.1017/S0263574708004694

Loutfi, A., Coradeschi, S., Mani, G. K., Shankar, P., & Balaguru Rayappan, J. B. (2015). Electronic noses for food quality: A review. *Journal of Food Engineering*, *144*, 103–111. doi:10.1016/j.jfoodeng.2014.07.019

Luo, D., Chen, J., Gao, L., Liu, Y., & Wu, J. (2017). Geographical origin identification and quality ontrol of Chinese chrysanthemum flower teas using gas chromatography–mass spectrometry and olfactometry and electronic nose combined with principal component analysis. *International Journal of Food Science & Technology*, *52*(3), 714–723. doi:10.1111/ijfs.13326

Luo, D., Hosseini, H. G., & Stewart, J. R. (2004). Application of ANN with extracted parameters from an electronic nose in cigarette brand identification. *Sensors and Actuators. B, Chemical*, *99*(2), 253–257. doi:10.1016/j.snb.2003.11.022

Lupascu, A., Gabrielli, M., Lauritano, E. C., Scarpellini, E., Santoliquido, A., Cammarota, G., ... Gasbarrini, A. (2005). Hydrogen glucose breath test to detect small intestinal bacterial overgrowth: A prevalence casecontrol study in irritable bowel syndrome. *Alimentary Pharmacology & Therapeutics*, *22*(11-12), 1157–1160. doi:10.1111/j.1365-2036.2005.02690.x PMID:16305730

Lv, P., Tang, Z., Wei, G., Yu, J., & Huang, Z. (2007). Recognizing indoor formaldehyde in binary gas mixtures with a micro gas sensor array and a neural network. *Measurement Science & Technology*, *18*(9), 2997–3004. doi:10.1088/0957-0233/18/9/034

Maamor, H. N., Rashid, F. N. A., Zakaria, N. Z. I., Zakaria, A., Kamarudin, L. M., Jaafar, M. N., . . . Adnan, K. N. A. K. (2014, August). *Bio-inspired taste assessment of pure and adulterated honey using multi-sensing technique*. Paper presented at the 2nd International Conference on Electronic Design, Penang, Malaysia. doi:10.1109/ICED.2014.7015812

Ma, C., Forbes, A., Llano, D., Berger-Wolf, T., & Kenyon, R. (2016). Swordplots: Exploring neuron behavior within dynamic communities of brain networks. *Electronic Imaging*, *60*(1), 1–13.

Madan, M., & Dandina, H. (1994). *Neuro-Control Systems: Theory and Applications*. IEEE Press.

Mahat, N. I., Zakaria, A., & Shakaff, A. Y. M. (2015). Comparisons of parametric and non-parametric classification rules for e-nose and e-tongue. In AIP Conference Proceedings (vol. 1691(1)). doi:10.1063/1.4937017

Mahat, N. I., Krzanowski, W. J., & Hernandez, A. (2007). Variable selection in discriminant analysis based on the location model for mixed variables. *Advances in Data Analysis and Classification*, *1*(2), 105–122. doi:10.1007/s11634-007-0009-9

Mahin, T. D. (2001). Comparison of different approaches used to regulate odours around the world. *Water Science and Technology*, *44*(9), 87–102. PMID:11762488

Majcher, M. A., Kaczmarek, A., Klensporf-Pawlik, D., Pikul, J., & Jeleń, H. H. (2015). SPME-MS-Based Electronic Nose as a Tool for Determination of Authenticity of PDO Cheese, Oscypek. *Food Analytical Methods*, *8*(9), 2211–2217. doi:10.1007/s12161-015-0114-x

Mallat, S. G. (1989). Multiresolution approximations and wavelet orthonormal bases of 2. *Transactions of the American Mathematical Society*, *315*(1), 69–87.

Mandow, A., de Gabriel, J. G., Martlnez, J., Mufioz, V., Ollero, A., & Garcla-Cerezo, A. (1996). The Autonomous Mobile Robot AURORA for Greenhouse Operation. *IEEE Robotics & Automation Magazine*, *3*(4), 18–28. doi:10.1109/100.556479

Marcelloni, F. (2001). Recognition of olfactory signals based on supervised fuzzy c-means and knn algorithms. *Pattern Recognition Letters*, *22*(9), 1007–1019. doi:10.1016/S0167-8655(01)00040-X

Marć, M., Śmiełowska, M., & Zabiegała, B. (2016). Concentrations of monoaromatic hydrocarbons in the air of the underground car park and individual garages attached to residential buildings. *Science of the Total Environment Journal*, *573*, 767–777. doi:10.1016/j.scitotenv.2016.08.173 PMID:27591527

Marco, S., Pardo, A., Ortega, A., & Samitier, J. (1997, May). Gas identification with tin oxide sensor array and self organizing maps: adaptive correction of sensor drifts. In Instrumentation and measurement technology conference, 1997. imtc/97. proceedings. sensing, processing, networking (Vol. 2, pp. 904-907). IEEE. doi:10.1109/IMTC.1997.610256

Marjovi, A., & Marques, L. (2014). Multi-Robot Odor Distribution Mapping in Realistic Time-Variant Conditions. In *IEEE International Conference on Robotics and Automation (ICRA 2014)* (pp. 3720–3727). IEEE. doi:10.1109/ICRA.2014.6907398

Markom, M. A., Shakaff, A. Y. M., Adom, A. H., Ahmad, M. N., Hidayat, W., Abdullah, A. H., & Fikri, N. A. (2009). Intelligent electronic nose system for basal stem rot disease detection. *Journal of Computers and Electronics in Agriculture*, *66*(2), 140–146. doi:10.1016/j.compag.2009.01.006

Marques, L., Martins, A., & de Almeida, A. T. (2005). Environmental monitoring with mobile robots. In *Intelligent Robots and Systems, 2005 (IROS 2005). 2005 IEEE/RSJ International Conference on* (pp. 3624–3629). IEEE. doi:10.1109/IROS.2005.1545133

Marques, L., Almeida, N., & de Almeida, A. T. (2003). *Olfactory sensory system for odour-plume tracking and localization* (Vol. 1). IEEE. doi:10.1109/ICSENS.2003.1278971

Marques, L., Nunes, U., & de Almeida, A. T. (2002). Olfaction-based mobile robot navigation. *Thin Solid Films*, *418*(1), 51–58. doi:10.1016/S0040-6090(02)00593-X

Marques, L., Nunes, U., & De Almeida, A. T. (2003). Odour searching with autonomous mobile robots: An evolutionary-based approach. In *Proceedings of the IEEE Int. Conf. on Advanced Robotics* (pp. 494–500). IEEE.

Marsili, R. T. (2000). Shelf-life prediction of processed milk by solid-phase microextraction, mass pectrometry, and multivariate analysis. *Journal of Agricultural and Food Chemistry*, *48*(8), 3470–3475. doi:10.1021/jf000177c PMID:10956135

Martinez, A. M., & Kak, A. C. (2001). PCA versus LDA. *IEEE Transactions on Pattern Analysis and Machine Intelligence*, *23*(2), 228–233. doi:10.1109/34.908974

Masnan, M. J., Mahat, N. I., Zakaria, A., Shakaff, A. Y. M., Adom, A. H., & Sa'ad, F. S. A. (2012). Enhancing classification performance of multisensory data trough extraction and selection of features. *Procedia Chemistry*, *6*, 132–140. doi:10.1016/j.proche.2012.10.139

Masnan, M. J., Zakaria, A., Shakaff, A. Y. M., Mahat, N. I., Hamid, H., Subari, N., & Saleh, J. M. (2012). Principal Component Analysis – A Realization of Classification Success in Multi Sensor Data Fusion. In P. Sanguansat (Ed.), *Principal Component Analysis/Book 2*. Rijeka, Croatia: InTech-Open Access Publisher.

Mathworks. (2017). Discriminant Analysis. In *MATLAB Documentation*. Retrieved April 3, 2017, from https://de.mathworks.com/help/stats/discriminant-analysis.html

McGinley, C. M., & McGinley, M. (2000). A Field odour monitoring and enforcement. *Proceedings of the 22nd Annual Hawaii Water Environment Association Conference.*

Melucci, D., Bendini, A., Tesini, F., Barbieri, S., Zappi, A., Vichi, S., ... Gallina Toschi, T. (2016). Rapid direct analysis to discriminate geographic origin of extra virgin olive oils by flash gas chromatography electronic nose and chemometrics. *Food Chemistry*, *204*, 263–273. doi:10.1016/j.foodchem.2016.02.131 PMID:26988501

Mendes, A., Bonassi, S., Aguiar, L., Pereira, C., Neves, P., Silva, S., ... Teixeira, J. P. (2015). Indoor air quality and thermal comfort in elderly care centers. *Urban Climate*, *14*, 486–501. doi:10.1016/j.uclim.2014.07.005

Mendoza, E., Crismatt, C., & Matos, R. (2007). Diagnosis of small intestinal bacterial overgrowth in children: The use of lactulose in the breath hydrogen test as a screening test. *Biomedica*, *27*(3), 325–332. doi:10.7705/biomedica.v27i3.195 PMID:18320098

Menzel, R., & Goschnick, J. (2000). Gradient gas sensor microarrays for on-line process control — a new dynamic classification model for fast and reliable air quality assessment. *Sensors and Actuators. B, Chemical*, *68*(1–3), 115–122. doi:10.1016/S0925-4005(00)00470-6

Micone, P. G., & Guy, C. (2007). Odour quantification by a sensor array: An application to landfill gas odours from two different municipal waste treatment works. *Sensors and Actuators. B, Chemical*, *120*(2), 628–637. doi:10.1016/j.snb.2006.03.026

Mildner-Szkudlarz, S., Jeleń, H. H., & Zawirska-Wojtasiak, R. (2008). The use of electronic and human nose for monitoring rapeseed oil autoxidation. *European Journal of Lipid Science and Technology*, *110*(1), 61–72. doi:10.1002/ejlt.200700009

Millward, D. J. (1999). The nutritional value of plant-based diets in relation to human amino acid and protein requirements. *The Proceedings of the Nutrition Society*, *58*(02), 249–260. doi:10.1017/S0029665199000348 PMID:10466163

Mohamed, E. I., Linder, R., Perriello, G., Di Daniele, N., Pöppl, S. J., & De Lorenzo, A. (2002). Predicting Type 2 diabetes using an electronic nose-based artificial neural network analysis. *Diabetes, Nutrition & Metabolism*, *15*(4), 215–221. PMID:12416658

Moncrieff, R. (1961). An instrument for measuring and classifying odours. *Journal of Applied Physiology*, *16*, 742–749. PMID:13771984

Monroy, J., & Gonzalez-Jimenez, J. (2015). Real-Time Odor Classification Through Sequential Bayesian Filtering. *16th International Symposium on Olfaction and Electronic Nose (ISOEN)*. Retrieved from http://mapir.isa.uma.es/mapirwebsite/index.php/mapir-downloads/papers/210

Monroy, J., Gonzalez-Jimenez, J., & Sanchez-Garrido, C. (2014). Monitoring Household Garbage Odors in Urban Areas Through Distribution Maps. In IEEE Sensors (pp. 1364–1367). IEEE. doi:10.1109/ICSENS.2014.6985265

Monroy, J., Blanco, J.-L., & Gonzalez-Jimenez, J. (2013). An Open Source Framework for Simulating Mobile Robotics Olfaction (Abstract only). *15th International Symposium on Olfaction and Electronic Nose (ISOEN).*

Monroy, J., Blanco, J.-L., & Gonzalez-Jimenez, J. (2016). Time-Variant Gas Distribution Mapping with Obstacle Information. *Autonomous Robots*, *40*(1), 1–16. doi:10.1007/s10514-015-9437-0

Monroy, J., & Gonzalez-Jimenez, J. (2017). Gas Classification in Motion: An Experimental Analysis. *Sensors and Actuators. B, Chemical*, *240*, 1205–1215. doi:10.1016/j.snb.2016.09.013

Monroy, J., Gonzalez-Jimenez, J., & Blanco, J.-L. (2012). Overcoming the slow recovery of MOX gas sensors through a system modeling approach. *Sensors (Basel)*, *12*(10), 13664–13680. doi:10.3390/s121013664 PMID:23202015

Monroy, J., Hernandez-Bennetts, V., Fan, H., Lilienthal, A., & Gonzalez-Jimenez, J. (2017). GADEN: A 3D gas dispersion simulator for mobile robot olfaction in realistic environments. *MDPI Sensors*, *17*(7), 1–16. doi:10.3390/s17071479

Monroy, J., Jaimez, M., & Gonzalez-Jimenez, J. (2017). Online Estimation of 2D Wind Maps for Olfactory Robots. *International Symposium on Olfaction and Electronic Nose (ISOEN).* doi:10.1109/ISOEN.2017.7968883

Monroy, J., Palomo, E., Lopez-Rubio, E., & Gonzalez-Jimenez, J. (2016). Continuous Chemical Classification in Uncontrolled Environments with Sliding Windows. *Chemometrics and Intelligent Laboratory Systems*, *158*, 117–129. doi:10.1016/j.chemolab.2016.08.011

Moody, J., & Darken, C. (1988). *Learning with localized receptive fields.* Yale Univ., Department of Computer Science.

Moon, J. Y., Han, H. J., Park, S. W., Dong, H., Han, K.-Y., Kim, H. J., ... Noh, B. S. (2014). Discrimination of the origin of commercial red ginseng concentrates using LC-MS/MS and electronic nose analysis based on a mass spectrometer. *Food Science and Biotechnology*, *23*(5), 1433–1440. doi:10.1007/s10068-014-0196-z

Morvan, M., Talou, T., & Beziau, J.-F. (2003). MOS–MOSFET gas sensors array measurements versus sensory and chemical characterisation of VOC's emissions from car seat foams. *Sensors and Actuators. B, Chemical*, *95*(1-3), 212–223. doi:10.1016/S0925-4005(03)00425-8

Mottram, D. S. (1998). Flavor formation in meat and meat a review. *Food Chemistry*, *62*(4), 415–424. doi:10.1016/S0308-8146(98)00076-4

Mukhtar, S., Ullman, J., Carey, J., & Lacey, R. (2004). A review of literature concerning odors, ammonia, and dust from broiler production facilities: 3. Land application, processing, and storage of broiler litter. *Journal of Applied Poultry Research*, *13*(3), 514–520. doi:10.1093/japr/13.3.514

Naddeo, V., Zarra, T., Giuliani, S., & Belgiorno, V. (2012). Odour Impact Assessment in Industrial Areas. *Chemical Engineering Transactions*, *30*, 85–90.

Nagarajan, N., & Stevens, C. (2008). How does the speed of thought compare for brains and digital computers? *Current Biology*, *18*(17), 756–758. doi:10.1016/j.cub.2008.06.043 PMID:18786380

Nagata, Y. (2003). Measurement of Odor Threshold by Triangle Odor Bag Method. *Odor Measurement Review*, 118–127.

Nagle, H. T., Gutierrez-Osuna, R., & Schiffman, S. S. (1998). The How and Why of Electronic Noses. *IEEE Spectrum*, *35*(9), 22–24. doi:10.1109/6.715180

Nakamoto, T., Ishida, H., & Moriizumi, T. (1997). Active odor sensing system. *Proceedings of the IEEE International Symposium on Industrial Electronics*, 128-133.

Nake, A., Dubreuil, B., Raynaud, C., & Talou, T. (2005). Outdoor *in situ* monitoring of volatile emissions from wastewater treatment plants with two portable technologies of electronic noses. *Sensors and Actuators. B, Chemical*, *106*(1), 36–39. doi:10.1016/j.snb.2004.05.034

National Consortium for the Study of Terrorism and Responses to Terrorism. (2016). *Global Terrorism Database*. Retrieved April 10, 2017, from http://www.start.umd.edu/gtd

Neumann, P. P., Bartholmai, M., Schiller, J. H., Wiggerich, B., & Manolov, M. (2010). Micro-drone for the characterization and self-optimizing search of hazardous gaseous substance sources: A new approach to determine wind speed and direction. In *Robotic and Sensors Environments (ROSE), 2010 IEEE International Workshop on* (pp. 1–6). IEEE. doi:10.1109/ROSE.2010.5675265

Neumann, C., Harris, D. M., & Rogers, L. M. (2002). Contribution of animal source foods in improving diet quality and function in children in the developing world. *Nutrition Research (New York, N.Y.)*, *22*(1-2), 193–220. doi:10.1016/S0271-5317(01)00374-8

Neumann, P. P., Hernandez-Bennetts, V., Lilienthal, A., Bartholmai, M., & Schiller, J. H. (2013). Gas source localization with a micro-drone using bio-inspired and particle filter-based algorithms. *Advanced Robotics*, *27*(9), 725–738. doi:10.1080/01691864.2013.779052

Nguyen, L. V, Kodagoda, S., Ranasinghe, R., & Dissanayake, G. (2014). *Mobile Robotic Wireless Sensor Networks for Efficient Spatial Prediction*. Academic Press.

Nicell, J. A. (2009). Assessment and regulation of odour impacts. *Atmospheric Environment*, *43*(1), 196–206. doi:10.1016/j.atmosenv.2008.09.033

Nicolas, J., Cerisier, C., & Delva, J. (2012). Potential of a Network of Electronic Noses to Assess in Real Time the Odour Annoyance in the Environment of a Compost Facility. *Chemical Engineering Transactions*, *30*, 133–138.

Nicolas, J., Romain, A.-C., & Ledent, C. (2006). The electronic nose as a warning device of the odour emergence in a compost hall. *Sensors and Actuators. B, Chemical*, *116*(1-2), 95–99. doi:10.1016/j.snb.2005.11.085

Nielsen, S. (Ed.). (2010). *Food Analysis*. Springer. doi:10.1007/978-1-4419-1478-1

Nimsuk, N. (2014, May). *Neural network based electronic nose for discrimination of fish sauces*. Paper presented at the 11th International Conference on Electrical Engineering/Electronics, Computer, Telecommunications and Information Technology, Nakhon Ratchasima, Thailand. doi:10.1109/ECTICon.2014.6839743

Nurjuliana, M., Che Man, Y. B., Mat Hashim, D., & Mohamed, A. K. S. (2011). Rapid identification of pork for halal authentication using the electronic nose and gas chromatography mass spectrometer with headspace analyzer. *Meat Science*, *88*(4), 638–644. doi:10.1016/j.meatsci.2011.02.022 PMID:21420795

Nurmaini, S., Tutuko, B., & Thoharsin, A. R. (2013). Intelligent Mobile Olfaction of Swarm Robots. *International Journal of Robotics and Automation*, *2*(4), 189–198.

Ocaño-Higuera, V. M., Maeda-Martinez, A. N., Marquez-Rios, E., Canizales-Rodriguez, D. F., Castillo-Yañez, F. J., Ruiz-Bustos, E., ... Plascencia-Jatomea, M. (2011). Freshness assessment of ray fish stored in ice by biochemical, chemical and physical methods. *Food Chemistry*, *125*(1), 49–54. doi:10.1016/j.foodchem.2010.08.034

Ojha, K. S., Tiwari, B. K., Kerry, J. P., & Cullen, P. J. (2016). Online meat quality and compositional assessment techniques. In E. J. Cummins & G. L. Lyng (Eds.), *Emerging Technologies in Meat Processing: Production, Processing and Technology* (pp. 375–390). Chichester, UK: John Wiley & Sons. doi:10.1002/9781118350676.ch14

Omatu, S., & Yano, M. (2014). Electronic nose System by Using Neural Networks. *Neurocomputing*, *172*(C), 311–318.

Onkal-Engin, G., Demir, I., & Engin, S. N. (2005). Determination of the relationship between sewage odour and BOD by neural networks. *Environmental Modelling & Software*, *20*(7), 843–850. doi:10.1016/j.envsoft.2004.04.012

Oroian, M., Amariei, S., Leahu, A., & Gutt, G. (2015). Multi-element composition of honey as a suitable tool for its authenticity analysis. *Polish Journal of Food and Nutrition Sciences*, *65*(29), 93–100.

Padilla, M., Perera, A., Montoliu, I., Chaudry, A., Persaud, K., & Marco, S. (2010). Drift compensation of gas sensor array data by orthogonal signal correction. *Chemometrics and Intelligent Laboratory Systems, 100*(1), 28 - 35. Retrieved from http://www.sciencedirect .com/science/article/pii/S0169743909001877

Pais, V. F., Oliveira, J. A., & Gomes, M. T. (2012). An electronic nose based on coated piezoelectric quartz crystals to certify ewes' cheese and to discriminate between cheese varieties. *Sensors (Basel)*, *12*(2), 1422–1436. doi:10.3390/s120201422 PMID:22438717

Pan, D. D., Wu, Z., Peng, T., Zeng, X. Q., & Li, H. (2014). Volatile organic compounds profile during milk fermentation by *Lactobacillus pentosus* and correlations between volatiles flavor and carbohydrate metabolism. *Journal of Dairy Science*, *97*(2), 624–631. doi:10.3168/jds.2013-7131 PMID:24359834

Pan, L., Yang, S. X., & DeBruyn, J. (2007). Factor Analysis of Downwind Odours from Livestock Farms. *Biosystems Engineering*, *96*(3), 387–397. doi:10.1016/j.biosystemseng.2006.10.017

Papadopoulou, O., Panagou, E. Z., Mohareb, F., & Nychas, G.-J. (2013). Sensory and microbiological quality assessment of beef fillets, using a portable electronic nose in tandem with support vector machine analysis. *Food Research International*, *50*(1), 241–249. doi:10.1016/j.foodres.2012.10.020

Park, K., Lim, J. H., Jin, H. J., Namgung, S., Lee, S. H., Park, T. H., & Hong, S. (2012b). A bioelectronic sensor based on canine olfactory nanovesicle–carbon nanotube hybrid structures for the fast assessment of food quality. *Analyst (London)*, *137*(14), 3249–3254. doi:10.1039/c2an16274a PMID:22497005

Park, S. J., Kwon, O. S., Lee, S. H., Song, H. S., Park, T. H., & Jang, J. (2012a). Ultrasensitive flexible graphene based field-effect transistor (FET)-type bioelectronic nose. *Nano Letters*, *12*(10), 5082–5090. doi:10.1021/nl301714x PMID:22962838

Pashami, S., Asadi, S., & Lilienthal, A. (2010). Integration of OpenFOAM flow simulation and filament-based gas propagation models for gas dispersion simulation. *Open Source CFD International Conference*. Retrieved from http://scholar.google.com/scholar?hl=en&btnG=Search&q=intitle:Integration+of+OpenFOAM+Flow+Simulation+and+Filament-Based+Gas+Propagation+Models+for+Gas+Dispersion+Simulation#0

Pearce, T. C., Schiffman, S. S., Nagle, H. T., & Gardner, J. W. (Eds.). (2003). Handbook of Machine Olfaction, Electronic Nose Technology. Wiley – VCH, Weinheim (FDR).

Pearce, T. C., Schiffman, S. S., Nagle, H. T., & Gardner, J. W. (2006). *Handbook of machine olfaction: electronic nose technology*. John Wiley & Sons.

Pearson, K. (1901). On lines and planes of closest fit to systems of points in space. *Philosophical Magazine*, *2*(11), 559–572.

Pennazza, G., Fanali, C., Santonico, M., Dugo, L., Cucchiarini, L., Dachà, M., ... Mondello, L. (2013). Electronic nose and GC-MS analysis of volatile compounds in Tuber magnatum Pico: Evaluation of different storage conditions. *Food Chemistry*, *136*(2), 668–674. doi:10.1016/j.foodchem.2012.08.086 PMID:23122112

Penza, M., Cassano, G., Tortorella, F., & Zaccaria, G. (2001). Classification of food, beverages and perfumes by WO 3 thin-film sensors array and pattern recognition techniques. *Sensors and Actuators. B, Chemical*, *73*(1), 76–87. doi:10.1016/S0925-4005(00)00687-0

Peris, M., & Escuder-Gilabert, L. (2013). On-line monitoring of food fermentation processes using electronic noses and electronic tongues: A review. *Analytica Chimica Acta*, *804*, 29–36. doi:10.1016/j.aca.2013.09.048 PMID:24267060

Perna, A., Simonetti, A., Intaglietta, I., Sofo, A., & Gambacorta, E. (2012). Metal content of southern Italy honey of different botanical origins and its correlation with polyphenol content and antioxidant activity. *International Journal of Food Science & Technology*, *47*(9), 1909–1917. doi:10.1111/j.1365-2621.2012.03050.x

Perrin, J. (1924). Loi de décroissance du pouvoir fluorescent en fonction de la concentration. *Comptes rendus de l'Academie des sciences, 178*, 1978-1980.

Persaud, K., & Dodd, G. (1982). Analysis of discrimination mechanisms in the mammalian olfactory system using a model nose. *Nature*, *299*(5881), 352–355. doi:10.1038/299352a0 PMID:7110356

Peveler, W. J., Roldan, A., Hollingsworth, N., Porter, M. J., & Parkin, I. P. (2016). Multichannel detection and differentiation of explosives with a quantum dot array. *ACS Nano*, *10*(1), 1139–1146. doi:10.1021/acsnano.5b06433 PMID:26579950

Piana, M. L., Persano Oddo, L., Bentabol, A., Bruneau, E., Bogdanov, S., & Guyot Declerck, C. (2004). Sensory analysis applied to honey: State of the art. *Apidologie*, *35*(Suppl. 1), 26–37. doi:10.1051/apido:2004048

Pimentel, M., Chow, E. J., & Lin, H. C. (2003). Normalization of lactulose breath testing correlates with symptom improvement in irritable bowel syndrome. a double-blind, randomized, placebo-controlled study. *The American Journal of Gastroenterology*, *98*(2), 412–419. PMID:12591062

Ping, J., Xiao-fang, H., & Ai-dong, G. E. (2014). Mobile Robot Gas Source Localization Based on Behavior Strategies. *Proceedings of The 33rd Chinese Control Conference*, 8304–8308.

Pobkrut, T., Eamsa-ard, T., & Kerdcharoen, T. (2016). Sensor drone for aerial odor mapping for agriculture and security services. In *Electrical Engineering/Electronics, Computer, Telecommunications and Information Technology (ECTI-CON), 2016 13th International Conference on* (pp. 1–5). Academic Press. doi:10.1109/ECTICon.2016.7561340

Politechnika Wrocławska. (2012). Semiconductor gas sensors for mobile robot navigation. *Electrical Review*, *88*(11b), 281-284.

Powers, W. (2004). *The science of smell, part 3: odor detection and measurement*. Iowa State University Extension PM.

Purnamadjaja, A., & Russell, R. (2004). Pheromone communication: implementation of necrophoric bee behaviour in a robot swarm. *Proceedings of the 2004 IEEE Conference on Robotics, Automation and Mechatronics,* 2, 638– 643.

Purnamadjaja, A. H., & Russell, R. A. (2005). Pheromone communication in a robot swarm: Necrophoric bee behaviour and its replication. *Robotica*, *23*(06), 731–742. doi:10.1017/S0263574704001225

Pyk, P., Badia, S. B., Bernardet, U., Knüsel, P., Carlsson, M., Gu, J., … Verschure, P. F. M. J. (2006). An artificial moth: Chemical source localization using a robot based neuronal model of moth optomotor anemotactic search. *Autonomous Robots*, *20*(3), 197–213. doi:10.1007/s10514-006-7101-4

Qiu, S., & Wang, J. (2017). The prediction of food additives in the fruit juice based on electronic nose with chemometrics. *Food Chemistry*, *230*, 208–214. doi:10.1016/j.foodchem.2017.03.011 PMID:28407902

Ramalho, O. (2000). Correspondences between olfactometry, analytical and electronic nose data for 10 indoor paints. *Analusis*, *28*(3), 207–215. doi:10.1051/analusis:2000280207

Randy, R. (2013). *These Machines Will Be Able to Detect Smells Your Own Nose Cannot*. Retrieved from https://www.smithsonianmag.com/innovation/these-machines-will-be-able-to-detect-smells-your-own-nose-cannot-8790482/

Reggente, M., & Lilienthal, A. (2010). The 3D-Kernel DM+V/W algorithm: Using wind information in three dimensional gas distribution modelling with a mobile robot. *2010 IEEE Sensors*, 999–1004. 10.1109/ICSENS.2010.5690924

Reggente, M., Lilienthal, A., Pardo, M., & Sberveglieri, G. (2009). Three-Dimensional Statistical Gas Distribution Mapping in an Uncontrolled Indoor Environment. *AIP Conference Proceedings*, *109–112*, 109–112. doi:10.1063/1.3156484

Reis, J. C., Morais, M. B., & Oliva, C. A. G. (2007). Breath hydrogen test in the diagnosis of environmental enteropathy in children living in an urban slum. *Digestive Diseases and Sciences*, *52*(5), 1253–1258. doi:10.1007/s10620-006-9288-9 PMID:17372830

Rencher, A. C. (1998). *Multivariate statistical inference and applications*. New York: Wiley.

Renna, M., Cornale, P., Lussiana, C., Giordano, M., Belviso, S., Zeppa, G., & Battaglini, L. M. (2012). Efficacy of fatty acids and terpenoids and weakness of electronic nose response as tracers of Asiago d'Allevo PDO cheese produced in different seasons. *Dairy Science & Technology*, *92*(3), 203–218. doi:10.1007/s13594-012-0056-7

Report, E. (2009). Robotic Ferret Will Detect Hidden Drugs and Weapons. *Engineering and Physical Sciences Research Council, Science Daily*. Retrieved from www.sciencedaily.com/releases/2009/06/090612115531.htm

Richard, M. D., & Lippmann, R. P. (1991). Neural network classifiers estimate bayesian a posteriori probabilities. *Neural Computation*, *3*(4), 461–483. doi:10.1162/neco.1991.3.4.461

Riordan, S. M., Mciver, C. J., Duncombe, V. M., & Bolin, T. D. (1995). Bacteriologic analysis of mucosal biopsy specimens for detecting small-intestinal bacterial overgrowth. *Scandinavian Journal of Gastroenterology*, *30*(7), 681–685. doi:10.3109/00365529509096313 PMID:7481532

Ripley, B. D. (2007). *Pattern recognition and neural networks*. Cambridge University Press.

Ritvay, D., & Kovács, D. (2006). The international and Hungarian practice of odor emission control (in Hungarian). National Environmental Conference, Siófok, 19-21 September 2006, Romain, A.C., Delva, J., & Nicolas J. (2008). Complementary approaches to measure environmental odours emitted by landfill areas. *Sensors and Actuators. B, Chemical*, *131*, 18–23.

RoboticsS. (2015). Retrieved from https://phys.org/news/2015-11-scientists-breakthrough-pheromone-based-swarm-robots.html#jCp

Rochel, O., & Martinez, D. (2002). Stereo-olfaction with a sniffing neuromorphic robot using spiking neurons. *16th European Conference on Solid State Transducers - EUROSENSORS*.

Rodriguez-Mendez, M. L., Arrieta, A. A., Parra, V., Bernal, A., Vegas, A., Villanueva, S., ... de Saja, J. A. (2004). Fusion of three sensory modalities for the multimodal characterization of red wines. *IEEE Sensors Journal*, *4*(3), 348–354. doi:10.1109/JSEN.2004.824236

Roinestad, K. S., Louis, J. B., & Rosen, J. D. (1993). Determination of pesticides in indoor air and dust. *Journal of AOAC International*, *76*(5), 1121–1126. PMID:8241815

Romain, A., Molitor, N., Adam, G., & Bietlot, E. (2016). Comparison " Low Cost Chemical Sensors – Analytical Instruments " for Odour Monitoring in a Municipal Waste Plant. *Chemical Engineering Transactions*, *54*, 211–216.

Rosenblith, W. A. (Ed.). (1961). *Sensory Communication*. New York: John Wiley.

Rozas, R., Morales, J., & Vega, D. (1991). Artificial smell detection for robotic navigation. *Advanced Robotics. 'Robots in Unstructured Environments', 91 ICAR., Fifth International Conference on*, 1730-1733. doi:10.1109/ICAR.1991.240354

Russell, R. A. (2006). Tracking chemical plumes in 3-dimensions. In *Robotics and Biomimetics, 2006. ROBIO'06. IEEE International Conference on* (pp. 31–36). IEEE.

Russell, R. A. (2014). Article. Journal of Bionic Engineering, 11(1), 82-89.

Russell, R. A. (2003). Chemical source location and the robomole project. *Proceedings Australian Conference on Robotics and Automation.*

Russell, R. A., Kleeman, L., & Kennedy, S. (2000). Using Volatile Chemicals to Help Locate Targets in Complex Environments. *Australian Conference on Robotics and Automation*, 87-91.

Russo, M., Serra, D., Suraci, F., Di Sanzo, R., Fuda, S., & Postorino, S. (2014). The potential of e-nose aroma profiling for identifying the geographical origin of licorice (Glycyrrhiza glabra L.) roots. *Food Chemistry*, *165*, 467–474. doi:10.1016/j.foodchem.2014.05.142 PMID:25038700

Ryusuke Fujisawa. (2008). Communication Using Pheromone Field for Multiple Robots. *2008 IEEE/RSJ International Conference on Intelligent Robots and Systems*, 1391-1396.

Saad, R. J., & Chey, W. D. (2014). Breath testing for small intestinal bacterial overgrowth: Maximizing test accuracy. *Clinical Gastroenterology and Hepatology*, *12*(12), 1964–1972. doi:10.1016/j.cgh.2013.09.055 PMID:24095975

Saad, S. M., Shakaff, A. Y. M., Saad, A. R. M., Yusof, A. M., Andrew, A. M., Zakaria, A., & Adom, A. H. (2017). Analysis of feature selection with Probabilistic Neural Network (PNN) to classify sources influencing indoor air quality. *AIP Conference Proceedings*, 1808.

Saevels, S., Lammertyn, J., Berna, A. Z., Veraverbeke, E. A., Di Natale, C., & Nicolaï, B. M. (2004). An electronic nose and a mass spectrometry-based electronic nose for assessing apple quality during shelf life. *Postharvest Biology and Technology*, *31*(1), 9–19. doi:10.1016/S0925-5214(03)00129-7

Safety Report. (1991). Safety aspects of nuclear power plant automation and robotics. *The International Atomic Energy Agency*. Retrieved from http://www.iaea.org/inis/collection/NCLCollectionStore/_Public/24/007/24007130.pdf

Sanaeifar, A., Mohtasebi, S.S., Ghasemi-Varnamkhasti, M., Ahmadi, H., (2016). Application of MOS based electronic nose for the prediction of banana quality properties. *Measurement: Journal of the International Measurement Confederation,* (82), 105-114.

Sanchez-Garrido, C., Monroy, J., & Gonzalez-Jimenez, J. (2014). A Configurable Smart E-Nose for Spatio-Temporal Olfactory Analysis. In IEEE Sensors (pp. 1968–1971). IEEE. doi:10.1109/ICSENS.2014.6985418

Sberveglieri, V. (2016). Validation of Parmigiano Reggiano cheese aroma authenticity, categorized through the use of an array of semiconductors nanowire device. *Materials (Basel)*, *9*(2), 81. doi:10.3390/ma9020081 PMID:28787881

Schaller, E., Bosset, J. O., & Escher, F. (1998). 'Electronic Noses' and Their Application to Food. *Lebensmittel-Wissenschaft + Technologie*, *31*(4), 305–316. doi:10.1006/fstl.1998.0376

Schieweck, A., Lohrengel, B., Siwinski, N., Genning, C., & Salthammer, T. (2005). Organic and inorganic pollutants in storage rooms of the Lower Saxony State Museum Hanover, Germany. *Atmospheric Environment*, *39*(33), 6098–6108. doi:10.1016/j.atmosenv.2005.06.047

Schleif, F.-M., Hammer, B., Monroy, J., Gonzalez-Jimenez, J., Blanco, J.-L., Biehl, M., & Petkov, N. (2016). Odor recognition in robotics applications by discriminative time series modeling. *Pattern Analysis & Applications*, *19*(1), 207–220. doi:10.1007/s10044-014-0442-2

Schutz, S., Schoning, M. J., Schroth, P., Malkoc, U., Weissbecker, B., Kordos, P., ... Hummel, H. E. (2000). An insect-based BioFET as a bioelectronic nose. *Sensors and Actuators. B, Chemical*, *65*(1-3), 291–295. doi:10.1016/S0925-4005(99)00325-1

Scorsone, E., Pisanelli, A. M., & Persaud, K. C. (2006). Development of an electronic nose for fire detection. *Sensors and Actuators. B, Chemical*, *116*(1-2), 55–61. doi:10.1016/j.snb.2005.12.059

Senter, S. D., Arnold, J. W., & Chew, V. (2000). APC values and volatile compounds formed in commercially processed, raw chicken parts during storage at 4 and 13 C and under simulated temperature abuse conditions. *Journal of the Science of Food and Agriculture*, *80*(10), 1559–1564. doi:10.1002/1097-0010(200008)80:10<1559::AID-JSFA686>3.0.CO;2-8

Serby, M. J., & Chobar, K. L. (1992). *Science of Olfaction*. New York: Springer Verlag. doi:10.1007/978-1-4612-2836-3

Seyama, M., Sugimoto, I., & Miyagi, T. (2002). Application of an array sensor based on plasma-deposited organic film coated quartz crystal resonators to monitoring indoor volatile compounds. *IEEE Sensors Journal*, *2*(5), 422–427. doi:10.1109/JSEN.2002.804576

Shahin, M., Maier, H., & Jaksa, M. (2004). Data division for developing neural networks applied to geotechnical engineering. *Journal of Computing in Civil Engineering*, *18*(2), 105–114. doi:10.1061/(ASCE)0887-3801(2004)18:2(105)

Sharma, P., Ghosh, A., Tudu, B., Sabhapondit, S., Baruah, B. D., Tamuly, P., ... Bandyopadhyay, R. (2015). Monitoring the fermentation process of black tea using QCM sensor based electronic nose. *Sensors and Actuators. B, Chemical*, *219*, 146–157. doi:10.1016/j.snb.2015.05.013

Shen, N., Moizuddin, S., Wilson, L., Duvick, S., White, P., & Pollak, L. (2001). Relationship of electronic nose analyses and sensory evaluation of vegetable oils during storage. *Journal of the American Oil Chemists' Society*, *78*(9), 937–940. doi:10.1007/s11746-001-0367-z

Shinyei. (n.d.). *Handheld Odor Meter OMX-GR Operation Manual*. Kyomachi: Shinyei.

Shraiman, B. I., & Siggia, E. D. (2000). Scalar Turbulence. *Nature*, *405*(6787), 639–646. doi:10.1038/35015000 PMID:10864314

Shurmer, H., Gardner, J., & Corcoran, P. (1990). Intelligent vapour discrimination using a composite 12-element sensor array. *Sensors and Actuators B: Chemical*, *1*(16), 256 - 260. Retrieved from http://www.sciencedirect.com/science/article/pii/

Siegel, M. W. (1990). Olfaction metal oxide semiconductor gas sensors and neural networks. In T. C. Henderson (Ed.), *Traditional and Non-Traditional Robotic Sensors, NATO ASI Series* (Vol. F63, pp. 143–157). Berlin: Springer-Verlag. doi:10.1007/978-3-642-75984-0_11

Sinanoglou, V. J., Strati, I. F., & Miniadis-Meimaroglou, S. (2011). Lipid, fatty acids and carotenoid content of edible egg yolks from avian species: A comparative study. *Food Chemistry*, *124*(3), 971–977. doi:10.1016/j.foodchem.2010.07.037

Singh, H., Raj, V. B., Kumar, J., Durani, F., Mishra, M., Nimal, A. T., & Sharma, M. U. (2016). SAW mono sensor for identification of harmful vapors using PCA and ANN. *Process Safety and Environmental Protection*, *102*, 577–588. doi:10.1016/j.psep.2016.05.014

Singh, S. (1999). A single nearest neighbor fuzzy approach for pattern recognition. *International Journal of Pattern Recognition and Artificial Intelligence*, *13*(01), 49–64. doi:10.1142/S0218001499000045

Singh, T. K., Drake, M. A., & Cadwallader, K. R. (2003). Flavour of Cheddar cheese: A chemical and sensory perspective. *Comprehensive Reviews in Food Science and Food Safety*, *2*(4), 139–162. doi:10.1111/j.1541-4337.2003.tb00021.x

Sklavounos, S., & Rigas, F. (2004). Validation of turbulence models in heavy gas dispersion over obstacles. *Journal of Hazardous Materials*, *108*(1–2), 9–20. doi:10.1016/j.jhazmat.2004.01.005 PMID:15081160

Śliwińska, M., Wiśniewska, P., Dymerski, T., Wardencki, W., & Namieśnik, J. (2015). The flavour of fruit spirits and fruit liqueurs: A review. *Flavour and Fragrance Journal*, *30*(3), 197–207. doi:10.1002/ffj.3237

Śmiełowska, M., Marć, M., & Zabiegała, B. (2017). Indoor air quality in public utility environments—a review. *Environmental Science and Pollution Research International*, 1–11. PMID:28236201

Smyth, H., & Cozzolino, D. (2013). Instrumental methods (spectroscopy, electronic nose, and tongue) as tools to predict taste and aroma in beverages: Advantages and limitations. *Chemical Reviews*, *113*(3), 1429–1440. doi:10.1021/cr300076c PMID:23256680

Sohn, J. H., Hudson, N., Gallagher, E., Dunlop, M., Zeller, L., & Atzeni, M. (2008). Implementation of an electronic nose for continuous odour monitoring in poultry shed. *Sensors and Actuators. B, Chemical*, *133*(1), 60–69. doi:10.1016/j.snb.2008.01.053

Solomon, D. A. (2015). Neuron the Memory Unit of the Brain. *IOSR Journal of Computer Engineering*, *17*(14), 48–61.

Song, K., Liu, Q., & Wang, Q. (2011). Olfaction and Hearing Based Mobile Robot Navigation for Odor/Sound Source Search. *Sensors (Basel)*, *11*(2), 2129–2154. doi:10.3390/s110202129 PMID:22319401

Song, S., Yuan, L., Zhang, X., Hayat, K., Chen, H., Liu, F., ... Niu, Y. (2013). Rapid measuring and modelling flavour quality changes of oxidised chicken fat by electronic nose profiles through the partial least squares regression analysis. *Food Chemistry*, *141*(4), 4278–4288. doi:10.1016/j.foodchem.2013.07.009 PMID:23993616

Son, M., Lee, J. Y., Ko, H. J., & Park, T. H. (2017). Bioelectronic nose: An emerging tool for odour standardization. *Trends in Biotechnology*, *35*(4), 301–307. doi:10.1016/j.tibtech.2016.12.007 PMID:28089199

Sorrenti, V., Di Giacomo, C., Acquaviva, R., Barbagallo, I., Bognanno, M., & Galvano, F. (2013). Toxicity of ochratoxin A and its modulation by andioxidants: A review. *Toxins*, *5*(10), 1742–1766. doi:10.3390/toxins5101742 PMID:24152986

Sousa, P., Marques, L., & De Almeida, A. (2008). Toward chemical-trail following robots. *Seventh International Conference on Machine Learning and Applications (ICMLA08)*, 489–494. doi:10.1109/ICMLA.2008.133

Spiteri, M., Jamin, E., Thomas, F., Rebours, A., Lees, M., Rogers, K. M., & Rutledge, D. N. (2015). Fast and global authenticity screening of honey using ^{1H}-NMR profiling. *Food Chemistry*, *189*, 60–66. doi:10.1016/j.foodchem.2014.11.099 PMID:26190601

Spitzer, M. (2006). Brain research and learning over the life cycle. In Schooling for tomorrow: Personalising education (pp. 47-62). Academic Press. doi:10.1787/9789264036604-4-en

Stachniss, C., Plagemann, C., & Lilienthal, A. (2009). Learning gas distribution models using sparse Gaussian process mixtures. *Autonomous Robots*, *26*(2–3), 187–202. doi:10.1007/s10514-009-9111-5

Stella, E., Musio, F., Vasanelli, L., & Distante, A. (1995). Goal-oriented mobile robot navigation using an odour sensor. *Proceedings of the Intelligent Vehicles Symposium*, 147–151. doi:10.1109/IVS.1995.528273

Stern, O., & Volmer, M. (1919). Über die Abklingungszeit der Fluoreszenz. *Physikalische Zeitschrift*, *20*, 193–188.

Stuetz, R. M., Fenner, R. A., & Engin, G. (1999). Assessment of odours from sewage treatment works by an electronic nose, H2S analysis and olfactometry. *Water Research*, *33*(2), 453–461. doi:10.1016/S0043-1354(98)00246-2

Subari, N., Saleh, J. M., & Shakaff, A. Y. M. (2014, September). *Fusion technique for honey purity estimation using artificial neural network*. Paper presented at the 2nd International Conference on Advances in Intelligent Systems in Bioinformatics, Chem-Informatics, Business Intelligence, Social Media and Cybernetics (IntelSys), Jakarta, Indonesia. doi:10.2495/Intelsys130071

Subari, N., Saleh, J. M., Shakaff, A. Y. M., & Zakaria, A. (2012). A hybrid sensing approach for pure and adulterated honey classification. *Sensors (Basel)*, *12*(10), 14022–14040. doi:10.3390/s121014022 PMID:23202033

Su, M., Zhang, B., Ye, Z., Chen, K., Guo, J., Gu, X., & Shen, J. (2013). Pulp volatiles measured by an electronic nose are related to harvest season, TSS concentration and TSS/TA ratio among 39 peaches and nectarines. *Scientia Horticulturae*, *150*, 146–153. doi:10.1016/j.scienta.2012.10.020

Suman, M., Riani, G., & Dalcanale, E. (2007). MOS-based artificial olfacotry system for the assessment of egg products freshness. *Sensors and Actuators. B, Chemical*, *125*(1), 40–47. doi:10.1016/j.snb.2007.01.031

Sung, H. J., Paik, C. N., & Chunget, W. C. (2015). Small Intestinal Bacterial Overgrowth Diagnosed by Glucose Hydrogen Breath Test in Postcholecystectomy Patients. *Journal of Neurogastroenterology and Motility*, *21*(4), 545–551. doi:10.5056/jnm15020 PMID:26351251

Sun, J., Wang, Q.-J., Huang, J., Hou, Y.-D., Chen, Y.-F., & Su, X.-R. (2013). Influence of heating temperature on the development of volatile compounds in bigeye tuna meat (*Thunnus obesus*) as assessed by E-nose and SPME-GC/MS. *International Food Research Journal*, *20*(6), 3077–3083.

Sun, J., Yu, X., Fang, B., Ma, L., Xue, C., Zhang, Z., & Mao, X. (2015). Effect of fermentation by *Aspergillus oryzae* on the biochemical and sensory properties of anchovy (*Engraulis japonicus*) fish sauce. *International Journal of Food Science & Technology*, *51*(1), 133–141. doi:10.1111/ijfs.12981

Sun, X., Wang, Y., & Lei, Y. (2015). Fluorescence based explosive detection: From mechanisms to sensory materials. *Chemical Society Reviews*, *44*(22), 8019–8061. doi:10.1039/C5CS00496A PMID:26335504

Surai, P. E., & Sparks, N. H. C. (2001). Designer eggs: From improvement of egg composition to functional food. *Trends in Food Science & Technology*, *12*(1), 7–16. doi:10.1016/S0924-2244(01)00048-6

Sutton, R., Naeem, W., & Chudley, J. (2007). Chemical plume tracing and odour source localization by autonomous vehicles. *Journal of Navigation*, *60*(2), 173–190. doi:10.1017/S0373463307004183

Szczurek, A., & Maciejewska, M. (2015). Classification of air quality inside car cabin using sensor system. Sensornets.

Tahri, K., Tiebe, C., Bougrini, M., Saidi, T., El Alami El Hassani, N., El Bari, N., ... Bouchikhi, B. (2015). Characterization and discrimination of saffron by multisensory systems, SPME-GC-MS and UV-Vis spectrophotometry. *Analytical Methods*, *7*(24), 10328–10338. doi:10.1039/C5AY01986A

Takeoka, G., Perrino, C., & Buttery, R. (1996). Volatile Constituents of Used Frying Oils. *Journal of Agricultural and Food Chemistry*, *44*(3), 654–660. doi:10.1021/jf950430m

Talaiekhozani, A., Bagheri, M., Goli, A., & Talaei Khoozani, M. R. (2016). An overview of principles of odor production, emission, and control methods in wastewater collection and treatment systems. *Journal of Environmental Management*, *170*, 186–206. doi:10.1016/j.jenvman.2016.01.021 PMID:26829452

Tang, K. T., Chiu, S. W., Pan, C. H., Hsieh, H. Y., Liang, Y. S., & Liu, S. C. (2010). Development of a Portable Electronic Nose System for the Detection and Classification of Fruity Odors. *Sensors (Basel)*, *10*(10), 9179–9193. doi:10.3390/s101009179 PMID:22163403

Taurino, A., Distante, C., Siciliano, P., & Vasanelli, L. (2003). Quantitative and qualitative analysis of vocs mixtures by means of a microsensors array and different evaluation methods. *Sensors and Actuators. B, Chemical*, *93*(1), 117–125. doi:10.1016/S0925-4005(03)00241-7

Thaler, E. R., & Hanson, C. W. (2006). Use of an electronic nose to diagnose bacterial sinusitis. *American Journal of Rhinology*, *20*(2), 170–172. PMID:16686381

Tian, X., Wang, J., & Cui, S. (2013). Analysis of pork adulteration in minced mutton using electronic nose of metal oxide sensors. *Journal of Food Engineering, 119*(4), 744–749. doi:10.1016/j.jfoodeng.2013.07.004

Tiwari, K., Tudu, B., Bandyopadhyay, R., & Chatterjee, A. (2013). Identification of monofloral honey using voltammetric electronic tongue. *Journal of Food Engineering, 117*(2), 205–210. doi:10.1016/j.jfoodeng.2013.02.023

Toal, S. J., & Trogler, W. C. (2006). Polymer sensors for nitroaromatic explosives detection. *Journal of Materials Chemistry, 16*(28), 2871–2883. doi:10.1039/b517953j

Torri, L., Sinelli, N., & Limbo, S. (2010). Shelf life evaluation of fresh-cut pineapple by using an electronic nose. *Postharvest Biology and Technology, 56*(3), 239–245. doi:10.1016/j.postharvbio.2010.01.012

Tran, P. D., Van de Walle, D., De Clercq, N., Winne, A. D., Kadow, D., Lieberei, R., ... Van Durme, J. (2015). Assessing cocoa aroma quality by multiple analytical approaches. *Food Research International, 77*, 657–669. doi:10.1016/j.foodres.2015.09.019

Trincavelli, M., & Loutfi, A. (2010). Feature selection for gas identification with a mobile robot. In *Robotics and Automation (ICRA), 2010 IEEE International Conference on* (pp. 2852–2857). IEEE. doi:10.1109/ROBOT.2010.5509617

Trincavelli, M., Lout, A., & Coradeschi, S. (2009). Classification of Odours for Mobile Robots Using an Ensemble of Linear Classifiers. In AIP Conference Proceedings (pp. 475–478). AIP. doi:10.1063/1.3156587

Trincavelli, M., Coradeschi, S., & Loutfi, A. (2009). Odour classification system for continuous monitoring applications. *Sensors and Actuators. B, Chemical, 139*(2), 265–273. doi:10.1016/j.snb.2009.03.018

Tsujita, W., Yoshino, A., Ishida, H., & Moriizumi, T. (2005). Gas sensor network for air-pollution monitoring. *Sensors and Actuators. B, Chemical, 110*(2), 304–311. doi:10.1016/j.snb.2005.02.008

Tudu, B., Jana, A., Metla, A., Ghosh, D., Bhattacharyya, N., & Bandyopadhyay, R. (2009). Electronic nose for black tea quality evaluation by an incremental fRBFg network. *Sensors and Actuators. B, Chemical, 138*(1), 90–95. doi:10.1016/j.snb.2009.02.025

Ullah, M., Tandy, K., Yambem, S. D., Aljada, M., Burn, P. L., Meredith, P., & Namdas, E. B. (2013). Simultaneous enhancement of brightness, efficiency, and switching in RGB organic light emitting transistors. *Advanced Materials, 25*(43), 6213–6218. doi:10.1002/adma.201302649 PMID:23963863

Ulloa, P. A., Guerra, R., Cavaco, A. M., da Costa, A. M. R., Figuera, A. C., & Brigas, A. F. (2013). Determination of the botanical origin of honey by sensor fusion of impedance e-tongue and optical spectroscopy. *Computers and Electronics in Agriculture, 94*, 1–11. doi:10.1016/j.compag.2013.03.001

Urbach, G., & Milne, T. (1988). The concentration of volatiles in pasteurized milk as a function of storage time and storage temperaturesa possible indicator of keeping quality. *Australian Journal of Dairy Technology, 43*, 53–58.

Vainionpaa, J., Smolander, M., Alakomi, H. L., Ritvanen, T., Rajamäki, T., Rokka, M., & Ahvenainen, R. (2004). Comparison of different analytical methods in the monitoring of the quality of modified atmosphere packaged broiler chicken cuts using principal component analysis. *Journal of Food Engineering, 65*(2), 273–280. doi:10.1016/j.jfoodeng.2004.01.025

Vallejo-Cordoba, B., & Nakai, S. (1994). Keeping-quality assessment of pasteurized milk by multivariate analysis of dynamic headspace gas chromatographic data. 1. Shelf life prediction by principal component regression. *Journal of Agricultural and Food Chemistry, 42*(4), 989–993. doi:10.1021/jf00040a029

Vergara, A., Vembu, S., Ayhan, T., Ryan, M. A., Homer, M. L., & Huerta, R. (2012). Chemical gas sensor drift compensation using classifier ensembles. *Sensors and Actuators B: Chemical, 166–167*(0), 320–329. https://doi.org/10.1016/j.snb.2012.01.074

Vergara, A., Fonollosa, J., Mahiques, J., Trincavelli, M., Rulkov, N., & Huerta, R. (2013). On the performance of gas sensor arrays in open sampling systems using Inhibitory Support Vector Machines. *Sensors and Actuators. B, Chemical, 185*(0), 462–477. doi:10.1016/j.snb.2013.05.027

Vergara, A., Vembu, S., Ayhan, T., Ryan, M. A., Homer, M. L., & Huerta, R. (2012). Chemical gas sensor drift compensation using classifier ensembles. *Sensors and Actuators. B, Chemical, 166–167*(0), 320–329. doi:10.1016/j.snb.2012.01.074

Vergassola, M., Villermaux, E., & Shraiman, B. I. (2007). "Infotaxis" as a strategy for searching without gradients. *Nature, 445*(7126), 406–409. doi:10.1038/nature05464 PMID:17251974

Vestergaard, J. S., Haugen, J.-E., & Byrne, D. V. (2006). Application of an electronic nose for measurements of boar taint in entire male pigs. *Meat Science, 74*(3), 564–577. doi:10.1016/j.meatsci.2006.05.005 PMID:22063061

Visser, F. R., & Taylor, M. (1998). Improved performance of the Aromascan A32S Electronic Nose and its potential for detecting aroma differences in dairy products. *Journal of Sensory Studies, 13*(1), 95–120. doi:10.1111/j.1745-459X.1998.tb00077.x

Wan, E. (1993). *Finite impulse response neural networks with applications in time series prediction* (Doctoral dissertation). Stanford University, Stanford, CA.

Wang, D., Yu, K., & Wang, Y. (2011). An Electronic Nose Based on Hybrid MOS-SAW Sensors for Detection of Different Biomarkers of Lung Cancer. American Institute of Physics.

Wang, Y., Meng, Q. H., Luo, B., Zeng, M., Xue, L., & Ma, S. G. (2016). Multi-robot based odor source declaration in 3D airflow fields. In *2016 12th World Congress on Intelligent Control and Automation (WCICA)* (pp. 384–388). Academic Press. doi:10.1109/WCICA.2016.7578671

Wang, J., Kliks, M. M., Jun, S., Jackson, M., & Li, Q. X. (2010). Rapid analysis of glucose, fructose, sucrose, and maltose in honeys from different geographic regions using Fourier transform infrared spectroscopy and multivariate analysis. *Journal of Food Science, 75*(2), 208–214. doi:10.1111/j.1750-3841.2009.01504.x PMID:20492227

Wang, Q., Jin, G., Ma, M., Wang, N., Liu, C., & He, L. (2014). Discriminating eggs from different poultry species by fatty acids and volatiles profiling: Comparison of SPME-GC/MS, electronic nose, and principal component analysis method. *European Journal of Lipid Science and Technology, 116*(8), 1044–1053. doi:10.1002/ejlt.201400016

Wang, X. R., Lizier, J. T., Nowotny, T., Berna, A. Z., Prokopenko, M., & Trowell, S. C. (2014). Feature selection for chemical sensor arrays using mutual information. *PLoS One, 89840*. doi:10.1371/journal.pone.0089840 PMID:24595058

Waphare, S., Botre, B., & Sadistap, S. (2009). Mobile odor tracking robot based on embedded technology. *International Conference on Emerging Trends in Electronic and Photonic Devices Systems*, 108-111.

Wardencki, W., Chmiel, T., Dymerski, T., & Biernacka, P. (2009). Instrumental techniques used for assessment of food quality. *Proceedings of ECOpole, 3*(2), 273.

Warren, M. W., Larick, D. K., & Ball, H. R. (1995). Volatiles and sensory characteristics of cooked egg yolk, white and their combinations. *Journal of Food Science, 60*(1), 79–84. doi:10.1111/j.1365-2621.1995.tb05611.x

Wei, L., Yuanyuan, H., Yanping, C., Jiaojiao, J., & Guohua, H. (2015). *Penaeus orientolis* prawn freshness rapid determination method based on electronic nose and non-linear stochastic resonance technique. *Bioengineered, 6*(1), 45–52. doi:10.4161/21655979.2014.993294 PMID:25551520

Weissburg, M. J., Dusenbery, D. B., Ishida, H., Janata, J., Keller, T., Roberts, P. J. W., & Webster, D. R. (2002). A multidisciplinary study of spatial and temporal scales containing information in turbulent chemical plume tracking. *Environmental Fluid Mechanics, 2*(1–2), 65–94. doi:10.1023/A:1016223500111

Wenjing, L., Zhenhan, D., Dong, L., Jimenez, L. M. C., Yanjun, L., Hanwen, G., & Hongtao, W. (2015). Characterization of odor emission on the working face of landfill and establishing of odorous compounds index. *Waste Management (New York, N.Y.)*, *42*, 74–81. doi:10.1016/j.wasman.2015.04.030 PMID:25997990

Wentzell, P. D., & Montoto, L. V. (2003). Comparison of principal components regression and partial least squares regression through generic simulations of complex mixtures. *Chemometrics and Intelligent Laboratory Systems*, *65*(2), 257–279. doi:10.1016/S0169-7439(02)00138-7

Werle, P., Slemr, F., Maurer, K., Kormann, R., Mücke, R., & Jänker, B. (2002). Near-and mid-infrared laser-optical sensors for gas analysis. *Optics and Lasers in Engineering*, *37*(2), 101–114. doi:10.1016/S0143-8166(01)00092-6

Wilkens, W., & Hatman, A. (1964). An electronic analog for olfactory processes. *Journal of Food Science*, *29*(3), 372–378. doi:10.1111/j.1365-2621.1964.tb01746.x PMID:14220555

Wilson, A. D. (2012). Review of Electronic-nose Technologies and Algorithms to Detect Hazardous Chemicals in the Environment. *First World Conference on Innovation and Computer Sciences (INSODE 2011), Procedia Technology*, 453-463. doi:10.1016/j.protcy.2012.02.101

Wilson, A. D., Oberle, C. S., & Oberle, D. F. (2013). Detection of off-flavor in catfish using a conducting polymer electronic-nose technology. *Sensors (Basel)*, *13*(12), 15968–15984. doi:10.3390/s131215968 PMID:24287526

Wilson, A., & Baietto, M. (2009). Applications and Advances in Electronic-Nose technologies. *Sensors (Basel)*, *9*(7), 5099–5148. doi:10.3390/s90705099 PMID:22346690

Wold, S., Antti, H., Lindgren, F., & Hman, J. (1998). Orthogonal signal correction of near-infrared spectra. *Chemometrics and Intelligent Laboratory Systems, 44*(12), 175 - 185. Retrieved from http://www.sciencedirect.com/science/article/pii/

Wongchoosuk, C., Khunarak, C., Lutz, M., & Kerdcharoen, T. (2012). WiFi electronic nose for indoor air monitoring. *9th International Conference on Electrical Engineering/Electronics, Computer, Telecommunications and Information Technology*, 1–4. doi:10.1109/ECTICon.2012.6254166

Worobiec, A., Samek, L., Spolnik, Z., Kontozova, V., Stefaniak, E., & Van Grieken, R. (2006). Study of the winter and summer changes of the air composition in the church of Szalowa, Poland, related to conservation. *Mikrochimica Acta*, *156*(3–4), 253–261. doi:10.1007/s00604-006-0619-5

Wright, R. H. (1966). Odor and molecular vibration. *Nature*, *209*(5023), 571–572. doi:10.1038/209571a0 PMID:5921179

Wu, C., Li, J., Zhao, P., Piringer, M., & Schauberger, G. (2016). Conversion of the chemical concentration of odorous mixtures into odour concentration and odour intensity: A comparison of methods. *Atmospheric Environment*, *127*, 283–292. doi:10.1016/j.atmosenv.2015.12.051

Wu, Z., Zhuang, B., Weng, P., & Zhang, X. (2016). Fermentation Quality Characteristics and Flavor Formation Changes during the Process of Pickled Wax Gourd in Eastern Zhejiang. *International Journal of Food Properties*, *19*(2), 409–419. doi:10.1080/10942912.2015.1027775

Xu, L., He, J., Duan, S., Wu, X., & Wang, Q. (2016). Comparison of machine learning algorithms for concentration detection and prediction of formaldehyde based on electronic nose. *Sensor Review*, *36*(2), 207–216. doi:10.1108/SR-07-2015-0104

Xu, L., Yu, X., Liu, L., & Zhang, R. (2016). A novel method for qualitative analysis of edible oil oxidation using an electronic nose. *Food Chemistry*, *202*, 229–235. doi:10.1016/j.foodchem.2016.01.144 PMID:26920289

Yang, C. Y., Chang, C. S., & Chen, G. H. (1998). Small-intestinal bacterial overgrowth in patients with liver cirrhosis, diagnosed with glucose H2 or CH4 breath tests. *Scandinavian Journal of Gastroenterology*, *33*(8), 867–871. doi:10.1080/00365529850171549 PMID:9754736

Yao, D. (2009). A gas sensing system for indoor air quality control and polluted environmental monitoring. In *9th IEEE Conference on Nanotechnology 2009* (Vol. 8, pp. 806–811). IEEE.

Yongwei, W., Wang, J., Zhou, B., & Lu, Q. (2009). Monitoring storage time and quality attribute of egg based on electronic nose. *Analytica Chimica Acta*, *650*(2), 183–188. doi:10.1016/j.aca.2009.07.049 PMID:19720190

Yoon, H., Lee, S. H., Kwon, O. S., Song, H. S., Oh, E. H., Park, T. H., & Jang, J. (2009). Polypyrrole nanotubes conjugated with human olfactory receptors: High-performance transducers for FET-type bioelectronic noses. *Angewandte Chemie*, *48*(15), 2755–2758. doi:10.1002/anie.200805171 PMID:19274689

Young, R. C., Buttner, W. J., Linnell, B. R., & Ramesham, R. (2003). Electronic nose for space program applications. *Sensors and Actuators. B, Chemical*, *93*(1-3), 7–16. doi:10.1016/S0925-4005(03)00338-1 PMID:14584511

Yuan, J., Oswald, D., & Li, W. (2015). Autonomous tracking of chemical plumes developed in both diffusive and turbulent airflow environments using Petri nets. *Expert Systems with Applications*, *42*(1), 527–538. doi:10.1016/j.eswa.2014.08.005

Yu, H., Wang, J., Xiao, H., & Liu, M. (2009). Quality grade identification of green tea using the eigenvalues of PCA based on E-nose signals. *Sensors and Actuators. B, Chemical*, *140*(2), 378–382. doi:10.1016/j.snb.2009.05.008

Yusuf, N., Omar, M. I., Zakaria, A., Abdullah, A. A., Kamarudin, L. M., Shakaff, A. Y. M., ... Yassin, M. S. (2013). Diagnosis of bacteria for diabetic foot infection using electronic nose technology. *IEEE Conference on Wireless Sensors 2013*. doi:10.1109/ICWISE.2013.6728791

Zakaria, A., Shakaff, A. Y., Masnan, M. J., Ahmad, M. N., Adom, A. H., Jaafar, M. N., ... Fikri, N. A. (2011). A biomimetic sensor for the classification of honeys of different floral origin and the detection of adulteration. *Sensors (Basel)*, *11*(8), 7799–7822. doi:10.3390/s110807799 PMID:22164046

Zampolli, S., Elmi, I., Ahmed, F., Passini, M., Cardinali, G. C., Nicoletti, S., & Dori, L. (2004). An electronic nose based on solid state sensor arrays for low-cost indoor air quality monitoring applications. *Sensors and Actuators. B, Chemical*, *101*(1), 39–46. doi:10.1016/j.snb.2004.02.024

Zappi, P., Bales, E., Park, J. H., Griswold, W., & Šimuni, T. (2012). The CitiSense Air Quality Monitoring Mobile Sensor Node. *Proceedings of the 11th ACM/IEEE Conference on Information Processing in Sensor Networks.*

Zarei, M., Najafzadeh, H., Eskandari, M. H., Pashmforoush, M., Enayati, A., Gharibi, D., & Fazlara, A. (2012). Chemical and microbial properties of mahyaveh, a traditional Iranian fish sauce. *Food Control*, *23*(2), 511–514. doi:10.1016/j.foodcont.2011.08.023

Zarzhitsky, D., Spears, D., Thayer, D., & Spears, W. (2004). Agent-based chemical plume tracing using fluid dynamics. In *International Workshop on Formal Approaches to Agent-Based Systems* (pp. 146–160). Academic Press. doi:10.1007/978-3-540-30960-4_10

Zeiger, E., Gollapudi, B., & Spencer, P. (2005). Genetic toxicity and carcinogenicity studies of glutaraldehyde—a review. *Mutation Research/Reviews in Mutation Research*, *589*(2), 136–151. doi:10.1016/j.mrrev.2005.01.001 PMID:15795166

Zhang, H., Balaban, M., & Principe, J. C. (2003). Improving pattern recognition of electronic nose data with time-delay neural networks. *Sensors and Actuators B: Chemical*, *96*(12), 385 - 389. Retrieved from http://www.sciencedirect.com/science/article/pii/S0925400503005744

Zhang, G., Patuwo, E., & Hu, Y., M. (1998). Forecasting with artificial neural networks: The state of the art. *International Journal of Forecasting*, *14*, 35–60. doi:10.1016/S0169-2070(97)00044-7

Zhang, H., Wang, J., & Ye, S. (2008). Predictions of acidity, soluble solids and firmness of pear using electronic nose technique. *Journal of Food Engineering*, *86*(3), 370–378. doi:10.1016/j.jfoodeng.2007.08.026

Zhang, J., Chen, W., Li, J., Yu, S., & Zhao, W. (2015). VOCs and Particulate Pollution due to Incense Burning in Temples, China. *Procedia Engineering*, *121*, 992–1000. doi:10.1016/j.proeng.2015.09.067

Zhang, M., Wang, X., Liu, Y., Xu, X., & Zhou, G. (2012). Species discrimination among three kinds of puffer fish using an electronic nose combined with olfactory sensory evaluation. *Sensors (Basel)*, *12*(9), 12562–12571. doi:10.3390/s120912562 PMID:23112731

Zhu, J., Chen, F., Wang, L., Niu, Y., & Xiao, Z. (2017). Evaluation of the synergism among volatile compounds in Oolong tea infusion by odour threshold with sensory analysis and E-nose. *Food Chemistry*, *221*, 1484–1490. doi:10.1016/j.foodchem.2016.11.002 PMID:27979119

Ziyatdinov, A., Marco, S., Chaudry, A., Persaud, K., Caminal, P., & Perera, A. (2010). Drift compensation of gas sensor array data by common principal component analysis. *Sensors and Actuators. B, Chemical*, *146*(2), 460–465. doi:10.1016/j.snb.2009.11.034

Zuppa, M., Distante, C., Persaud, K. C., & Siciliano, P. (2007). Recovery of drifting sensor responses by means of fDWTg analysis. *Sensors and Actuators B: Chemical, 120*(2), 411 - 416. Retrieved from http://www.sciencedirect.com/science/article/pii/

Zuppa, M., Distante, C., Siciliano, P., & Persaud, K. C. (2004). Drift counteraction with multiple self-organising maps for an electronic nose. *Sensors and Actuators B: Chemical, 98*(23), 305 - 317. Retrieved from http://www.sciencedirect.com/science/article/pii/

About the Contributors

Fatema A. Albalooshi is an Assistant Professor at the Computer Engineering Department, College of IT, University of Bahrain. She graduated from the Electrical and Computer Engineering (ECE) department, University of Dayton (USA). As a graduate research assistant, Dr. Albalooshi contributed an active member of the Computer Vision and Wide Area Surveillance Laboratory (UD Vision Lab) at the University of Dayton. Dr. Albalooshi had completed her Master of Science degree at the University of Nottingham (United Kingdom) in Electronic communications and computer engineering. Specializing in computer vision research, her primary research focus is in the field of image processing and object segmentation where she has completed several publications in the related areas. Dr. Albalooshi's other research interests include cybersecurity and authentication, object recognition and tracking, image enhancement, neural networks, medical imagery segmentation, e-noses, and 3D reconstruction.

* * *

Muhd Khairulzaman Abdul Kadir starts off his first degree in BEng (Hons) in Electrical and Electronics at Universiti Tenaga Nasional. First job works as Electronics R&D Engineer in Asia R&D Center for Matsushita Television Network Co. which currently known as Panasonic Audio Video Company in Shah Alam, Malaysia. After few years, then, a change career from engineer to academics in British Malaysian Institute, Gombak, Malaysia which currently known as Universiti Kuala Lumpur – British Malaysian Institute. Started to do research in embedded system, image processing and continue to enhance knowledge in intelligent system till pursue study in PhD Engineering at University of Warwick, UK in the same area. Main interest of research area are Data Mining, Machine Learning, Big Data Analysis, Intelligent System Design, any related Electronics and Electrical project, image processing, embedded system and food security. He really would like to work and collaborate with anybody that involves on the application of artificial intelligence techniques in any discipline (multidisciplinary) areas or any of the interest area.

Sanad Al Maskari is a PhD student in School of Information Technology and Electrical engineering, The University of Queensland, Australia. His major research interest is currently in the field of Data mining, machine learning specifically in sensor data and Big Data. His other research area of interest is in the field of information security, and artificial intelligence.

Yousif Al-Bastaki received a BSc. degree from University of Bahrain, Msc from University of Leeds, UK and a PhD degree from University of Nottingham, UK. Recently he has been appointed as an IT advisor at the Deputy Prime Minster at the Kingdom of Bahrain and previously worked as the Dean of College of IT at the University of Bahrain. Currently he is an associate professor and the Dean of Admission and Registration at the University of Bahrain. His research interests are Neural Networks, genetic algorithms E-Learning, Distance Education and e-government strategies and implementation.

Rajib Bandyopadhyay received the Ph.D. degree from Jadavpur University, Kolkata, India, in 2001. He is currently a Professor with the Department of Instrumentation and Electronics Engineering, Jadavpur University, Kolkata. His research interests include the fields of machine olfaction, intelligent control, and NIR-based instrumentation systems. Dr. Bandyopadhyay is a fellow of the Institution of Electronics and Telecommunication Engineers, India and member of ISA.

Runu Banerjee Roy received her Ph.D. degree from Jadavpur University, Kolkata, India, in 2015. She is currently an Associate Professor with the Department of Instrumentation and Electronics Engineering, Jadavpur University. Her main research interest includes pattern recognition, artificial intelligence, machine olfaction and the electronic tongue.

Nabarun Bhattacharyya received the Ph.D. degree from Jadavpur University, Kolkata, India, in 2008. He is the Additional Director with the Centre for the Development of Advanced Computing (C-DAC), Kolkata, which is a premier R&D Institute under Department of Information Technology, Government of India. His research areas include agrielectronics, machine olfaction, soft computing and pattern recognition. Dr. Bhattacharyya is a member of IEEE.

Nico Bolse received the Bachelor of Science (2011) and the Master of Science (2013) in electrical engineering and information technologies with a focus on micro-, nano-, and optoelectronics from the Karlsruhe Institute of Technology (KIT). Since late 2013 he has been with the Light Technology Institute (LTI) at KIT as a Ph.D. student. His research focus is on the design and development of fluorescence quenching sensors, arrays and systems.

Ambra Di Rosa received the Ph.D. degree on "Animal-Source Food Production and Safety" in 2007. Currently she is a contract professor at the University of Messina (Italy) and her research interests concern the sensory evaluation of animal source food by means of chemical sensors and innovative technologies. Currently she collaborates with the instrumental sensory analysis laboratory of the University of Messina.

Tomasz Dymerski received his MSc degree in Chemical Technology in 2008 and obtained his PhD degree in 2013 from the Gdańsk University of Technology. His main attention is focused on electronic noses, two-dimensional gas chromatography, mass spectrometry applied in food, environmental and medical studies.

Carsten Eschenbaum received the diploma degree in microsystems technology from the University of Applied Sciences in Kaiserslautern in 2006. Afterwards he became a research assistant and developed piezoresistive materials based on amorphous carbon. In parallel to his work, he received the Master of Engineering in 2009. In 2010 he joined the Light Technology Institute (LTI) at the Karlsruhe Institute of Technology (KIT) as a Ph.D. student and graduated in 2016. Since 2015, he has been heading the sensor systems group at the LTI. His research focus is on the development of printed optical sensor systems.

Fan Gao is currently educated at Zhejiang University in China and working for his doctorate. His main research fields are electronic nose, biosensors and bioelectronics.

Jacek Gębicki received his MSc degree in Chemical Technology in 1997 and obtained his PhD degree in 2008 from the Gdańsk University of Technology. His main attention is focused on chemical sensors and electronic noses applied in environmental, food and medical studies.

Javier Gonzalez-Jimenez is the head of the MAPIR group and full professor at the University of Malaga. Prof. Gonzalez-Jimenez received his B.S. degree in Electrical Engineering from the University of Seville in 1987. He joined the Department of "Ingenieria de Sistemas y Automatica" at the University of Malaga in 1988 and received his Ph.D. from this University in 1993. In 1990-1991 he was at the Field Robotics Center, Robotics Institute, Carnegie Mellon University (USA) working on mobile robots as part of his PhD. Since 1996 he has been leading Spanish and European projects on mobile robotics and perception. In these areas he is co-author of 3 books and more than 200 peer-reviewed papers.

Anne Habermehl received the Master of Science (2014) in electrical engineering and information technologies with a focus on micro-, nano-, and optoelectronics from the Karlsruhe Institute of Technology (KIT) and the Diplôme d'ingénieur (2014) from Phelma/Grenoble INP. Since 2014 she has been with the Light Technology Institute (LTI) at KIT as a Ph.D. student. Her research focus is on optical sensors with a strong interest in Surface-enhanced Raman Spectroscopy.

Mudassir Ismail received his BSc degree in Computer Science from the University of Bahrain with first class honors in 2017. During his BSc, with the help of his team, he succeeded in securing the Best Senior Project award.

Paulina Kolasińska is a PhD student at Gdańsk University of Technology.

Uli Lemmer received the diploma degree in physics from RWTH Aachen University in 1990 and a Ph.D. from the University of Marburg in 1995. From 1995 to 1996, he held a postdoctoral position with the University of California at Santa Barbara. He was with the University of Munich from 1996 to 2002. In 2002, he was appointed a full Professor and director of the Light Technology Institute, Karlsruhe Institute of Technology (KIT). His research interests are in the technology and the applications of printable organic and inorganic semiconductors for various optoelectronic applications such as photovoltaics, lighting and sensing.

Francesco Leone received his Master's Degree in Chemistry from Messina University, Italy. His research interests include chemical sensors and measurement techniques.

Xue Li is a professor in School of Information Technology and Electrical engineering, The University of Queensland, Australia. His research interests are in data fusion, data analytics, and data mining.

Nor Idayu Mahat is the Director at the Center for Testing, Measurement and Appraisal, Universiti Utara Malaysia. She is an Associate Professor of Mathematics and Statistics at the same university and actively involved in research activities, business consulting, and community services. Her major researcher interests are on classification and pattern recognition, statistical computation, and mixed variable problems. Her recent work is enhancing statistical computing with Big Data.

Tomasz Majchrzak received his degree in Chemical Technology in 2016 from the Gdańsk University of Technology. He is currently pursuing a PhD degree at the Gdańsk University of Technology, where his research is mostly focused on electronic olfaction.

Maz Jamilah Masnan is a Senior Lecturer at the Institute for Engineering Mathematics, Universiti Malaysia Perlis since 2006. Her research interest includes discriminant analysis and feature selection approach. She is currently investigating the potential of feature selection which is more flexible and definite compared to feature extraction. She has explored the capability of the unbounded and bounded Mahalanobis distance as the distance-based criteria for identifying and selecting influential features for good classification performance. Her explorations on the distance-based criteria were applied for multi sensor data fusion of low and intermediate level data fusion involving electronic nose and tongue data. Despite her publication in the current study, most of her publications were also related to different classification techniques of sensor data which concerns with food and beverages, herb and diabetes foot infection. Currently, she is working on a fundamental research grant scheme using medical data of diabetic patient.

Javier Monroy is a postdoc researcher associated with the Machine Perception and Intelligent Robotics group (MAPIR) at the University of Málaga (Spain). He received his B.Sc (2007), and M.Sc (2010) in Electrical Engineering from this University, and his Ph.D, focusing on the integration between artificial olfaction and mobile robotics, in 2013. During spring 2011 he was a guest researcher at the Applied Autonomous Sensor Systems, Örebro University (Sweden), collaborating with the MR&O group. His research interest includes mobile robotics, machine learning and electronic noses. In these fields, he has authored more than 25 scientific works.

Jacek Namieśnik received his MSc degree in Inorganic chemistry in 1972, obtained his PhD degree in 1978 and defend his DSc degree in 1985 from the Gdańsk University of Technology, he obtained scientific title of professor conferred by President of Poland in 1996. All his areas of interest is related with widely understood analytical chemistry.

Ali Yeon Md. Shakaff is a Full Professor and Director at the Centre of Excellence for Advanced Sensor Technology (CEASTech), Universiti Malaysia Perlis (UniMAP). He is also a Professor at the School of Mechatronic Engineering, UniMAP. He earned his Ph.D. in Electronic Engineering from the University of New Castle-upon-Tyne, England in 1987. His specialization includes electronics, embedded systems design, sensor applications, machine olfaction and gustation. His vast experience of being the research grant leader of enormous research, expert of his area of interest as well as in research publication has help many master and Ph.D degree candidates to graduate successfully.

B. Szulczyński (born in 1991) received his B.Sc. degree in chemical technology from Gdansk University of Technology in 2014. Currently, he's studying chemical technology on M.Sc. level and power systems in B.Sc. level. He's interested in electronics. His main research interests are electronic noses applications, data processing algorithms and chemical detectors construction.

Bipan Tudu received the Ph.D. degree from Jadavpur University, Kolkata, India, in 2011. He is currently a Professor with the Department of Instrumentation and Electronics Engineering, Jadavpur University. His main research interest includes pattern recognition, artificial intelligence, machine olfaction and the electronic tongue.

Wojciech Wojnowski received his degree in Chemical Technology in 2013 from the Gdańsk University of Technology, and also completed the China Studies course at Nanjing University. He is currently pursuing a PhD degree at the Gdańsk University of Technology, where his research is mostly focused on electronic olfaction.

Ammar Zakaria is a Senior Lecturer at the School of Mechatronic Engineering, UniMAP. He is also the Manager at the Centre of Excellence for Advanced Sensor Technology (CEASTech), UniMAP. He earned his Ph.D. in Mechatronic Engineering from UniMAP in 2013. He is a member of the International Society for Olfaction and Chemical Sensing (ISOCS) and IEEE Electronic. He has extensive experience of being the research grant leader for various research grants.

Index

G

H

I

L

M

N

O

P

Q

R

S

T

U

V

www.ingramcontent.com/pod-product-compliance
Lightning Source LLC
LaVergne TN
LVHW060304011026
849161LV00011B/2771